Andreas Maehly Lars Strömberg

Chemical Criminalistics

With 70 Figures and 65 Tables

Springer-Verlag
Berlin Heidelberg New York 1981

Andreas Maehly Lars Strömberg
The Swedish National Laboratory of Forensic Science
Statens Kriminaltekniska Laboratorium
S-581 01 Linköping

ISBN-13: 978-3-642-68063-2 e-ISBN-13: 978-3-642-68061-8
DOI: 10.1007/978-3-642-68061-8

Library of Congress Cataloging in Publication Data. Maehly, Andreas Christian, 1917–. Chemical criminalistics. Bibliography: p. Includes index. 1. Chemistry, Forensic. 2. Crime laboratories. 3. Forensic scientists. I. Strömberg, Lars, 1934–. II. Title. HV8073.M224. 363.2'56. 81-5313. ISBN 0-387-10723-1 (U.S.). AACR2.

Softcover reprint of the hardcover 1st edition 1981

Preface

In recent years, a number of textbooks on forensic science have been published, most of them directed to two groups, viz. the students of forensic science, and the customers so to say, (prosecutors, police officers, judges, defense lawyers).

In this book, while covering fundamental concepts, we try to go a little further and address also active workers in the field of forensic chemistry. This is mainly achieved by relatively numerous literature references. We hope that they may assist the forensic chemist in penetrating further into the subjects covered in this volume.

At the end of most chapters there are examples of actual cases handled at the Swedish National Laboratory of Forensic Science. Many of these cases could, no doubt, have been investigated in greater detail, but they reflect the compromises often necessary for achieving a reasonable turnover.

Some parts of the book are quite strongly colored by the personal opinions of the authors. We felt that these passages will give a little more life to the text than in other treatises of a more objective, but possibly duller character.

The authors welcome all constructive criticism which will help to improve the book, should there be a second edition.

This reminds us of a well-known exchange of messages between Winston Churchill and George Bernard Shaw, which roughly went thus: Shaw sent two tickets for the first performance of one of his plays to W. C. with a note "... one ticket for you, and one for a friend – if you have any!" The tickets came back with an answer: "Unfortunately, I am engaged that evening, but will be delighted to attend the second performance – if there is any."

Linköping, Fall 1981

Andreas Maehly
Lars Strömberg

Acknowledgements

This book has been born with certain difficulties and would not have succeeded without the help of many people.

The difficulties were mostly concerned with the lack of time. Most of the book was written on weekends and holidays, and we want to thank our families for the understanding and patience they showed us.

The whole text was read and corrected for contents, style and language by Xenia Maehly who sacrificed many hours of piano studies and picture painting for this task.

Ms. Birgit Selldén, secretary to one of us (A. M.) typed the whole manuscript – not only once, but often twice or thrice. In her capacity as main librarian, she ordered all the necessary literature not represented in our own library.

The chapter on "automobile painting" was written by engineer Inge Horkeby, AB Volvo. The chapter on soil was contributed by Dr. Jan Andrasko of this laboratory.

Several drawings were executed by Mr. Lars Lindström, and numerous photographs by our photographers, all of this laboratory.

The authors want to thank all of these coworkers for their contributions.

Table of Contents

A. General Introduction

I. Historical Notes

In the history of human civilization, there was an early urge for thought and experimentation in the fields of religion, philosophy, medicine and law. This was of course due to the basic needs of man, beyond the "natural" needs for food and propagation. From these roots, the tree of knowledge grew and branched. Some twigs from these branches eventually contributed to the field of forensic sciences. A very recent development, from the historical point of view, was the emergence of police forces about a century ago. A schematic diagram (Fig. A.1) depicts these relationships.

Since there is no established academic training in forensic science as such, the workers in this field – bordering on law, medicine, philosophy and police operations naturally have their basic training as lawyers, doctors, natural scientists or police

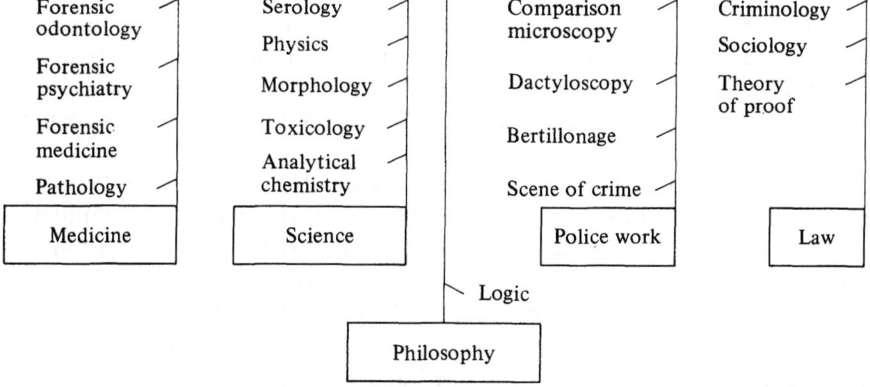

Fig. A.1. Schematic and simplified view of the growth of forensic sciences from the roots of human endeavor – medicine, philosophy, and law-as well as the additional fields of science and police work. On the top of the figure some of the professionals working in forensic science are listed: the medical doctor, the natural scientist, the sociologist, the police officer, and the member of a legal profession

officers. They have to acquire some knowledge of each others' fields and have to learn to understand each others way of reasoning (Fig. A. 1).

The judicial systems vary widely in space and time, and it would be a major task indeed to follow their development within each cultural region. It has become costumary to treat historical developments in the western world alone, beginning, say, with Egypt, and continuing through Greece, ancient Rome, medieval Europe and North America. Within this limited region one might observe a general trend toward carrying out criminal court procedures. The earliest way for obtaining proof of guilt was often semi-religious, where god(s) were supposed to influence certain tests. Later, this system was modified or supplemented by the instrument of torture. Whereas torture unfortunately still is in inofficial practice, it has been largely replaced by other evidence, foremost the hearing of witnesses under oath.

Eventually, a special kind of "witness" began to enter the scene, the *professional expert*. The expert was only exceptionally a witness proper of a crime, but had to testify on an expert opinion, based in turn on observations and experimentation. The first experts to be heard by courts were almost always doctors or surgeons. These medical experts gave their opinion on wounds and other trauma inflicted in fights or accidents, and before all, on causes of death in capital crimes.

It is thus not surprising that the medical profession was the pioneer in the field of forensic science. Departments of forensic medicine emerged in many Western Countries in the 19th century and became centers for research and teaching, initially almost exclusively in pathology.

At the beginning of that century, in 1813, the Spanish-born Mathieu Joseph Bonaventure Orfila, working in France, published a 2-volume textbook on toxicology. Orfila was also one of the first to use animals for toxicity studies. The first application of chemical analysis to a forensic case was probably the isolation and identification of nicotine in the digestive tract of a murder victim by Jean Servais Stas in Belgium 1851. The murderer, a count Bocarmé, had forced his brother-in-law to drink a strong solution of the alkaloid prepared from tobacco.

It was not until the end of the 19th century that other professions than medical doctors and chemists began playing a role in court. Of course, in special cases – such as accidents – building engineers, manufacturers of various machines etc. were heard as expert witnesses. But the specialists in providing criminal evidence emerged quite late. It is hard in this context not to mention Conan Doyle (1859–1930) whose novels describe the work of a personality we today would call a criminalist.

The time around the turn of the century saw the emergence of the pioneers in criminalistics in France, Austria, Germany, Switzerland, the United States, Argentina and India. The first main field of activity was the establishment of the identity of an individual. This was to take two different paths. One approach was the comparison of fingerprints, first suggested by Henry Faulds in 1880, and put into practical use by two schools of workers, i. e. Francis Galton in England (1892) and Edward R. Henry in India on the one hand, and Juan Vucetich in the Argentina (1891) who had read Galton's earlier work on the other hand. The two systems, with numerous variations, still exist side by side.

The other approach to the identification of criminals was devised in France by Alphonse Bertillon. In 1885 he described his system of anthropometrics (measurements of various parts of the human body), later expanded by the "portrait parlé", describing numerous features of the body. These procedures became known as "bertillonage" and competed for several years with the fingerprint method.

The turn of the century saw the development of criminalistic theory and practice in many other fields, such as ballistics, tool mark investigation, and eventually the analysis of handwriting. The pioneers of the emerging new science were inter alia H. Gross (Linz), R. A. Reiss (Lausanne), R. Heindl (Berlin), and E. Locard (Lyon). These scientists had an enormous influence on the development of forensic science and educated a new generation of criminalists who were active between the two world wars. The methods used were mainly optical observation by sophisticated microscopical and photographic procedures.

During this period, the first forensic science laboratories were organized, often against great opposition by the policy makers.

A new, almost explosive development of criminalistics took place after the Second World War and was largely due to the introduction of modern instrumentation in analytical chemistry. Thus we could date the birth of forensic chemistry at around 1950.

Among the chemists who got interested in this field we find one outstanding personality, the American Paul L. Kirk (1902–1970) who had the gift of being an outstanding researcher as well as an inspiring teacher. At the University of California in Berkley he established a school of criminalistics where – for the first time – degrees in criminalistics could be obtained. Many of his pupils are still concentrated in California and make this state a center for criminalistics in the United States.

Kirks most active period occurred at a time when it was still possible for one individual to master the whole field of criminalistics. He established the concept of the "generalist", the criminalist who could, and was supposed to, tackle any problem and investigation in the field.

Table A. 1 (p. 6) gives a short list of the major historical events in criminalistics and forensic chemistry.

The International Reference Organization in Forensic Medicine (INFORM)*) has compiled a bibliography on the history of criminalistics with 295 references prior to 1974.

---

*) W. G. Eckert (ed.), St. Francis Hospital Wichita, Kansas 67214, USA

Table A.1. Some historical dates on the development of criminalistics and forensic chemistry

1880	Henry Faulds (England) suggests fingerprinting in search for criminals
1885	Alphonse Bertillon (France) publishes his first work on anthropometrics ("Bertillonage")
1891	Juan Vucetich (Argentina) devises his fingerprint system, giving credit to Galton's early work
1892	Francis Galton (England) publishes his treatise "Finger Prints"
1893	Alphonse Bertillon describes his system of the "portrait parlé"
1893	Hans Gross (Austria) publishes his renowned "Handbuch für Untersuchungsrichter" – Start of criminalistics in the modern sense
1900	A. L. Hall (USA) publishes his book "The Missile and the Weapon" – Start of scientific ballistics
1901	The Galton fingerprint system, perfected by Edward R. Henry (India and England) is introduced in England
1910	Albert S. Osborn (USA) publishes his work "Questioned Documents" – Start of forensic document work
ca. 1910	The ABO blood group system, discovered by Landsteiner in 1901, begins to be used in forensic science
1911	R. A. Reiss (Switzerland) writes "Manuel de police scientifique"
1912	A "Criminalistics Institute" is opened by Hans Gross in Graz (Austria)
1920	Edmond Locard publishes his "L'Enquête criminelle et les méthodes scientifiques"
1925	The comparison microscope is introduced into criminalistics by Waite and Goddard (USA)
1926	A criminalistics laboratory is opened in Helsinki (Finland)
1935	The first Metropolitan Police Laboratory is opened at Hendon (London)
ca. 1950	Four methods in analytical chemistry are applied to forensic science: Ultraviolet and infrared spectrophotometry, X-ray diffraction, and paper chromatography
1953	Paul L. Kirk (USA) publishes "Crime Investigation"
1957	First triannual meeting of the International Association of Forensic Sciences in Ghent (Belgium)

II. Forensic Science Today

Forensic Chemistry is an integral part of a relatively new discipline, Forensic Science. The definition of this discipline is still under discussion and varies from country to country and even between local regions. Etymologically the word "forensic" is derived from latin "forensis" (public), which is related to "forum" (market place) where roman courts originally had their sessions. "Forensic" today can be defined as "related to court procedures". The second part of the expression "Forensic Science" is not as clearly defined. It is generally agreed on that legal medicine, criminalistics, and legal toxicology belong to its domain. The situation for legally oriented sociology (criminology) and the whole field of law is less clear and opinions about whether or not to regard these disciplines as parts of forensic science vary geographically and temporally.

In Figure A. 2, these authors have tried to place "Forensic Science" into the framework of related sciences and technologies. It is evident from this diagram that "Forensic Science" is a very complex conglomerate of bits of human knowledge and endeavour. It is obviously not possible for a single scientist to master all aspects of forensic science.

On the European continent, the most common model is a division into roughly three areas:

(i) the university departments of legal medicine and the medical examiners,
(ii) the forensic science laboratories,
(iii) the local police departments, primarily concerned with crime scene investigations.

By tradition and training, the legal medical expert encompasses most of the medico-legal disciplines such as pathology, histology and psychiatry – but he often also extends his work to forensic serology, toxicology, osteology and odontology.

The forensic science laboratories concern themselves with all or parts of criminalistics, but traditionally with the laboratory aspects of ballistics, tool mark investigation, questioned document work, arson and trace evidence investigation.

In recent years, the forensic science laboratories have also increasingly been concerned with the analysis of narcotics and with the determination of genetic markers in samples of blood, semen etc. Some laboratories assist also in crime scene work. This development has led to high demands of education and training of the criminalist. The senior staff members are expected to have Ph.D.'s in physics, chemistry or biology.

Local police departments are in charge of the investigation of scenes of crimes, accidents, and fires. The larger departments often run their own laboratories, staffed

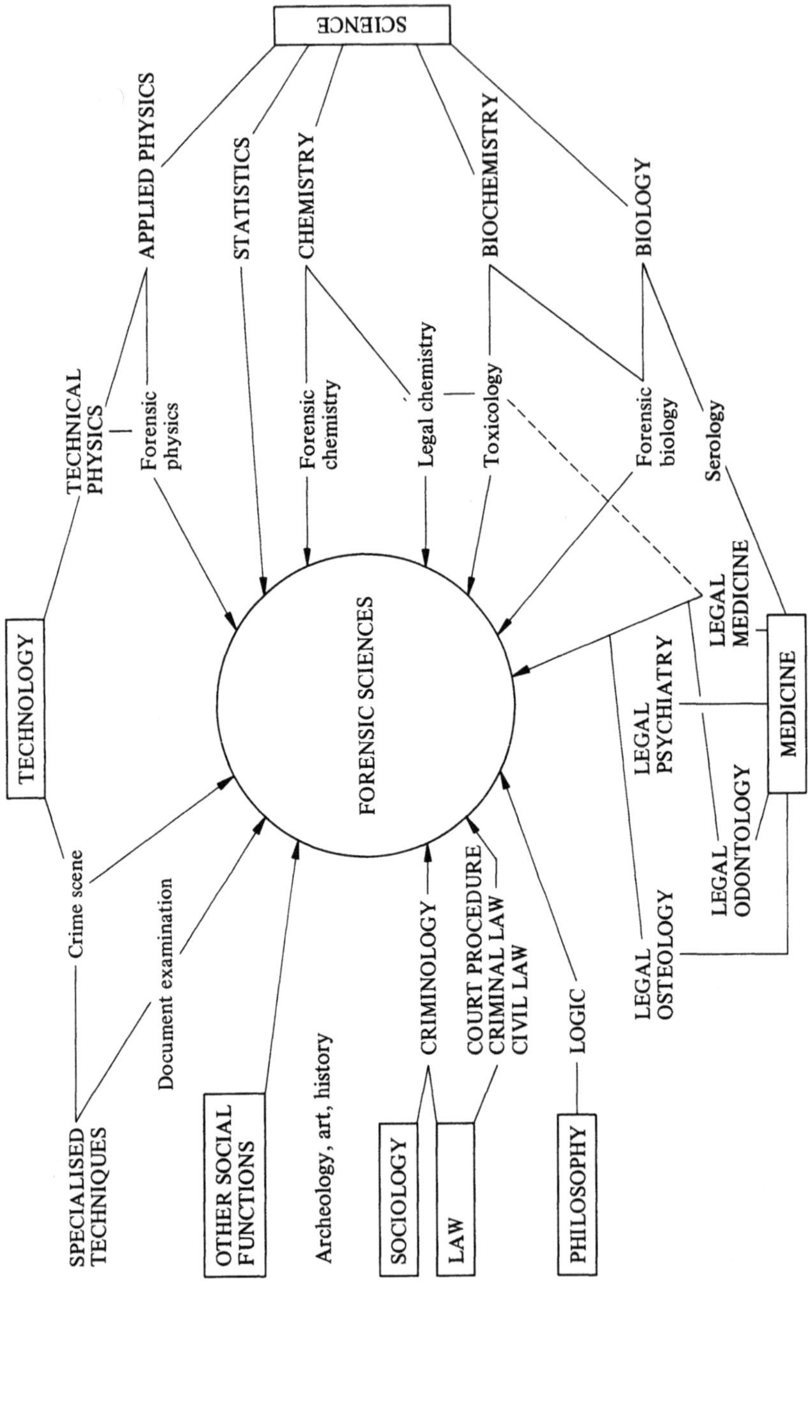

Fig. A.2. Branches and "twigs" of human endeavour connected to Forensic Sciences

either by specially trained police officers alone, or by a team of civilians and sworn personnel. Such laboratories' tasks sometimes overlap with the civilian staffed "proper" forensic science laboratories.

The division of forensic case work between the medical, the natural science, and the police laboratories in Sweden is schematically presented in Fig. A. 3.

In other parts of the Western World, but also including countries like India and Japan, the division of tasks is shifted somewhat. Forensic toxicology and serology, in particular, are more often parts of the activities of the forensic science laboratories.

Tables A. 2 (p. 10–13) and 3 (p. 13) give overviews of the laboratories in the major part of the world (China and the Soviet Union are notably absent). It is easily seen that, in continental Europe, the medical forensic institutes play a much greater role than the forensic science laboratories. In the United States, the situation is reversed. The authors realize of course that a mere counting of numbers cannot accurately reflect the actual situation; they know, however, from personal experience that the overall picture is correct.

Within the lego-medical and the forensic science laboratories, two main approaches have been chosen for meeting the increasing specialization in the field of forensic science, namely the "generalist" and the "specialist" approach. Traditionally, the generalist had been able to penetrate all of the subdisciplines of his field by constant self-studies and continuous application of the acquired

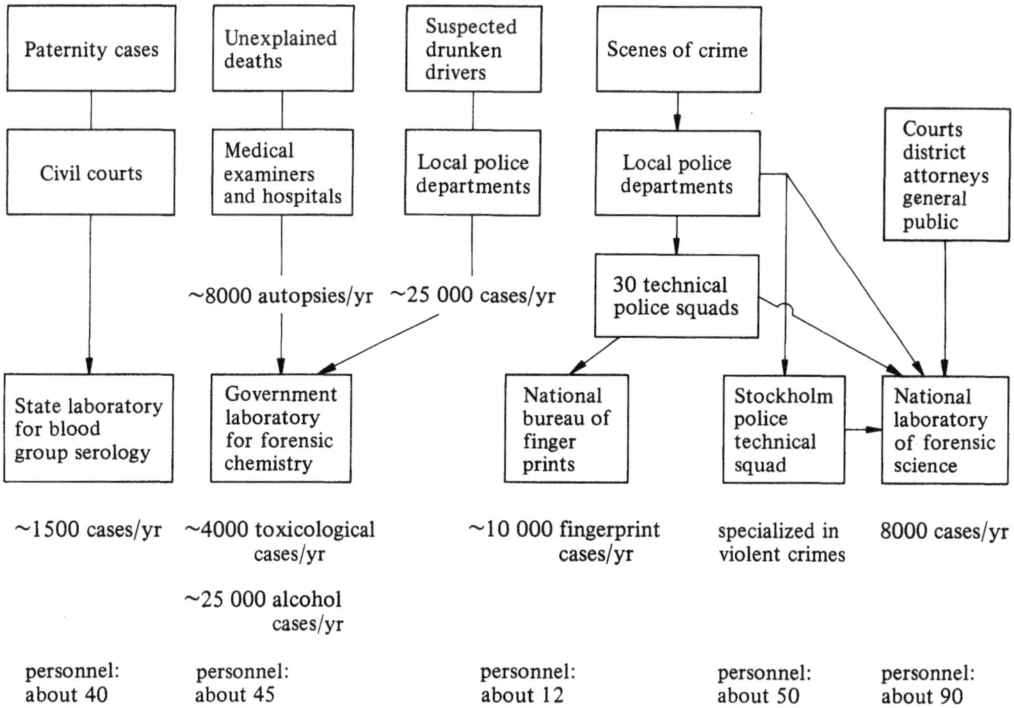

Fig. A. 3. The distribution of forensic casework among different agencies in Sweden

knowledge to actual case work. During the last two decades or so, however, the development in many fields of forensic science has been so rapid that it is now hardly possible for either a legal pathologist or a criminalist to keep abreast of his whole field. The most explosive development has taken place in three topics, i.e. toxicology, serology and chemical analysis.

Thus, compromises must be accepted. These compromises depend, inter alia on the tasks to be performed by the laboratory and on the size of the organization. Some aspects of this situation will be discussed in Chap. C. V of this volume.

As can be seen from Fig. A. 2, chemistry plays a great part in forensic science. Chemical methods are needed and used in forensic toxicology and in various fields of criminalistics. In this treatise, various regions of interest to the criminalist are discussed, mostly, but not exclusively from the point of view of analytical chemistry.

In Part C, some general aspects of forensic science, including criminalistics as a whole, are discussed.

Table A. 2. The distribution of major laboratories and private practitioners in forensic science in the world; acc. to the "World List of Forensic Science Laboratories, 3rd edition, The Forensic Science Society, Harrogate, N-Yorks., 1977

Africa (37)			
Algeria	1	Rhodesia	1
Ethiopia	1	Sierra Leone	1
Ghana	2	South Africa	9
Kenya	1	Sudan	3
Libya	1	Tanzania	1
Malagasy	1	Uganda	1
Malawi	1	United Arab Rep	3
Mauritius	1	Zaire	1
Nigeria	3	Zambia	5
			37
Middle East (14)			
Iran	2	Kuwait	1
Irak	3	Lebanon	2
Israel	3	Saudi Arabia	2
Jordan	1		
			14

Table A. 2 (continued)

Europe (386)			
Austria	9	Malta	1
Belgium	10	The Netherlands	6
Bulgaria	3	Norway	6
Cyprus	2	Poland	5
Czechoslovakia	7	Portugal	4
Denmark	8	Romania	3
Eire	4	Spain	11
Finland	4	Sweden	13
France	31	Switzerland	11
East Germany	9	Turkey	5
West Germany	33	England	130
Greece	3	Wales	8
Hungary	6	Channel Islands	1
Iceland	1	Northern Ireland	3
Italy	23	Scotland	20
Luxemburg	1	Yugoslavia	5
			386

North America I (212)

		Survey of criminalistics laboratories 1968[a]			Survey of criminalistics laboratories 1968[a]
Canada	31	(4)	D.C.	10	(2)
Mexico	5	–	Florida	18	(3)
Alabama	13	(2)	Georgia	9	(1)
Alaska	4	–	Guam	1	–
Arizona	9	–	Hawaii	2	(1)
Arkansas	2	(1)	Idaho	2	–
California	64	(21)	Illinois	19	(3)
Colorado	5	(1)	Indiana	7	(2)
Connecticutt	5	(2)	Iowa	4	(1)
Delaware	2	–			
					212

[a] According to a study report "John Jay College National Survey". Office of Law Enforcement Assistance (LEAA) project 013. US Department of Justice, 1968

Table A. 2 (continued)

North America II (158)

		Survey of criminalistics laboratories 1968[a]			Survey of criminalistics laboratories 1968[a]
Kansas	6	(1)	Missouri	10	(3)
Kentucky	5	(1)	Montana	2	–
Louisiana	8	(2)	Nebraska	4	–
Maine	3	–	Nevada	4	(1)
Maryland	10	(1)	New Hampshire	2	(1)
Massachusetts	8	(3)	New Jersey	10	(1)
Michigan	32	(6)	New Mexico	4	–
Minnesota	4	(3)	New York	36	(9)
Mississippi	3	(1)	North Carolina	7	(2)
					158

North America III (146)

		Survey of criminalistics laboratories 1968[a]			Survey of criminalistics laboratories 1968[a]
North Dakota	2	–	Tennessee	12	(1)
Ohio	21	(7)	Texas	28	(4)
Oklahoma	9	(3)	Utah	4	–
Oregon	8	(1)	Vermont	3	–
Pennsylvania	16	(3)	Virginia	10	(4)
Puerto Rico	4	(1)	Washington	9	(1)
Rhode Island	4	(2)	West Virginia	1	(1)
South Carolina	3	–	Wisconsin	8	(2)
South Dakota	1	(1)	Wyoming	3	–
					146
			North America total		516

Asia (97)

Burma	2	Malaysia	6
Hong Kong	3	Phillipines	2
India	30	Singapore	2
Indonesia	4	Sri Lanka	2
Japan	38	Taiwan	1
Korea	3	Thailand	4
			97

Australia and New Zealand (62)

NSW and A. C. T.	19	Tasmania	5
Northern Territory	2	Victoria	9
Queensland	4	Western Australia	4
South Australia	13	New Zealand	6
			62

Table A. 2 (continued)

South and Middle America (37)			
Argentina	4	Dominican Rep.	1
Bermuda	1	Ecuador	2
Brazil	10	Jamaica	3
Chile	2	Peru	3
Colombia	2	Trinidad	1
Costa Rica	1	Uruguay	1
Cuba	1	Venezuela	5
			37

Africa	37	Middle East	14
Asia	97	North America	516
Australia and New Zealand	62	South and Middle America	37
Europe	386		
			1149

Table A. 3. The distribution of forensic science laboratories according to the "World List of Forensic Science Laboratories" 3rd ed. (1977), published by the Forensic Science Society

C: Laboratories with predominantly criminalistic case work
M: Laboratories handling predominantly forensic cases in medicine, odontology, veterinary science and toxicology
O: Other laboratories and individuals, mainly in the consultant capacity

Geographical area	C	M	O
Africa	16	15	4
North America	287	166	81
Middle and South America	17	19	4
Asia	46	61	4
Australia and Oceania	25	15	26
Europe	80	217	80
Total	471	493	199
USA alone	273	146	76
United Kingdom alone	33	53	73
Europe minus UK	47	164	7
India	24	5	0
Japan	3	35	0

B. The State of the Art

I. Narcotics and Dangerous Drugs

1. Introduction

a. General Remarks on Drugs of Abuse

Man has used extracts and fumes from plants as stimulants for thousands of years and has also discovered the intoxicating effect of ethyl alcohol. This use (and abuse) is continuing also today, but a new situation has developed during the last 100 years or so due to the rapid development of synthetic organic chemistry. Synthetic drugs have created a much more serious scenario for drug abuse than ever before in human history. This has led to national and international efforts to stop or at least to limit this abuse and its damaging effect on individuals and indeed on major parts of human society.

b. The Abuse of Alcohol and of Traditional Drugs Obtained from Plants

The use of alcohol (ethanol) as an intoxicating agent has been accepted in many cultures for thousands of years. Significant exceptions are cultures strongly dominated by religions such as the Islam and Judaism, where the use of alcohol is forbidden in principle. While it is true that only a minor part of alcohol inbibers become psychologically and finally also physically dependent on this drug, the existence of many millions of alcoholics constitutes a major problem. However, no international agreement against the use of alcohol is likely ever to be reached.

In this book we obviously cannot cover the abuse of all major plant constituents in the world but we shall list some of the major examples of this kind.

One widely used drug is of course opium from the opium poppy (*Papaver somniferum* L.) which originally was mainly abused in the Far East but is now cultivated in vast regions of the tropical and subtropical countries.

Hemp (*Cannabis sativa* L.) is the other widespread plant of abuse. In countries where alcohol is banned for religious or other traditional reasons, cannabis is often abused as a kind of substitute.

The abuse of other natural drugs has been more limited to certain geographical regions. In the northwest of South America, indian tribes have been chewing the leaves of the coca bush (*Erythroxylon coca* L.) for probably thousands of years. The active principle, cocaine, is a relative to atropine as shown in Fig. B. 1.

In East Africa and southern Arabia, parts of the evergreen shrub khat (*Catha edulis*) are chewed for obtaining euphoria. That contains ψ-norephredine (Fig. B. 3, XV) as a stimulant. It may be noted here that ψ-norephedine is not taken up as an illicit drug on the presently accepted international lists.

In some regions of Mexico, psilocybe mushrooms (*Psilocybe mexicana*) and mescaline cactus (*Lopophora williamsii*) have been used as intoxicants, sometimes in connection with ritual dances. The active principles are psilocin (Fig. B.5, XXIII) and mescaline (Fig. B. 4, XIX) respectively.

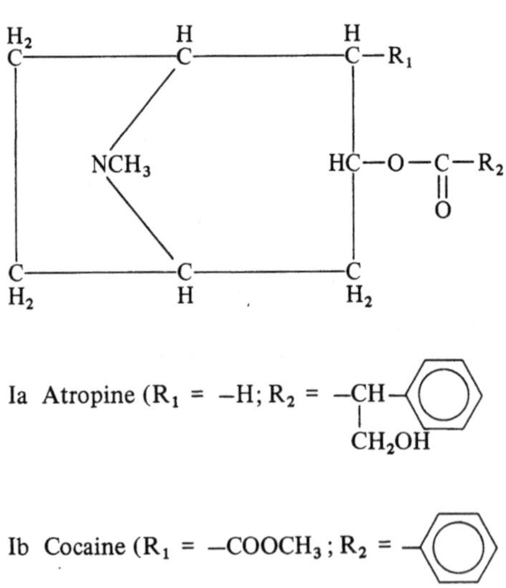

Fig. B. 1

In some instances, the thornapple or devil's apple (Datura stramonium) is abused due to its content of atropine (Fig. B. 1, Ia).

The seeds of the morning glory (Ipomoea purpurea and tricolor) as well as ergot (the dried sclerotia of the parasitic fungus Claviceps purpurea, growing on rye) contain lysergic acid which, in high doses, has a hallucinogenic effect.

On the bottom of the list, so to speak, are tobacco (containing not only nicotine but also carcinogenic aromatic compounds etc.), tea and coffee (containing in addition to caffeine also tannic acids etc.), maté and many other plant extracts of similar nature.

c. The Introduction of Semi-synthetic and Synthetic Drugs

Pure morphine was isolated from opium already in 1805 by the German pharmacist Friedrich Sertürner. In 1855, the injection syringe was introduced. The combination morphine/syringe was then intensively used in the civil war in the United States (1861–65) and in the Franco-German war (1870–71). The use became widespread and led soon to the abuse of morphine, mainly among medical personnel.

In 1874, diacetylmorphine (Fig. B. 2, VI) was synthesized by C. R. Wright in London. The German manufacturer Friedrich Bayer began to sell diacetylmorphine in 1898 under the name of "Heroin".

The name is derived from "heroic" (medicamentum heroicum) which is supposed to indicate a dramatically acting drug. It was introduced as an antitussic and analgesic drug and was even recommended as an antidote to morphine (!). It is interesting to note at this point that, according to Moore and Klein [11] the acetylation of morphine (Fig. B. 2, II) starts at the 3-position (yielding 3-monoacetylmorphine), and that hydrolysis of heroin also begins at the 3-position, yielding 6-monoacetylmorphine (Fig. B. 2, V).

The next drugs of widespread importance, the phenylalkylamines, were obtained by total synthesis (a large number of synthetic pathways are known today) and were claimed to be effective against obesity by "spoiling" the appetite. It soon became known, however, that some phenylalkylamines at the same time had a stimulating effect on the central nervous system. This property was e. g. used by students working for their examinations. Later, the stimulating effect was used by the armed forces in some countries. The German tank divisions invading Belgium and France in 1940 e. g. were given ample doses of methamphetamine to "keep them going" for several days in succession. After the Second World War, a major epidemic of amphetamine abuse hit Japan. By a concentrated effort of the police, the courts, and by educational measures, this epidemic was brought under control after several difficult years. A similar situation developed in the 1960's in Scandinavia and is only slowly ebbing out.

Other anti-obesity drugs were introduced into the official drug market, such as methyl phenidate (ritalin) and phenmetrazine, always in the hope of finding an effective appetite regulator without stimulating side effects and without habit-forming properties. At this point it may be pointed out that a variety of these CNS-stimulating amines are structurally related by sharing a phenethylamine moiety as illustrated in Fig. B. 3.

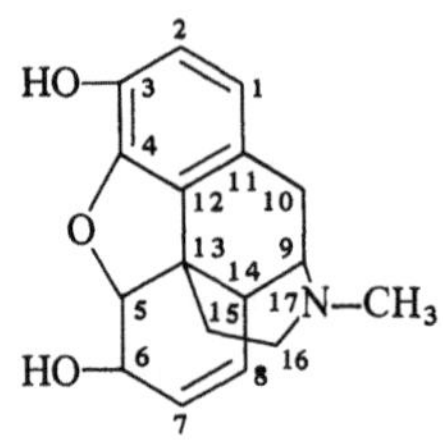

II Morphine

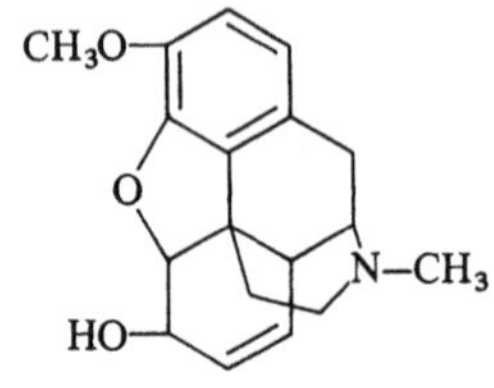

III Codeine

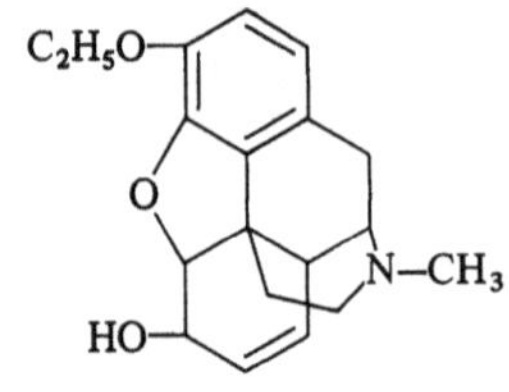

IV Ethylmorphine

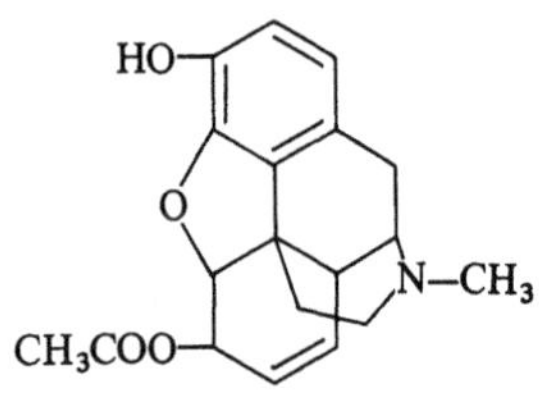

V 6–Monoacetylmorphine

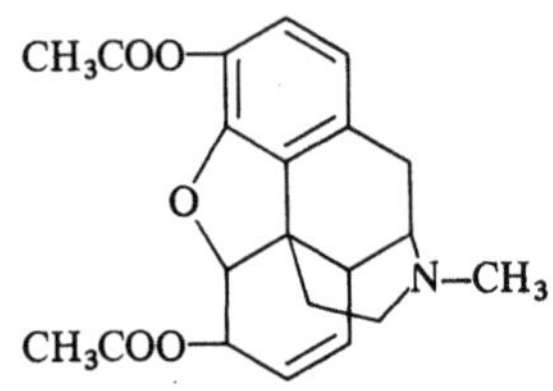

VI Diacetylmorphine (Heroin)

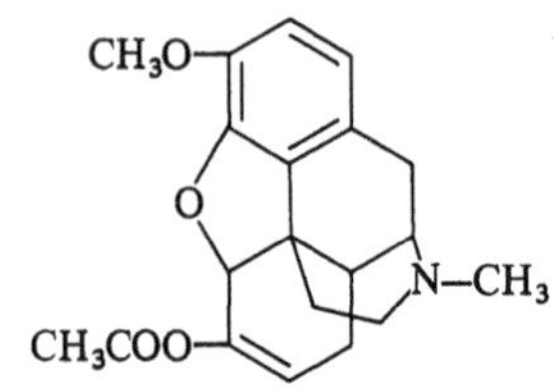

VII Thebacodone (Acetylcodeine)

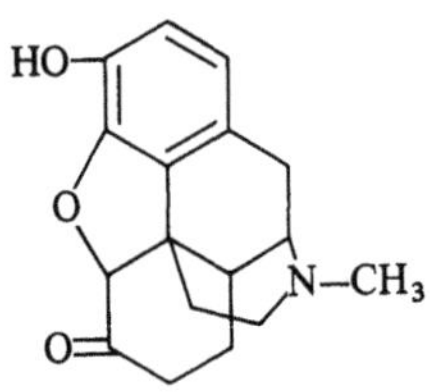

VIII Hydromorphone (Dilaudid)

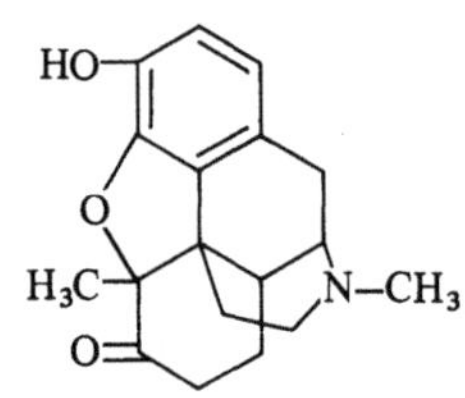

IX Methopone

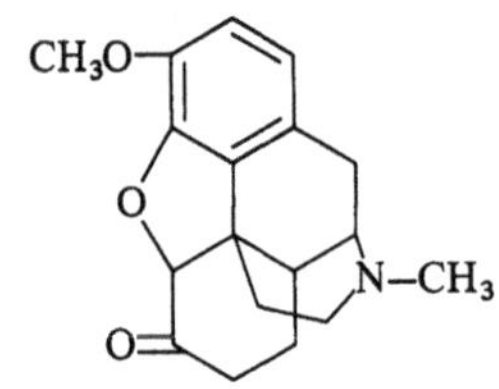

X Hydrocone (Dicodid)

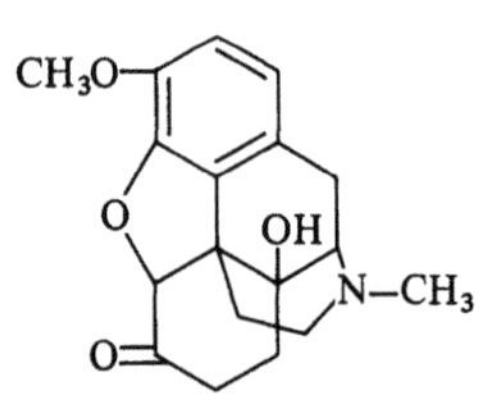

XI Hydroxycodone (Eukodal)

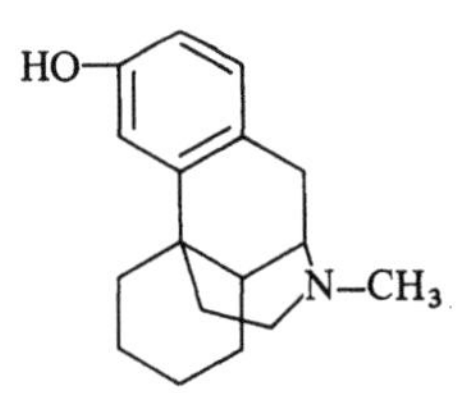

XII Levorphanole (Dromorane)

Fig. B. 2

The introduction of methoxy groups into the aromatic ring, especially in the m-position, introduces a strong hallucinogenic effect. The natural representative of this type of hallucinogen is mescaline (Fig. B. 4, XIX), but a number of drugs with similar effects have been obtained synthetically. Some of these are shown in Fig. B. 4, XX–XXII.

A quite different class of hallucinogenic drugs appears in the psilocybe mushroom, and also in this case, synthetic drugs with similar properties have been manufactured as shown in Fig. B. 5.

The Swiss pharmaceutical firm Sandoz was conducting intensive research on the ergot drugs occurring in spurred rye (cf. above). One of their research chemists,

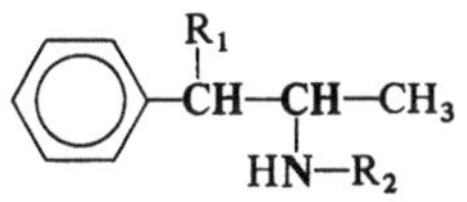

XIII Amphetamine (R_1 = –H; R_2 = –H)
XIV Methamphetamine (R_1 = –H; R_2 = $-CH_3$)
XV Norephedrine
and ψ–norephedrine (R_1 = –OH; R_2 = –H)

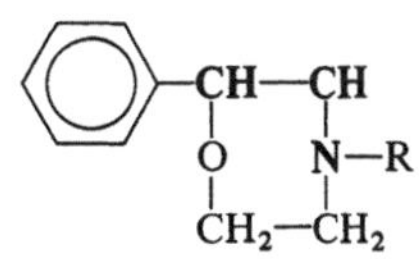

XVI Phenmetrazine (R = –H)
XVII Phendimetrazine (R = $-CH_3$)

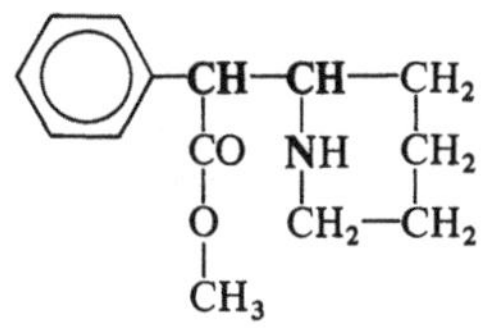

XVIII Methylphenidate

Fig. B. 3

A. Hofmann, while synthesizing a derivative of lysergic acid in 1943, suddenly began to suffer from severe hallucinations [16]. The substance was identified as lysergic acid diethylamide (Fig. B. 6, XXVII) and it was found that doses as small as 50 µg could lead to strong hallucinations. Its action is similar to that of mescaline (XIX) but it is about 10 000 times more potent. The effect usually lasts for 6–9 hours but can last much longer in some individuals. The drug was actually distributed under the name of lysergide or LSD-25 for experimental purposes in psychotherapy but was soon withdrawn.

The next decades saw the syntheses of numerous other synthetic drugs and of analogs and homologs of known drugs. Some of these were introduced into the

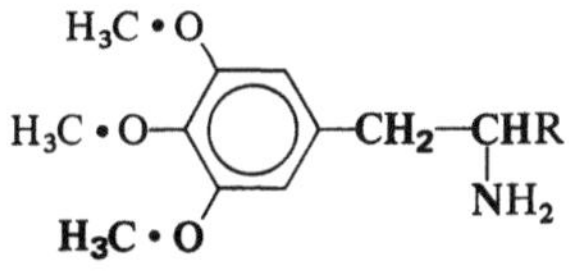

XIX Mescaline (R = H)
XX 3, 4, 5–Trimethoxy–amphetamine (TMA) (R = $-CH_3$)

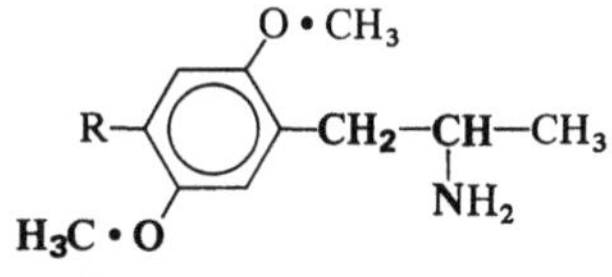

XXI 2, 5–Dimethoxyamphetamine (DMA) (R = –H)
XXII 2, 5–Dimethoxy–4–methylamphetamine (STP, DOM) (R = $-CH_3$)

Fig. B. 4

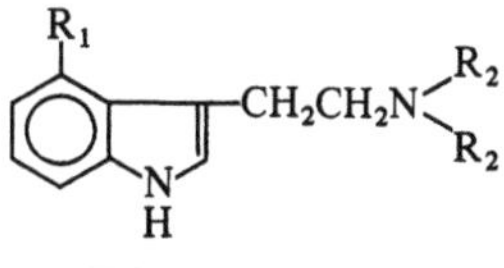

XXIII Psilocin (R_1 = –OH; R_2 = $-CH_3$)
XXIV Dimethyltryptamine, DMT (R_1 = –H; R_2 = $-CH_3$)
XXV Diethyltryptamine, DET (R_1 = –H; R_2 = $-C_2H_5$)

Fig. B. 5

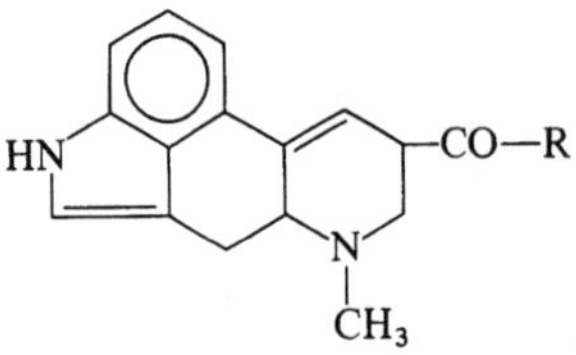

XXVI Lysergic acid (R = –OH)
XXVII Lysergic acid diethylamide,
lysergide, LSD–25 (R = $-N(C_2H_5)_2$) **Fig. B. 6**

market by established pharmaceutical firms for therapeutical purposes. Thus, dextromoramide (Palfium) was hoped to be a safer replacement for morphine (Fig. B. 7, XXVIII). Unfortunately, the users also became dependent on this drug.

Synthetic drugs were – and are – also manufactured in clandestine laboratories. Either illegal drugs (listed as prohibited substances) are made, or derivatives of such drugs, not found on these lists are synthesized. In this way, the race between illicit manufacturers and the authorities is constantly going on.

Only one of the numerous drugs of this kind will be mentioned here because of the alarming increase of its abuse. In 1958, a new anaesthetic was patented for use in veterinary medicine. It got the trade name of Sernyl and is chemically 1-(1-phenylcyclohexyl) piperidine or phencyclidine (PCP). Its structure is shown in Fig. B. 8, XXIX. It became soon evident that this compound had strong hallucinogenic properties and began to be abused sporadically under the slang names of "hog" or "angel dust". This endemic use suddenly exploded into a major epidemic of heavy abuse from about 1977 or 1978 on, predominantly in the United States. By 1980, the numberof PCP-addicts in the USA alone has been estimated to be several hundred thousand. In some individuals, the abuse of the drug leads to acts of extreme violence, and numerous deaths are reported from many parts of the United States.

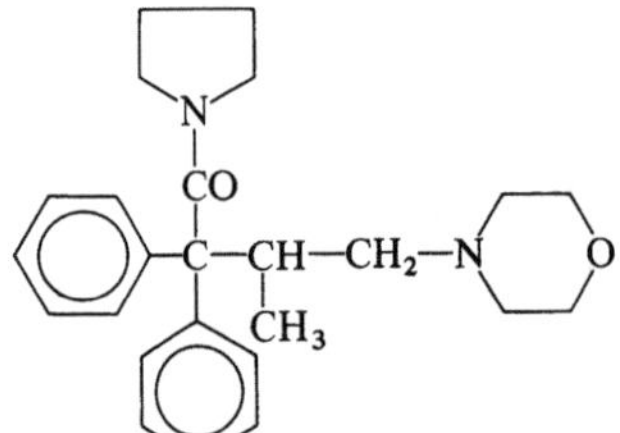

XXVIII Dextromoramide, 4–[2–methyl–4–oxo–3,3–
diphenyl–4–(1–pyrrolidinyl)butyl]morpholine **Fig. B. 7**

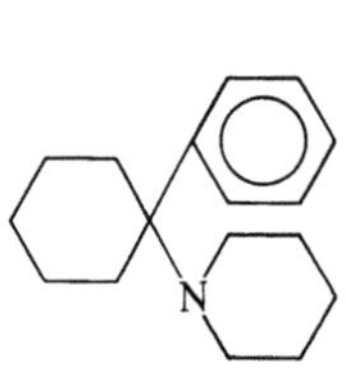

XXIX Phencyclidine (PCP),
1–(1–phenylcyclohexyl)piperidine **Fig. B. 8**

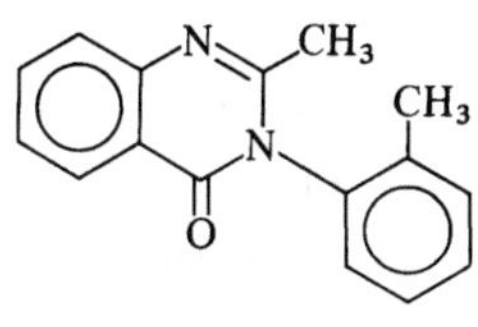

Fig. B. 9 XXX Methaqualone
3, 4–Dihydro–2–methyl–4–oxo–3–O–tolylquinazoline

In the 1950's and 60's it was realized that a number of therapeutic drugs could lead to dependence when taken for prolonged periods of time. The dependence could be psychological or even physical. As a consequence, many classes of these drugs were put on the list of "dangerous drugs". The majority of these compounds never became part of the international illegal drug market. One of the notable exceptions is methaqualone (Fig. B.9, XXX) which is widely abused due to organized drug peddling.

Among the classes of compounds regarded as "dangerous drugs" in many countries one can find e. g. barbituric acid derivatives (Fig. B. 10, with the general formula XXXI), and benzodiazepines (Fig. B. 10, general formula XXXII).

From the chemical point of view it is interesting to note that the vast majority of narcotics and dangerous drugs contain nitrogen. Thus, out of approximately 140 drugs on the Swedish list, only six lack nitrogen. These are (cf. Fig. B. 11):

1. The cannabinoids (cf. Sect. B. I. 5) (XXXVIII),
2. Chloral hydrate (XXXIII),
3. Chloralodol (XXXIV),

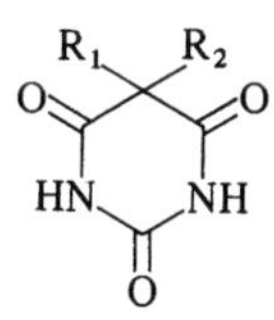

XXXI General formula for barbituric acic derivatives.
R_1 and R_2 can be $-H, -CH_3, -C_2H_5, -C_3H_7, -CH(CH_3)_2,$
$-CH_2-CH_2-CH(CH_3)_2, -CH_2-CH{=}CH_2, -CH_2-CBr{=}CH_2,$
–phenyl etc.
Methyl–barbiturates have a methyl group at one of the ring nitrogens.
In thiobarbiturates, the ketogroup between the two ring nitrogens is replaced by =C=S

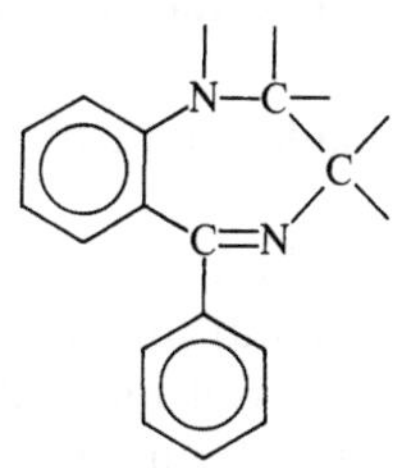

Fig. B. 10 XXXII General formula for benzodiazepines

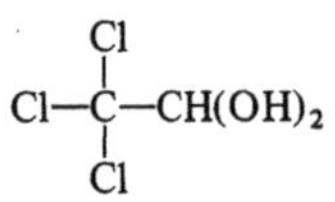

XXXIII Chloral hydrate, trichloro-acetaldehyde hydrate

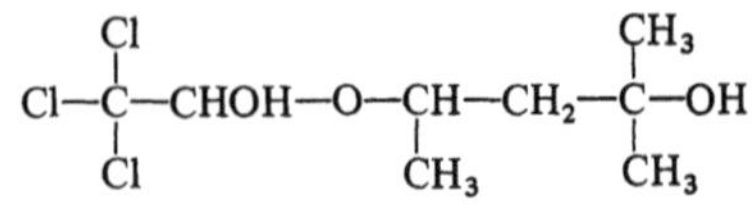

XXXIV Chloralodol

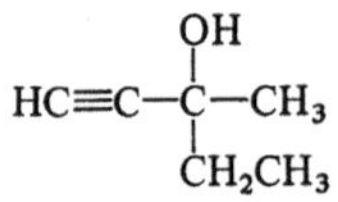

XXXV 3–Methyl–1–pentyn–3–ol

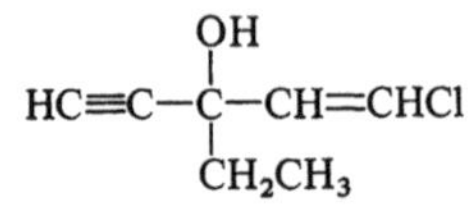

XXXVI Ethchlorvynol, 1–Chloro–3–ethyl–1–penten–4–yn–3–ol

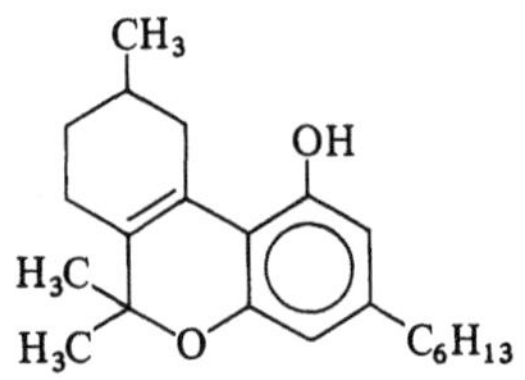

XXXVII Parahexyl, 1–Hydroxy–3–n–hexyl–6, 6, 9–trimethyl–7, 8, 9, 10–tetrahydro–6–dibenzopyran

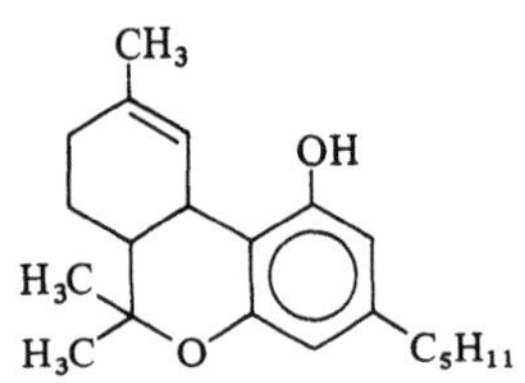

XXXVIII Δ^9–Tetrahydrocannabinol, THC

Fig. B. 11

4. Methylpentynol (XXXV),
5. Ethchlorvynol (XXXVI) and
6. Parahexyl (XXXVII).

d. The Origin and the Distribution of Narcotics and Illegal Drugs

In this book, only a few remarks can be made about the pathways along which illegal drugs reach their consumers (victims).

Opium and its refined product, heroin, originates predominantly in the Far East (the "Golden triangle": Burma, Laos, Thailand) and the Middle East (notably Turkey). From there, heroin is transported to Europe and North America.

Cocaine is mainly produced in Northwestern South America, notably Colombia and her neighboring states.

Cannabis grows practically everywhere on earth, but the genetic varieties giving high yields of cannabinoids are raised in tropical and subtropical countries. Three regions dominate the market: The Far East (Burma, Bangladesh, Pakistan), the Mediterranian region (Lebanon, Marocco, Algeria), and Mexico.

Synthetic drugs are often produced within the countries of sale or in their vicinities, thus avoiding border crossings. The synthesis of the amphetamines is relatively simple, and the majority of clandestine laboratories raided every year produce this kind of stimulant. The manufacture of phencyclidine is somewhat more complicated but requires no extraordinary skills. A much more difficult task is the synthesis of LSD; only a few laboratories have been found which produced this compound. The considerable risks for the manufacturer due to the potency of LSD are offset by the high profit since one "trip" usually only requires about 0.1 mg of the drug.

Considerable progress has been made in recent years in establishing the synthetic pathway of illicitly produced drugs and of detecting the common origin of peddled samples. This will be discussed in Sect. B. I. 4.

e. Some Remarks on Legal Actions Against the Narcotics Trade

Seventy-three countries of the United Nations met to discuss the international control of narcotic drugs in 1958. The result of these discussions was the "Single Convention on Narcotic Drugs, 1961". It contained a list of 96 substances grouped into four "schedules". The convention was eventually ratified by most countries in the world. The "Single Convention" was amended in 1972 under the name of "Single Convention on Narcotic Drugs, 1961 as amended by the 1972 Protocol Amending the Single Convention on Narcotic Drugs, 1961". Its latest edition was published by the United Nations in New York in 1977.

The national legislations often take up additional compounds not listed in the amendment. The Swedish list e.g. takes up at present 143 items, grouped into 5 schedules.

The seizures of narcotic drugs by the customs and police authorities of the world (except for the USSR, China, and some other countries) in 1978 are briefly listed in Table B. 1 (p. 50).

2. General Remarks on Drug Analysis

Three main objectives of forensic chemistry are identification, quantification, and comparison. Especially, this holds for the field of drug offences where the sentences heavily depend on the analytical results obtained in the laboratory. In this field, the forensic chemist also faces problems of organizational nature, that is, the investigation of large numbers of specimens, variations between pure drugs and multicomponent drug mixtures on the market and finally a varying case load depending on changes in the drug market or police resources available for drug intelligence and raiding. The problem of the varying case load is generally discussed in Sect. C.V.8. The other problems mentioned above connected with drug analysis will be discussed in the following part.

3. Identification

The materials sent to the drug section of a forensic laboratory may be divided into two classes. The first class comprises industrially manufactured legal preparations, usually tablets or capsules, suspected to be illegally possessed or misused. The second class contains materials which lack characteristic external properties, usually white powders and sometimes solutions. The most dangerous drugs generally belong to this class as well as "home-made" tablets and doped legal preparations.

The simplest drug seizure consists of a single *item* such as a paper bag containing white powder or a tablet. The situation is more complicated in the case of a number of items, displaying similar external properties. Such a group of items will henceforth be called *exhibit*. The most general case implies several different exhibits. Usually, the items of an exhibit have the same or similar chemical composition and roughly the same amount. The exhibit may then be said to be uniform chemically and with respect to quantity. Although this uniformity cannot be taken for granted, it seems quite obvious that apparently similar items containing different drug preparations would have a limited value on the illicit market. Exceptions from this "rule" show up on the street level but are very rare in cases of big exhibits in the earlier levels of the distribution pyramid. The uniformity of exhibits is a very important resource-saving factor in practical drug casework. The reason is that one typical analysis will reveal the chemical composition of many items.

The first step in the investigation of a drug seizure is grouping the items into exhibits displaying the same external properties and roughly the same amount. The exhibits are then weighed and their chemical uniformity is established. In cases of very high number of items, random sampling is done for reasons of expediency. For the first class of exhibits mentioned above which display characteristic external properties and match the laboratory's reference collection, the uniformity may be established by careful visual inspection at low magnification and/or in ultraviolet light. For the establishment of the uniformity of the second class of exhibits, especially when characteristic external properties are lacking, it is advisable to use a combination of visual inspection and a chemical method. In this way, a valuable first hint concerning the composition of the items is achieved as well. Since this method is used for the analysis of all the items, it should be fast, simple and cheap. The best technique for this purpose is thin-layer chromatography, although it should be pointed out that this method will not detect small quantitative differences in the compositions of the various items. If different compositions are detected within a group, a further division is subsequently made.

The final qualitative analysis should be representative of the whole exhibit. To this end two methods can be used, i. e. (i) random selection of a sufficient number of items and (ii) making a composite or pool from parts of all items of the exhibit. Random selection is recommended for the first class of exhibits which generally show good uniformity. As far as the second class of exhibits is concerned, pooling, although somewhat more laborious, is the method of choice since it is more representative of the exhibit. This is especially important in quantitative analysis which will be discussed in Sect. B. I. 6.

The final step in the identification of drugs is the qualitative analysis of the random samples and composites mentioned above. The identification of legal preparations which match the reference collection of the laboratory is generally supported by characteristic external properties. A rather simple analytical method will therefore suffice in these cases. The choice of this method depends on the composition of the samples. In some cases infrared or ultraviolet spectrophotometry may be used directly, whereas sometimes the excipients will have to be removed by solvent extraction of the characteristic constituents.

The aim of the analysis of samples which lack recognizable external properties is simply the general identification or organic compounds. The analytical methods used in organic chemistry may be divided into two types according to their selectivity, although there is no sharp border-line between them. The highly selective methods associate the unknown compound with a very small group of knowns (standards). Typical examples are infrared spectrophotometry (IR), combined gas chromatography – mass spectrometry (GCMS) and the determination of mixed melting points. The methods of medium to low selectivity may be called indicative and consequently associate the unknown with a larger group of knowns. The chromatographic methods (GC, TLC, HPLC) generally belong to this group. The selectivity of ultraviolet spectrophotometry is strongly structure-dependent, but as far as drug analysis is concerned, it is generally limited.

IR is a fast and extremely selective method. The cost of the equipment is moderate, around $ 20.000. The method is, however, not suited for drug mixtures because the spectra will depend on the quantitative composition. Furthermore, mixtures often give broad absorption peaks which reduces selectivity. This limitation is the main serious drawback of IR in routine drug analysis. Moderate sensitivity is a limitation in some cases. It should be pointed out, however, that some compounds, e. g. morphine and heroin are conveniently isolated from common diluents such as sugars and caffeine simply by precipitation in ammoniacal water solution.

GCMS is the drug analyst's dream method. It is very selective, rather fast and works well with drug mixtures. Because of its extreme sensitivity, it may also be used in drug trace analysis. In gas chromatography as well as in organic mass spectrometry, the analytes are processed in gaseous form. This means that certain requirements concerning volatility and thermal stability must be met. These requirements are much more stringent in gas chromatography. Consequently, some compounds which do not pass the gas chromatograph properly may be analyzed by direct inlet probe mass spectrometry, provided that they are in pure form or purified. A typical example of this is LSD. The main drawback of GCMS is the cost of the equipment, around $ 200.000, roughly equally divided between the spectrometer and the computer system which is quite necessary for the rational use of the method. There are at least three good reasons for the happy marriage between the MS and the GC. First, the fractions are delivered by the GC in gaseous form, as required by the MS. Second, the sensitivity of the MS happens to be well suited for the capacity of the GC.Third, the MS is fast enough to analyze the GC fractions "on line", that is, during the short time required for the fraction to leave the end of the GC column.

If a pure substance (standard) with a well-defined melting point is mixed with any other substance, its melting point will generally be lowered. This means that if the mixture and the standard have the same melting points (within error), the identity between the standard and the admixed unknown compound is very strongly indicated. On the other hand, different melting points indicate either non-identity or an impure unknown. The method of mixed melting point determination is very selective, fast and cheap but it is not suited for direct use on impure or mixed drugs. However, it works very well with individual drugs, purified as crystalline precipitates or derivatives.

Conventional GC, i.e. with packed column and a single detector is a method of moderate selectivity. There are two possibilities of improving the selectivity, the use of detector combinations and the introduction of capillary columns. In 1977 Meola [7] and Baker [1] described GC systems with two detectors, connected in parallel to a packed column by an effluent splitter. The detectors chosen were a flame ionization detector (FID) and a nitrogen-phosphorous detector (NPD) operated in the nitrogen mode. The use of an internal standard eliminated small changes in retention times as well as in NPD-responses. The drugs analyzed are characterized by relative retention times as well as relative detector response ratios. In the authors' laboratory a study is in progress for further improvement of the selectivity of such a system by the introduction of wide-application capillary columns. By feeding the detector signals to a computer system, relative retention times and relative peak area ratios are computed and automatically compared with preset values from the analysis of standard mixtures. The tolerances set for this comparison should be small enough to minimize the risk of false identifications (false positives) but on the other hand large enough to avoid frequent false exclusions (false negatives). A gas chromatographic system of the type described above has many advantages, some of them already mentioned. The cost of the equipment seems at present to be roughly $ 30.000. When equipped with an auto-injector, the system may handle sudden increases of the case load resulting from big drug raids.

The chromatographic methods commonly used in qualitative drug analysis are thin-layer chromatography (TLC), gas chromatography (GC) and high pressure liquid chromatography (HPLC). Although these methods are generally not highly selective, they play an important role in combination with the highly selective methods. TLC may be used as a valuable screening method for sorting out "uninteresting" materials such as sugar and aspirin at the beginning of an analytical scheme, for the establishment of uniformity of exhibits and for getting a valuable first hint concerning the possible qualitative composition of these materials.

GC is not as fast and cheap as TLC but has the advantage of higher selectivity which may be further enhanced as described above. The importance of GC in combination with MS has already been discussed. The overall high sensitivity of the flame ionization detector makes GC a valuable tool for drug trace screening. Compounds of high polarity, like carboxylic acids and phenols may be derivatized to meet the volatility requirements. The cost of a gas chromatograph varies roughly between $ 10.000 and 30.000 from the simplest manually controlled instruments to the micro-processor controlled ones with a data acquisition system.

HPLC gives virtually the same chromatographic resolution as GC. Fraction collection may be done very conveniently for further analysis, e.g. by IR or MS.

HPLC has also a wider application range than GC, because only solubility is required but not volatility and thermal stability. The detectors available for HPLC work on differences in UV-absorption, fluorescence or optical refraction between analyte and solvent (mobile phase). Although most drugs have structures which give UV-absorption, the lack of an overall high sensitive detector for HPLC is a limitation of the method. Such a detector will be developed, no doubt, in the near future as well as a HPLC-MS interface. In 1976 Wheals [20] reviewed the forensic applications of HPLC.

The analytical methods generally used in qualitative drug analysis have now been briefly discussed. The question then arises, how many and which methods should be used for drug identification. It should first be pointed out that this question does not only concern analytical chemistry but also legal security. Since the illegal handling of drugs is considered a severe crime in most countries, the analyses must be very reliable, at least for the identification of controlled substances.

An analytical method is related to a certain number, N, of distinguishable experimental results. This number reflects the selectivity of the method. The positive identification of all the individuals of a given population of M compounds is possible if, and only if, the method can discriminate any pair of these individuals. As pointed out by Duer et al. [3] a necessary condition for positive identification then is $N \geq M$. This condition is, however, not sufficient since the experimental data are generally not rectangularly distributed over the sites of N so that some of these are "overoccupied" and some are empty. N may be increased by combining methods to an analytical scheme. Ideally, the N of such a scheme is the product of the N : s of the individual methods. In practice, the ideal N will be reduced due to correlations between the methods. The concept of N serves as a valuable upper estimation of the selectivity of an analytical scheme in many cases.

In order to create a reliable scheme of methods for drug identification, the number of relevant compounds, i. e. M, should be estimated. Duer et al. [3] mention the figure 5.000.000. The N of the scheme should consequently exceed this figure. This may be achieved either by a long scheme of indicative methods or a shorter scheme including selective methods. Some examples of such schemes will be given at the end of this section.

A good measure of the selectivity of a scheme of indicative methods is the discrimination power, DP, which, according to Smalldon and Moffat [15], is defined as the probability that two randomly selected individuals from a given population could be discriminated. This definition is also valid for a particular method. The DP of an indicative method will depend on the distribution of the results over the population, of experimental errors and of M. A rectangular distribution and small errors will enhance DP. The DP of a scheme of k methods will depend on the k individual DP : s and on the correlations between the $k(k-1)/2$ combined pairs of the methods. The DP of a scheme may be determined experimentally on a group of standard compounds which is representative of the total population. The quality of this representation is of course an uncertainty. On the other hand, a greater part of individual compounds in the "total" population must be considered as extremely rare in the drug case load. Perfect identifications or analyses without uncertainty cannot be done in practice. The aim of the forensic analyst is the proof beyond reasonable doubt. According to the considerations above it seems to be reasonable and justified to use a scheme of one selective method and one or two suitable indicative methods (including TLC for screening and the establishment of uniformity of exhibits). As far as schemes of indicative methods are concerned, the required number of methods is difficult to specify but the DP of such schemes may be determined experimentally as mentioned above. For details, the reader is referred to a series of papers by Moffat et al. [8–10].

In the final choice of an analytical scheme for routine drug analysis, the management must consider not only the technical problems but also factors such as changes between single- and multicomponent drugs in the case load, legislation, economy and staff saving by automation.

This part of Chapter B.I will be concluded by giving some examples of analytical schemes used in the author's laboratory for the qualitative analysis of some different types of drug preparations. It is assumed that TLC has been used for screening etc.

Single component drugs:
UV + IR
GC + IR (when UV is uncharacteristic)

Mixtures of CNS-stimulants and diluents of medium molecular weight (caffeine, procaine) and/or sugar:
second TLC + FID/NPD-capillary GC
GC + MS (GCMS)

Morphine or heroin diluted with sugar or caffeine:
UV + IR (after precipitation as bases)
HPLC + UV (when coprecipitation occurs)

LSD:
HPLC + MS (direct inlet)
HPLC + UV

4. Chemical Signatures of Drugs

The investigation of relationships in the field of drug analysis is an important task in forensic chemistry for court evidence and drug intelligence. Common sources of drug seizures are typical examples of such relationships. The methods used for this purpose were reviewed by Sanger et al. in 1979 [14] and may be divided into groups according to the following scheme.

A. Physical or "ballistic" methods for comparing items having characteristic external properties, e. g. tablets, capsules and fractured hashish cakes.

B. Chemical methods for the analysis of items lacking characteristic external properties, such as white powders or solutions.
   I. Inorganic chemical methods for the display of trace element profiles.
   II. Organic chemical methods.
      1. Identification of impurities characteristic of the synthetic route (key impurities), which may establish relationships with precursor chemicals.

2. Comparison based on the analysis of the chemical composition for the establishment of common sources.
   a. Comparison based on chemical signatures (displayed impurity profiles).
   b. Comparison based on the analysis of known main constituents.

The discussion in this part will be limited to the establishment of common sources by using the methods under B.II.2. The common source for single-component drugs will generally be a batch resulting from a drug synthesis. In practice, the comparison of single-component drug items may start by identification of the main constituent. The organic impurities may then be separated from the main constituent by solvent extraction and analyzed using a gas chromatographic method involving a dual detector system. The resulting chromatograms are the chemical signatures to be compared.

For drug preparations containing several active constituents or diluents, the solvent-extraction method is often inferior. There are several explanations for this. Firstly, the possible common signature of a particular constituent may be obscured by the presence of other constituents and their impurities. Secondly, false positive, or rather trivial positive, results may occur if the mixed signature is strongly dominated by the impurities originating from a widely available diluent. Finally, since all the impurities in a drug preparation do not stem from the same batch of synthesis, the different impurities in the solvent extract may interact during the analytical procedure. Preliminary findings from experiments in the authors' laboratory support this conjecture. This effect may cause false negative results. Thus, drug preparations can be compared only with respect to their main compositions in many cases. The common source established in this way will not be a batch resulting from a synthesis, but a batch resulting from mixing the individual constituents. Relationships between such batches are generally less interesting from the point of view of drug enforcement than those between single constituents since they pertain to lower levels of the distribution pyramid where dilution and adulteration are generally made. Comparisons of "multiple"-drug preparations with respect to main composition are of "lower quality" for another reason as will be pointed out later. It should be mentioned here, however, that a mechanical isolation of a constituent having a characteristic crystal form or color may sometimes be done at low magnification using a needle, spatula or a suitable "micro-vacuum cleaning" technique.

For the discussion of comparisons according to B.II.2 some basic concepts should be introduced. These comparisons are based on concentration parameters. In analogy to qualitative organic analysis, the discrimination power (DP) of a comparative method depends on the distribution of the parameters over a population of non-connected objects (objects of different origin) as well as correlations between these parameters. The experimental errors in the measurement of these parameters will of course affect the DP. There is an important difference, however, between the comparison of single compounds and of mixtures of compounds. The properties of compounds depend only on natural laws whereas concentration parameters do not. A concentration parameter may show a variation over a population of connected objects (objects of common origin). This is called *intra-variation* of parameters. For example, the main composition of a drug preparation may show a pronounced variation due to the difficulty of preparing homogeneous mechanical mixtures of crystalline powders. This intra-variation implies a considerable uncertainty since it may be very different from case to case. Furthermore, the presumtive common batch is often not available for an experimental determination of the intra-variation.

The intra-variation of impurity profiles of single-component drugs is generally small due to the homogeneity of the reaction mixtures of the syntheses. Such intra-variation may also be caused by chemical instability of the constituents or by contamination. A large intra-variation may cause false negative results in comparative analysis.

The variation of a parameter over a population of non-connected objects is called *inter-variation*. A small inter-variation may cause false positive results in comparative analysis. As far as comparisons of the main compositions of drug preparations are concerned, it should be kept in mind that "traditional" mixtures may occur on the drug market. One example is brown sugar (heroin no. 3) which is a grained mixture of heroin hydrochloride and caffeine in the traditional proportion of about 2 : 3. This proportion therefore gives poor information about common origins of such heroin seizures.

For the comparison of concentration parameters of drugs and for the formulation of the subsequent statements concerning the likelihood of their common or different sources a probabilistic or a "non-probabilistic" approach can be used. In the probabilistic approach, the concentrations of a certain number of preselected constituents are considered. In this selection, those constituents should be preferred which show

(i) small correlations,
(ii) small intra-variations and
(iii) large inter-variations.

On the other hand, the number of constituents should be sufficiently large for minimizing the risk of coincidental agreements between compositions. The final selection may then have to be a compromise where "non-ideal" constituents are accepted. In a paper by Clark [2] the selection of comparison parameters for cocaine is discussed. The probabilistic approach is in a way analogous with the one described in the previous section for the identification of drugs where the concentration parameters for the comparison of compositions correspond to the methods in the analytical schemes for establishing the relationship between the unknowns and the standards. For the probabilistic comparison of drug compositions, background information concerning the inter- and intra-variations of the selected parameters as well as the correlations between them is needed. The problem of formulating statements from this background information the and experimental results is considered in two parts, i. e. similarity and identification [6]. The similarity problem involves the use of a similarity index for calculating the extent of similarity between two compositions. The identification problem concerns the rarity of the considered compositions. The strongest statements would then pertain to cases where rare compositions are in extreme agreement. The statement may be of the "classical" verbal type such as "the evidence is consistent with a common batch" or it may be a numerical "cold figure statement" such as "with regard to the observed similarity, the probability that the items stem from different batches is only 0.002".

In the "non-probabilistic" approach, the comparisons are based on a number of parameters which are not selected on statistical grounds, although obviously "improper" parameters are discarded. For the comparison of single-component drugs, the problem is to find the sufficient number of parameters which can discriminate drug samples from different batches of synthesis, irrespective of possible correlations between these parameters. This number may be estimated by a geometric interpretation. The chemical signatures, e. g. gas chromatograms, are then interpreted as vectors or points in a multidimensional Euclidean space. The peak areas or peak heights will then correspond to the components of the vectors and the number of peaks considered equals the number of dimensions of the space. If the number of parameters is sufficient for the purpose concerned, the points corresponding to signatures of different batches will generally form isolated clusters in the space. In the ideal case, any pair of clusters would be isolated from each other, i. e. the DP of the comparative method would equal unity. This also means that the identification problem mentioned in the paragraph above needs not to be considered and the statements given in such cases will depend only on the observed similarities. The geometric interpretation is quite illustrative – the sizes of the clusters demonstrate intra-variation whereas the distances between the clusters correspond to inter-variation. Several metrics for multidimensional distances have been described. The "classical" squared Euclidean distance has the disadvantage of being too sensitive to differences pertaining to single components. Among other metrics, the "Canberra Metric Similarity Coefficient" described by Thornton [19], is superior to the squared Euclidean distance in this respect. This is important in the comparative analysis of drugs, since the chemical signatures are generally strongly dominated by the peaks of a few main impurities (usually the route-specific key impurities).

The possibility of discriminating between different batches of a given drug synthesized by a given method depends on how strongly the conditions of the synthesis influence the chemical signature. The development of a comparative method based on the chemical signature of a drug should therefore include a study of this influence regarding for example reaction temperatures and reaction times of the different steps of the synthesis. Valuable information concerning the DP of a comparative method is also obtained from case-work experience. Here, a high DP is indicated by a pronounced dominance of cases of clear agreements and of clear disagreements over cases located between these extremes. If a high DP is found in the way described above, the risk of false positive results in the comparisons is negligible. To safegard against the risk of false negative results, the chemical stability of the main constituent under ordinary storage conditions should be studied since instability in this respect may cause considerable intra-variation.

The background information used in the non-probabilistic approach concerns the relation between inter- and intra-variation in the impurity profiles. Thus, the non-probabilistic approach is grounded on a simplified background information compared to the probabilistic approach where the background information concerns the individual components of the impurity profiles and the correlations between them. The resolution of the comparative process will therefore be lower in the non-probabilistic approach and the corresponding statement philosophy consequently coarser. If a series of verbal statements is used for this purpose, the number of statements in this series should reflect the resolution of the comparative process. For example, a series of five statements may be based on the following situations:

1. There is a strong indication that the items belong to a common batch.
2. There is an indication that the items belong to a common batch.
3. The result of the comparison is inconclusive.
4. There is an indication that the items belong to different batches.
5. There is a strong indication that the items belong to different batches.

As far as conclusions from dissimilarities are concerned, intra-variation must be considered as pointed out earlier. Comparative analyses of amphetamine sulfate were started in the authors' laboratory in 1973 [18]. At that time, the chemical stability of this drug under normal storage conditions had been confirmed experimentally but little was known about other causes of intra-variation such as heterogeneity and contamination. This situation has changed with time. Firstly, case-work experience has shown that intra-variation due to heterogeneity may generally be neglected in comparative analyses of single-component drugs such as amphetamine, methamphetamine and phenmetrazine. Secondly, the risk of false negative results due to contamination is drastically decreased by the introduction of highly selective gas chromatography. Under these circumstances, strong conclusions may be drawn from dissimilarities, especially when the chemical signatures show qualitative variations. In the comparison of two signatures, a contamination is indicated if additional peaks appear in *one* signature. These additional peaks may be regarded as a background signature which is characteristic of the contamination in question. One example is the contamination from gasoline, caused by smuggling of drugs in fuel tanks. In a signature recorded by the FID, such a contamination is easily recognized by a typical pattern of additional peaks. Furthermore, it is elegantly eliminated by the NPD which does not respond to hydrocarbons.

The formulations of the statements themselves is a linguistic rather than a scientific problem and such formulations will not be discussed here. The series of statements may be regarded as corresponding to a series of intervals on a scale. The end points of this scale then correspond to categoric statements used in qualitative chemical analysis, e.g. "the sample contained amphetamine".

The aim of the probabilistic approach is to make consistent decisions on a rational basis. For this purpose, the interpretation of the results of the comparisons are based on an "objective" formal probability model and not entirely on the "subjective" experience of the investigator. Nevertheless, forensic laboratories have been reluctant to use and courts have been reluctant to accept formal probability models, mainly because of their limited applicability to individual cases. These matters are discussed in the previously mentioned paper by Thornton [19]. In the comparative analysis of some single-component drugs, experimental results and case work experience have strongly indicated that batches from different sources can be individualized. In such cases, the non-probabilistic approach is well justified since only the similarity problem needs to be considered.

5. Cannabis

Three types of drug preparations from the hemp plant *Cannabis sativa* L. occur on the illicit market, i. e. the herbal type (marihuana, grass, pot), the resin type (hashish, charas) and cannabis extract. The herbal type is often in the form of small green flakes from manicuring the flowering tops of the plant. The resin type preparations have a more or less sticky consistence and occur often as flat cakes or "soles". Cannabis extracts are more or less viscous brown liquids from solvent extraction (usually with alcohol) of hemp plant material. In this part, all of these preparations will be referred to as *cannabis*.

Unlike the synthetic drugs, cannabis is a natural product of a very complex chemical composition. The plant material has characteristic morphological properties which are not always found in the cannabis extract. Furthermore, the active principle of cannabis. Δ^9-trans-tetrahydrocannabinol, does not contain nitrogen. Because of these circumstances, somewhat special methods are needed for the identification, quantification and comparison of cannabis items.

The reliable identification of cannabis is conveniently carried out by microscopy and thin-layer chromatography. The simultaneous presence of microscopic cystolith hairs and mushroom-shaped glandular hairs clearly indicates the presence of hemp plant material (cf. Fig. B. 12, 13). A thin-layer chromatogram visualized by Fast Blue B will then demonstrate the presence of the characteristic phenolic constituents (cannabinoids) of cannabis resin. As pointed out in Sect. B. I. 3, TLC is an indicative method in the qualitative analysis of single chemical compounds. In this application, however, a complex but very characteristic composition is demonstrated by the positions, the intensities and the hues of the spots. A properly filtered extract will not show the morphological characteristics. In such cases, a temperature programmed gas chromatographic analysis will strongly indicate the presence of cannabis resin (cf. Fig. B. 14). The reason for this is the complex but

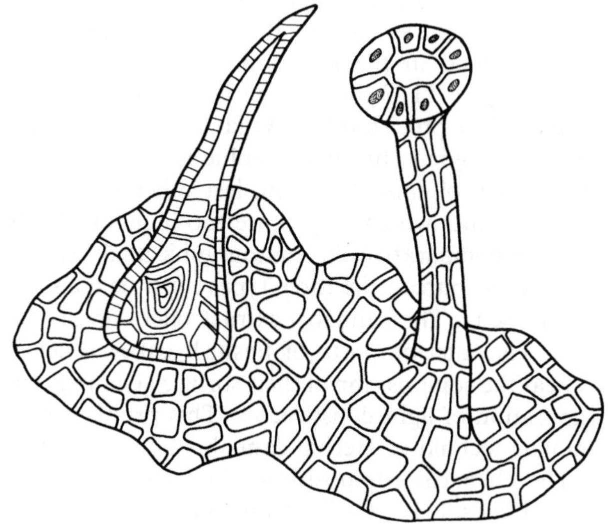

Fig. B. 12. Schematic view of a small part of the epidermis of a leaf of Cannabis sativa L. showing a cystolith hair to the left and a glandular hair to the right

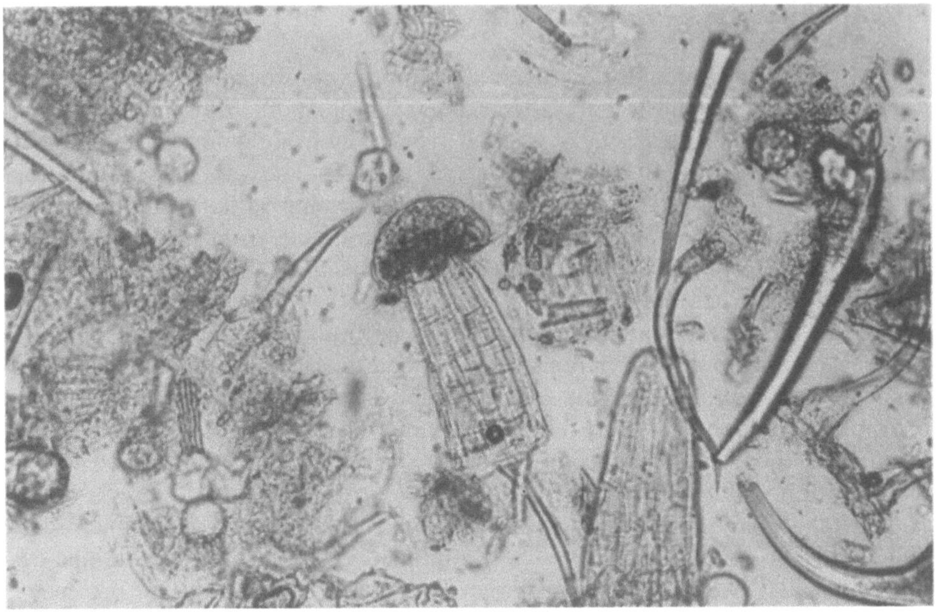

Fig. B. 13. Micrograph of isolated cystolith and glandular hairs from Cannabis sativa L. (Photograph by MA von Wachenfelt, The National Laboratory of Forensic Science, Sweden)

typical chemical composition of the resin which gives rise to characteristic peak clusters from the various types of constituents [17].

Quantification of cannabis is generally aimed at the determination of its main active constituent, Δ^9-trans-tetrahydrocannabinol (THC). THC occurs not only in its free (phenolic) form but also as carboxylic acid, especially in fresh plant material. THC acid is slowly decarboxylated to free THC during storage. When cannabis is smoked, this process is instantaneous. It may be said, therefore, that the "true" measure of the potency of cannabis is the combined content of THC and THC acid. This makes GC suited for the quantification of cannabis, since THC acid is decarboxylated in the injection region of the gas chromatograph. In the determination of THC by GC, the use of an internal standard is recommended for enhancing the precision [4]. As far as accuracy is concerned, it should be kept in mind that a good chromatographic resolution is needed because of the complex composition of cannabis resin, lest other constituents appear at closely similar retention times which may cause too high THC values.

In the comparative chromatographic analysis of cannabis for the establishment of common sources, the three types must be considered differently with regard to heterogeneity. The cannabis extract offers no problem since it is easily homogenized. The interior parts of hashish cakes generally show good homogeneity, whereas the composition of their surface layer may change due to air oxidation and evaporation of light constituents. This problem is easily overcome by a suitable sampling technique. Herbal cannabis often shows severe heterogeneity.

The chemical composition of cannabis resin is strongly dominated by the cannabinoids THC, CBD (cannabidiol), CBN (cannabinol) and their carboxylic

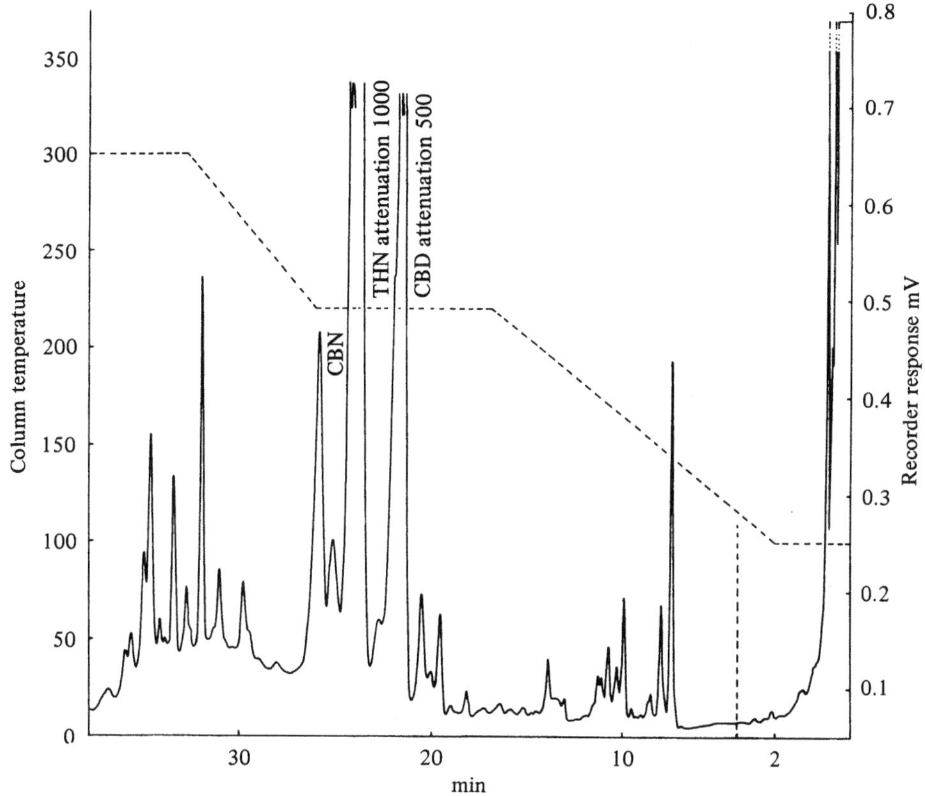

Fig. B. 14. Gas chromatogram of cannabis resin extract. The typical peak clusters are: sesquiterpenes (7–10 minutes), sesquiterpene alcohols (10–12 minutes), main cannabinoids (21–27 minutes) and "heavy constituents" (29–38 minutes). The broken line indicates the temperature program. From Strömberg [17]

acids. The concentrations of these constituents, however, are strongly correlated. For example, THC is converted to CBN by air oxidation and the carboxylic acids are converted to free phenols by decarboxylation [5]. This reduces the discrimination power of these concentration parameters which in turn may lead to coincidental agreements, i.e. false positive results if the comparisons were based only on these constituents. Therefore, comparative analyses of cannabis should also be based on minor constituents such as terpenes, terpene alcohols and side chain homologs of the cannabinoids. In such analyses, the sample size has to be large enough for displaying a sufficient number of minor peaks. On the other hand, the concentrations of the main constituents and the capacity of the chromatographic system will determine the upper limit of the sample size.

The chromatographic methods used for the comparative analysis of cannabis are GC [17], TLC and HPLC [20]. TLC is inferior to the other methods with respect to chromatographic resolution. The highest resolution is obtained by capillary gas chromatography. Such a method has been described by Novotny et al. [12]. These authors also introduced a partition step for the partial removal of the main cannabinoids, i.e. the samples were enriched with respect to minor constituents. The analytical results were excellent but the method seems somewhat laborious for case-work purposes. As far as comparison between HPLC and conventional GC (with packed column) is concerned, Wheals points out [20], that HPLC, unlike GC, gives an analysis of the cannabinoid acids. However, regarding the known instability of these constituents [5], the risk of false negative results cannot be neglected.

6. Quantification

There are several reasons why quantitative analyses of drugs are needed. Firstly, in some legislative systems, the penalty for drug crimes is affected by the concentration of narcotic constituents in the actual drug preparations. Secondly, sudden changes in the contents of the preparations of an illicit drug market may imply a serious threat to the users, especially as far as such highly toxic drugs as heroin and morphine are concerned. In order to monitor these changes, quantitative analyses of street drugs are needed. Thirdly, such analyses will also establish *possible* common origins of drug seizures. Matching seizures of special interest may then be subject to closer comparisons using the methods described in Sect. B. I. 4.

The basic demands of quantitative analysis are accuracy and precision. Accuracy concerns the agreement between the obtained value and the true (expected) value. The accuracy of a quantitative method may be checked by the analysis of known standard samples. A more realistic way of checking the accuracy is the method of standard addition which means the analysis of a case-work sample before and after the addition of a known amount of the constituent in question. The recovered fraction calculated (recovery percentage) will then be a direct measure of the accuracy. Precision concerns the reproducibility of the results, i. e. the repeated results of a precise method will fall within a narrow range. An accurate method of low precision will give a mean result close to the true value. However, a large number of determinations would be needed to achieve this and that is a disadvantage from the practical point of view. It may be said therefore, that good accuracy requires good precision. In this part, some methods for quantitative drug analysis will be briefly discussed with regard to the demands mentioned above. The discussion will be limited to methods used in practice in the authors' laboratory.

The first thing to consider prior to a quantitative analysis is the homogeneity of the item in question. Before homogenization of an item is carried out, it should be kept in mind that a heterogeneity may be of forensic interest per se. For example, the discoloration or size distribution of a crystal form in a drug item may give a valuable contribution to the establishment of a common source. The difficulty of homogenizing a mechanical mixture of crystalline constituents was pointed out in Sect. B. I. 3. This difficulty rapidly increases with the size of the item. Very large items are preferably sampled at random. A quantitative analysis of the joined (and homogenized) samples will then give the average composition of the item. Similar situations often occur in the analyses of groups of items with similar composition since an average value is sufficient in many such cases. Small items or representative samples of these are conveniently homogenized by dissolving them in a suitable solvent. In practice this is done in a volumetric flask.

The methods for quantitative drug analysis may be divided into two groups. In the first group measurements are made on the samples directly or after dissolving them. UV-, IR- and NMR- (nuclear magnetic resonance) spectroscopy belong to this group. In the second group of methods, the constituents are quantified after isolation, e. g. by a chromatographic procedure or by wet chemistry. UV-spectrophotometry is an important method for quantitative drug analysis because most drugs show absorption within this wavelength range. The method is very precise and also accurate provided that the analyte is the single absorbing constituent of the sample. The interference from a second known absorbing constituent may be compensated for by measuring absorbances at different wavelengths but the method is generally inferior as far as multicomponent drug preparations are concerned. Thus, amphetamine – sugar mixtures are conveniently analyzed by UV-spectrophotometry whereas amphetamine – methamphetamine – ephedrine – caffeine mixtures are not.

The most important chromatographic methods for quantitative drug analysis are GC and HPLC. Of these methods, HPLC is more precise due to the relatively mild conditions of the chromatographic process. HPLC has the further advantage over GC of being non-destructive. On the other hand, it may be pointed out that if a capillary gas chromatograph with a data acquisition system is used in qualitative drug analysis, quantitative data may be obtained without further analysis. The achieved precision may not be better than $\pm 5\%$ but this is quite sufficient for most forensic purposes.

By coincidence, two of the most dangerous drugs, morphine and heroin may be quantified by simple gravimetric methods, because the basic forms of these drugs are crystalline, non-volatile and sparingly soluble in water. In such analyses, co-precipitation of water-soluble diluents may occur but this is generally remedied by solvent extraction of the bases in question.

7. Practical Examples

a. A Typical Case with 5 Different Samples

The drug police group in a Swedish province sent in the following material:

1) White powder in a plastic bag.
2) White powder and empty capsules in a plastic bag.
3) A small ball of cotton and traces of a white substance in double plastic bags.
4) Traces of a white powder in double plastic bags.
5) A capsule with a white powder.

The analysis of all materials for illegal or dangerous drugs was requested. The following results were obtained:

Items 1 and 2: The weight of the powders was 0.72 and 0.48 g respectively. Neither thin-layer chromatography nor IR- or UV-spectrophotometry gave any indication of the presence of drugs. Thin-layer chromatography and reaction with Fehling's solution indicated the presence of glucose.

Item 3: The powder and cotton ball contained amphetamine as identified by mass spectrometry, supported by thin-layer- and gas chromatography. Sulfate ions were identified in the usual way ($BaSO_4$).

Item 4: The trace of powder was not weighed. It contained both methamphetamine and phendimetrazine as shown by mass spectrometry, supported by thin-layer chromatography and gas chromatography. The gas chromatogram showed also a small peak with the retention time for procaine.

Item 5: The powder weighed 0.07 g and contained methamphetamine and phendimetrazine (MS, TLC, GC). Gas chromatography further gave peaks with the same retention times as for procaine, ephedrine, caffeine, and phenazone. Figures B. 15 and 16 show the gas chromatograms obtained with a packed column and an open tubular capillary column respectively.

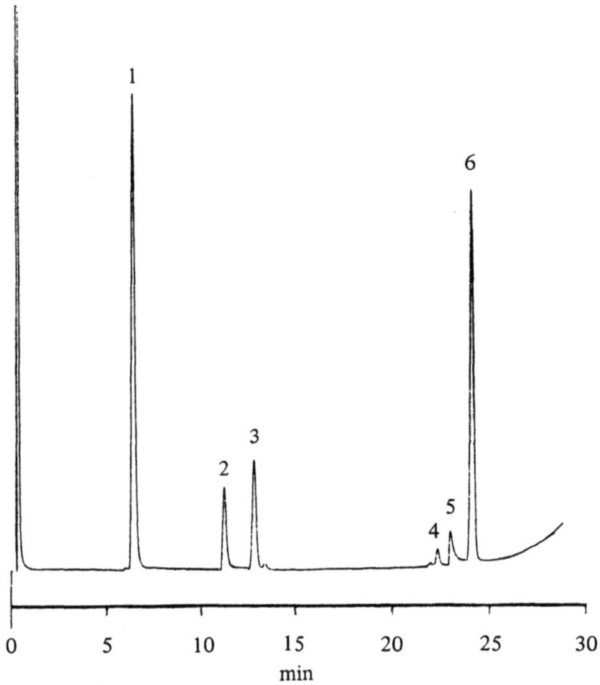

Fig. B. 15. Packed column gas chromatogram of case a. Column: 2 m × 2 mm i. d. glass. Filling: 3% OV-17 on Gaschrom. Q (80–100 mesh). Injector: 250°. Detector (FID): 280°. Oven temperature program: Start at 85°, increase 6°/min to 200°, then 8°/min to 300°. Total analysis time: 30 minutes

Compound	Retention time (min)
1. Methamphetamine	6.40
2. Ephedrine	11.32
3. Phendimetrazine	12.96
4. Caffeine	22.50
5. Phenazone	23.18
6. Procaine	24.22

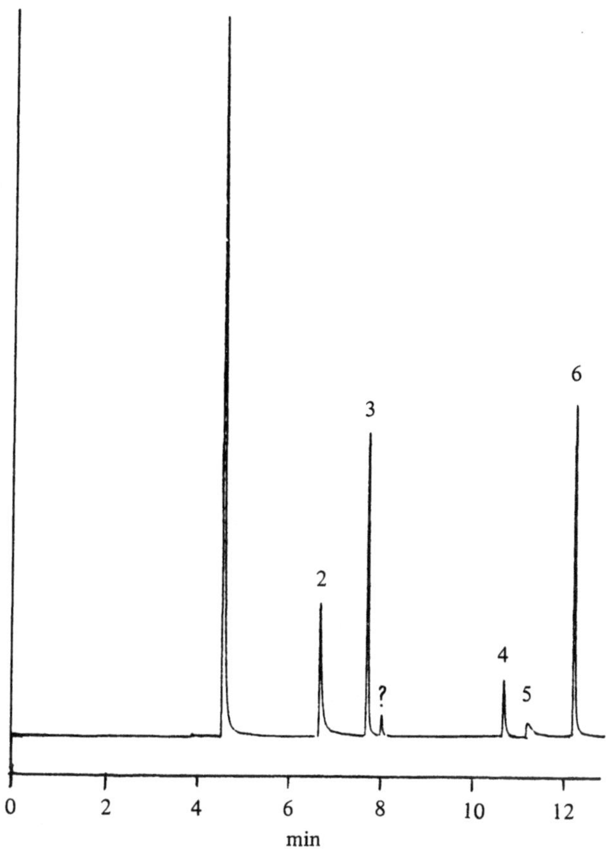

Fig. B. 16. Capillary gas chromatogram of case a. Column: 14.5 m × 0.22 mm i. d. Lining: Bonded methyl silicone gum. Injector: 280°. Detector (NPD): 300°. Oven temperature program: 115° for 3 min, increase 15°/min to 300°. Total analysis time: 13.4 min (stopped at 270°)

Compound	Retention time (min)
1. Methamphetamine	4.64
2. Ephedrine	6.80
3. Phendimetrazine	7.85
4. Caffeine	10.88
5. Phenazone	11.40
6. Procaine	12.45

Amphetamine, methamphetamine and phendimetrazine are illicit drugs on schedule II of the Swedish drug laws.

This example shows the average complexity of a typical drug case in the authors' laboratory. Often, one and the same sample contains 3–4 different compounds. In this case, item 5 contained no less than 7.

b. Analysis of a Multicomponent Street Drug

A tan-colored powder, packed into double plastic bags, was sent to the laboratory for drug analysis. The powder weighed 5.7 g and was shown to contain salts of amphetamine, methamphetamine, and phendimetrazine (MS, GC, TLC). According to GC it further probably contained ephedrine and procaine. Fehling's reagent showed the presence of a sugar. The identity of the three controlled drugs was established by mass spectrometry. The mass spectra obtained are shown in Figs. B. 17 a–c.

c. Variations in the Composition of Similar Looking Seizures

The national drug police squad sent the following seizures to the laboratory for analysis:

Group 1: Twenty-one gelatine capsules containing a white powder.
Group 2: White powder in a plastic bag.
Group 3: Two capsules containing a white powder.
Group 4: White powder and a piece of filter paper in a plastic bag.

1. According to the routine described in this chapter, each of the powders in the 21 capsules was analyzed by thin-layer chromatography. Twenty of these powders had apparently the same composition (further analysis showed that they consisted of a "health" preparation of a harmless kind). The twenty-first, however, weighing 0.18 g, contained amphetamine sulfate, adulterated with sugar. The amphetamine was identified by IR-spectrophotometry and confirmed by TLC and UV-spectrophotometry.

2. and 4. Both powders, weighing 0.32 g and 0.30 g respectively, contained sugar-adulterated amphetamine sulfate (IR, TLC, UV).

3. One of the capsules contained the "health" preparation mentioned above whereas the other contained 0.15 g of sugar-adulterated amphetamine sulfate (IR, TLC, UV).

This example shows the importance of testing each single item received.

d. A Major Clandestine Laboratory

In the autumn of 1978, a woman driver had a breakdown of her car on a country road in the middle of Sweden. She looked for a telephone and knocked at the door of an isolated house. The door was opened by a man wearing a gas mask! He told her that he had no telephone and slammed the door in her face.

This episode lead to the discovery of the largest clandestine laboratory in Northern Europe. The woman informed the police who surrepticiously garded the house for several days, waiting for would-be "customers". Finally, they entered and arrested the sole occupant. The basement and garage of the premise had been turned into a small factory for producing amphetamine by electrolytic reduction of

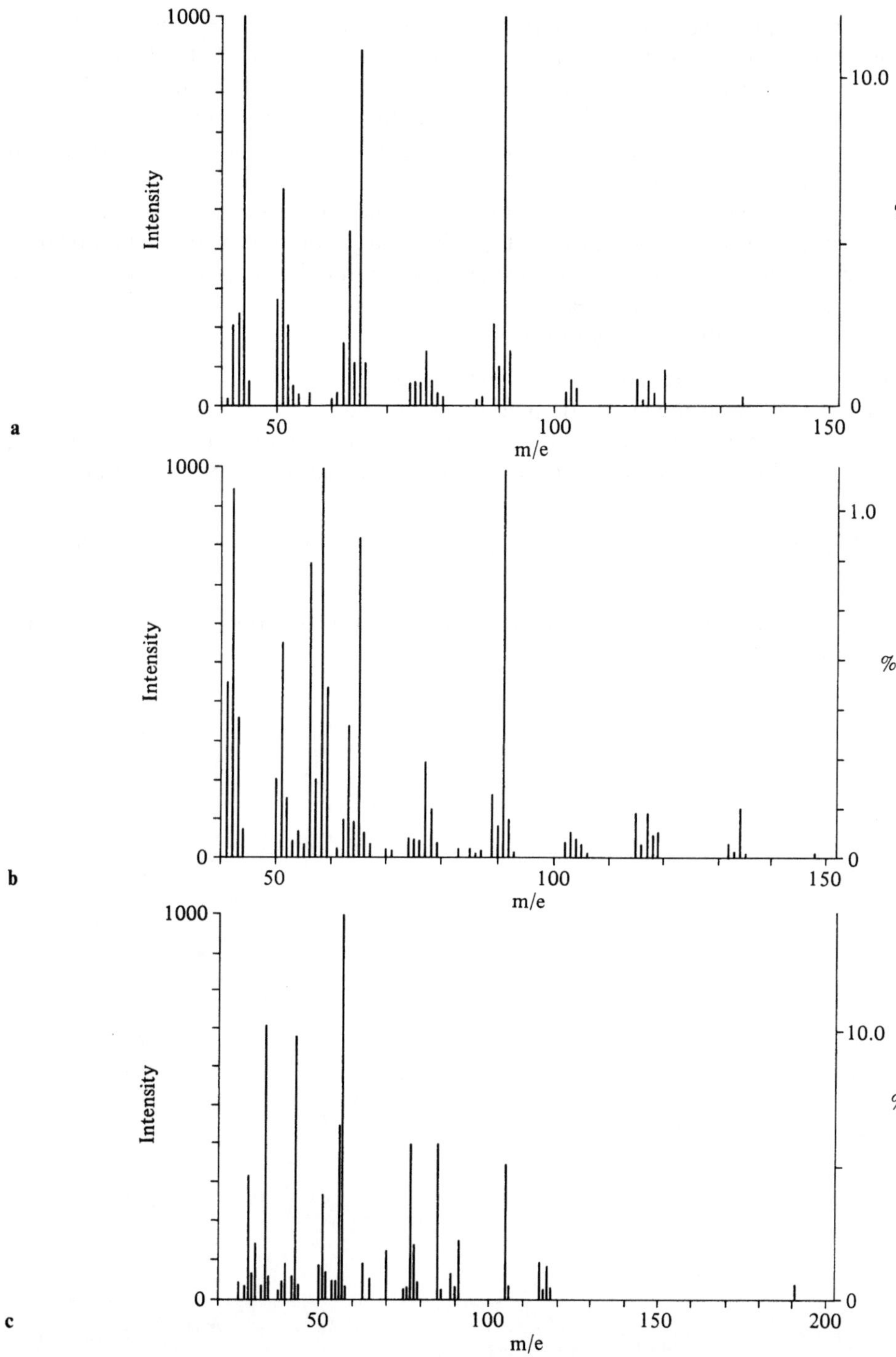

Fig. B. 17 a–c. Mass spectra of the illicit constituents in case b. **a** Mass spectrum of amphetamine. The base peak is situated at m/e = 44, but the spectrum is normalized for the peak with m/e = 91, **b** Mass spectrum of methamphetamine. The base peak lies at m/e = 58 in this case. The spectrum is normalized for the peak at m/e = 91 as above, **c** Mass spectrum of phendimetrazine. Base peak: m/e = 57

phenylnitropropene. The cost of the equipment, including a large rectifier, hoods, electrolysis and distillation units etc. was estimated at about $ 100.000.

The National Laboratory of Forensic Science was alerted, and two chemists joined the CID for a search of the premises. In this way, only the relevant items were brought to the laboratory. Even so, 107 different items had to be labeled and transported to the laboratory.

Figures B. 18 a, b show two interior views of the drug laboratory. It soon became evident that the timing of the raid was perfect. Large amounts of amphetamine base had been manufactured, but only a small amount had been used to make the final product, amphetamine sulfate.

The aims of the investigation were:

(i) to analyze the material for illegal drugs,
(ii) to establish the synthetic method used, and
(iii) to estimate the total yield if all material had been used up for the synthesis; for this purpose, the written instructions found on the premises were to be followed.

The investigation took more than 2 months and gave, in short, the following results:

1. The starting material, phenylnitropropene, was found in 45 items and had a total weight of about 45 kg.

Fig. B. 18 a, b. Two interior views of the clandestine laboratory described in case d. **a** Part of the ten electrolysis units for reducing phenylnitropropene

Fig. B. 18. b Part of the eight distillation units for the distilling of the amphetamine base after the electro lyte had been made alkaline

2. Amphetamine base was found in 26 items and had a total weight of about 7.9 kg.
3. Amphetamine sulfate was found in 20 items and had a total weight of 0.74 kg.
4. Traces of amphetamine were detected in 6 items.
5. The synthesis was based on the electrolytic reduction of phenylnitropropene.
6. The method used gave yields of 12–31% in four trials carried out in the laboratory. Thus the total yield could have been around 20 kg of amphetamine sulfate.

e. Establishment of a Common Source for Two Seizures of Cannabis Resin

In the summer of 1980, a person in Middle Sweden was found in the possession of 6 "cakes" of cannabis resin with a total weight of 0.38 kg, and also a package of marihuana weighing 0.21 kg.

In the same week, 9 packages of cannabis resin, weighing 9.78 kg were found in a forest about 200 km further north.

The police wanted to know if the seizures were indeed cannabis material, and if they had the same origin.

On arrival to the laboratory it was found that the two seizures were packaged in the same fashion and had the same trademark (Nipal) stamped into each "cake". The samples were identified as cannabis resin by microscopical observations and by thin-layer chromatography. One-gram samples from each seizure were extracted (soxhlet), and their chemical signatures determined by gas chromatography. The results are pictured in Fig. B.19.

Based on these findings it was stated that all material sent in consisted of cannabis products and that there is "cause to assume that the two seizures of cannabis resin had a common origin".

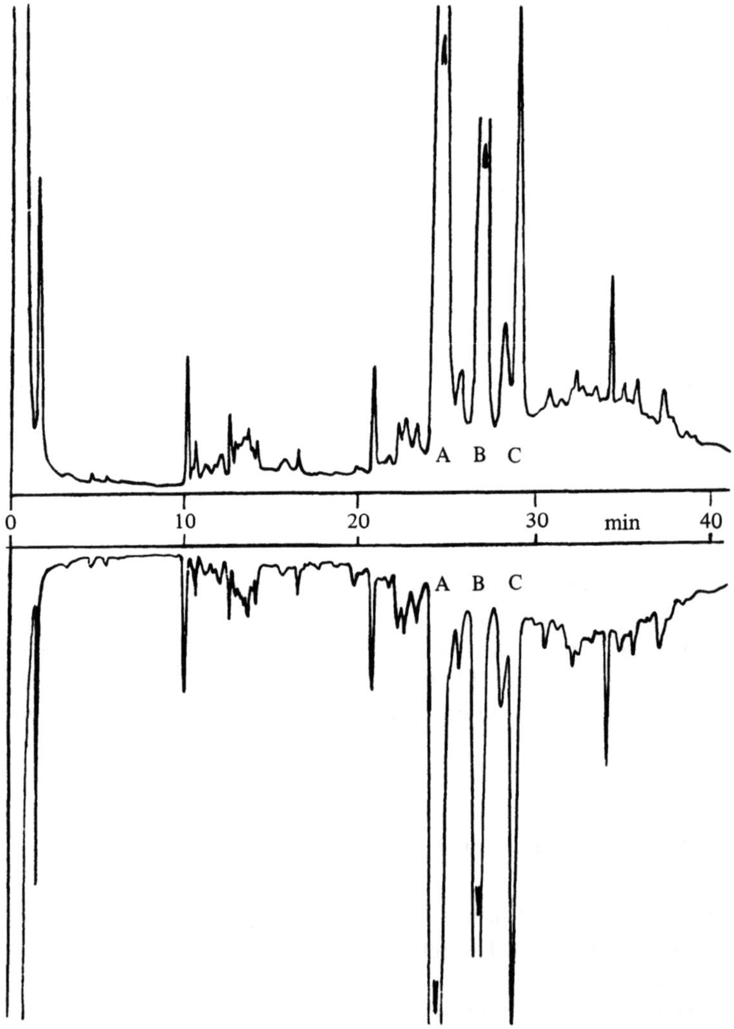

Fig. B. 19. The chemical signatures of the two seizures of cannabis resin mentioned in case e. Horizontal scale: time in minutes. The peaks denoted *A*, *B*, and *C* correspond to CBD, Δ^9-THC, and CBN respectively. The peaks *A* and *B* were attenuated 5 and 2.5 times respectively

f. Quantitative Determination of Amphetamine Sulfate

A Swedish police authority sent in the following material:

1) White powder (about 8 grams) in a plastic bag.
2) Thirteen hypodermic syringes containing liquids.
3) One hypodermic syringe, eight gelatin capsules, one plastic bag, one glass jar and one glass cup, all containing traces of white powder.

The analysis of all materials for illegal or dangerous drugs was requested.

Amphetamine was found in four of the items, including the white powder in the first one. This powder was subjected to quantitative gas chromatographic analysis, using N-ethylamphetamine neutral oxalate as internal standard. The procedure was as follows.

In the development of the procedure, a series of water solutions containing various proportions of amphetamine and N-ethylamphetamine were prepared. Four-milliliter portions of these solutions were made alkaline with KOH pellets and 2 ml of ether were added. After thorough shaking, 1 µl of the ether layer was injected into the gas chromatograph. For each analysis, the peak area ratio (amphetamine peak area: standard peak area) was calculated and plotted against the concentration ratio (calculated from the compositions of the 4-ml water solutions). The linearity was good in the range studied which covered two orders of magnitude in amphetamine concentration. The slope of the calibration line was calculated to be 0.781 by the least squares fit.

From the item in question, a sample weighing 1.005 g was transferred to a 100 ml volumetric flask. After the addition of water up to the mark, 2 ml of the solution were transferred to a graduated shaking cylinder using a pipet. Two ml of standard solution (containing 10 mg of standard substance), a KOH pellet and 2 ml of ether were added. GC analysis was then carried out as described above. The weight percentage of amphetamine sulfate in the sample is given by the expression

$$\frac{w_s \times A_x \times 100}{w_x \times A_s \times 0.781}$$

where w_s and w_x denote the amounts of standard substance and sample, respectively, transferred to the shaking cylinder. A_s and A_x, respectively, denote the areas of the standard and amphetamine peaks. The result in the present case was

$$\frac{0.0100 \times 193 \times 100}{0.0201 \times 662 \times 0.781} = 18.6\% \text{ (w/w)}$$

reported as $19 \pm 1\%$ (w/w) of amphetamine sulfate.

8. General References (In Chronological Order)

A) G. Schmidt: Detection and Estimation of Barbituric Acid Derivatives. In: Methods of Forensic Science, Vol. 1. Lundquist, F. (ed.). New York and London: Interscience 1962, pp. 373–496

B) I. Sunshine, S. R. Gerber: Spectrophotometric Analysis of Drugs Including Atlas of Spectra Springfield, Ill.: Ch. C. Thomas 1963

C) D. C. Garratt: The Quantitative Analysis of Drugs. 3rd edit. London: Chapman & Hall 1964
D) M. Gordon (ed.): Psychopharmacological Agents, Vols. I and II. New York and London: Academic Press 1964, 1967
E) M. Negwer: Organisch-chemische Arzneimittel und ihre Synonyma. Berlin: Akademie-Verlag 1966
F) H. Buzikiewicz, C. Djerassi, D. H. Williams: Mass Spectrometry of Organic Compounds. San Francisco, Cambridge, London and Amsterdam: Holden-Day 1967
G) B. J.: Gudzinowicz: Gas Chromatographic Analysis of Drugs and Pesticides. New York: Marcel Dekker 1967
H) E. Usdin, D. H. Efron: Psychotropic Drugs and Related Compounds. Washington D. C.: U. S. Dept. of Health, Education and Welfare 1967
I) J. H. Beynon, R. A. Saunders, A. E. Williams: The Mass Spectra of Organic Molecules. Amsterdam, London and New York: Elsevier 1968
J) P. C. Stecher (ed.): The Merck Index, 9th edit. Raway, N. Y. 1976
K) E. G. C. Clarke (ed.): Isolation and Identification of Drugs, Vol. 1. London: The Pharmaceutical Press 1969, reprinted 1971
L) E. Stahl: Thin-Layer Chromatography. A Laboratory Handbook. M. R. Ashworth (translator), 2nd edit. Berlin, Heidelberg, New York: Springer 1969
M) I. Sunshine (ed.): Handbook of Analytical Toxicology. Cleveland, Ohio: The Chemical Rubber Co. 1969
N) D. E. Efron: Psychotomimetic Drugs. New York: Raven Press 1970
O) C. R. B. Joyce, S. H. Curry (eds.): The Botany and Chemistry of Cannabis. London: J. & A. Churchill 1970
P) C. W. Waller, J. J. Denny: Annotated Bibliography of Marihuana (Cannabis sativa L.) 1964–1970. Mississippi 38677: The Research Institute of Pharmaceutical Sciences, University of Mississippi 1971
Q) C. W. Waller, J. J. Denny, M. A. Walz: Annotated Bibliography of Marihuana (Cannabis sativa L.) 1971 Supplement. Mississippi 38677: The Research Institute of Pharmaceutical Sciences, University of Mississippi 1972
R) Anonymous: Analysis of Drugs of Abuse. J. Chromatogr. Sci., Special Issue, May 1972
S) R. Mechoulam (ed.): Marijuana. New York and London: Academic Press 1973
T) L. J. Casarett, J. Doull (eds.): Toxicology. The Basic Science of Poisons. New York, Toronto, London: Macmillan Publishing Co., Inc. 1975
U) E. G. C. Clarke (ed.): Isolation and Identification of Drugs, Vol. 2. London: The Pharmaceutical Press 1975
V) A. Cornu, R. Massot: Compilation of Mass Spectral Data, Vols. 1 and 2. London, New York, Rheine: Heyden 1975
W) G. G. Nahas (ed.): Marihuana – Chemistry, Biochemistry, and Cellular Effects. New York, Heidelberg, Berlin: Springer 1976
X) W. E. Simons (ed.): The Sadtler Handbook of Infrared Spectra. Philadelphia: Sadtler Res. Labs. 1978. Abridged edition with approx. 3000 spectra
Y) N. H. Choulis: Identification Procedures of Drugs of Abuse. Ghent (Belgium): European Press 1977
Z) H. W. Dibbern (ed.): UV and IR Spectra of Some Important Drugs. Aulendorf (West Germany): Editio Cantor 1978. UV and IR spectra of 491 pharmaceuticals

9. Methodology References

It is not the purpose of this book to describe analytical methods in any detail. For drug analysis, a great variety of techniques is available and used in forensic laboratories.

In connection with a proficiency testing program in North America in 1977/78 [13], about 200 laboratories reported on the kinds of analytical methods used for drug analysis. Their reports are summarized in Tables B. 2 (p. 51) and 3 (p. 51). These data show that color tests were used by over 80% of these laboratories, and microcrystalline tests by about half of them. Some chemists have developed considerable experience with such tests, which, however, will always depend on subjective interpretations. Moreover, these tests are of very limited value for the examination of multicomponent drug mixtures. Color and microcrystalline tests will not be discussed in this book.

The most important instrumental techniques as well as thin-layer chromatography are described in numerous publications. We have chosen to list some of the more recent ones in the form of tables. Table B. 4 (p. 52 ff.) lists papers dealing with illegal drugs in general. Table B. 5 (p. 56 ff.) includes methods for the analysis of opium alkaloids, cocaine, central nervous system stimulating amines, LSD, phencyclidine, other hallucinogens, and cannabis.

10. Specific References

1) J. K. Baker: Identification and Chemical Classification of Drugs Based on the Relative Response of a Nitrogen Selective Detector and a Flame lonization Detector in Gas Chromatographic Analysis. Anal. Chem. *49*, 906 (1977)
2) C. C. Clark: Cocaine Comparison Parameters. Microgram *11*, 184 (1978)
3) W. C. Duer, J. de Kanel, T. D. Hall: A Necessary Condition for Drug Identification. Microgram *13*, 23 (1980)
4) N.-Å. Ek, E. Lönberg, A. C. Maehly, L. Strömberg: Cannabinoid Content of Fifty Seized Hashish Samples. J. Forens. Sci. *17*, 456 (1972)
5) M. J. de Faubert Maunder: The Forensic Significance of the Age and Origin of Cannabis. Med. Sci. Law *16*, 78 (1976)
6) K. W. Kemp, R. J. Rowlands, I. W. Evett: Some Statistical Aspects of Similarity and Identification. J. Forens. Sci. Soc. *14*, 335 (1974)
7) J. M. Meola: Rapid Procedure for Routine Drug Overdose Screening Using GC and the Nitrogen-selective and Flame Ionization Detector. Chromatogr. Newslett. (Perkin-Elmer) *5*, 1 (1977)
8) A. C. Moffat, K. W. Smalldon, C. Brown: Optimum Use of Paper, Thin-layer and Gas-liquid Chromatography for the Identification of Basic Drugs. I. Determination of Effectiveness for a Series of Chromatographic Systems. J. Chromatogr. *90*, 1 (1974)
9) A. C. Moffat, K. W. Smalldon: Optimum Use of Paper, Thin-layer and Gas Chromatography for the Identification of Basic Drugs. II. Paper and Thin-layer Chromatography. J. Chromatogr. *90*, 9 (1974)
10) A.C. Moffat, A.H. Stead, K.W. Smalldon: Optimum Use of Paper Thin-layer and Gas Chromatography for the Identification of Basic Drugs. III. Gas-liquid Chromatography. J. Chromatogr. *90*, 19 (1974)
11) J. M. Moore, M. Klein: Identification of O^3-Monoacetylmorphine in Illicit Heroin by Using Gas Chromatography-electron-capture Detection and Mass Spectrometry. J. Chromatogr. *154*, 76 (1978)
12) M. Novotny et al.: Analysis of Marijuana Samples from Different Origins by High-Resolution Gas-Liquid Chromatography for Forensic Application. Anal. Chem. *48*, 24 (1976)
13) J. L. Peterson et al.: Crime Laboratory Proficiency Testing Research Program. Washington: National Institute of Law Enforcement and Criminal Justice 1978
14) D. G. Sanger, I. J. Humphreys, J. R. Joyce: A Review of Analytical Techniques for the Comparison and Characterization of Illicit Drugs. J. Forens. Sci. Soc. *19*, 65 (1979)
15) K. W. Smalldon, A. C. Moffat: The Calculation of Discriminating Power for a Series of Correlated Attributes: J. Forens. Sci. Soc. *13*, 291 (1973)
16) W. A. Stoll: 11. Lysergsäurediäthylamid, ein Phantastikum aus der Mutterkorngruppe. Schweiz. Arch. Neurol. Psychiat. *60*, 279 (1947)
17) L. Strömberg: Minor Components of Cannabis Resin. III. Comparative Gas Chromatographic Analysis of Hashish. J. Chromatogr. *68*, 253 (1972)
18) L. Strömberg: Comparative Gas Chromatographic Analysis of Narcotics. II. Amphetamine Sulphate. J. Chromatogr. *106*, 335 (1975)
19) J. I. Thornton: The Use of an Agglomerative Numerical Technique in Physical Evidence Comparisons. J. Forens. Sci. *20*, 693 (1975)
20) B. B. Wheals: Forensic Aspects of High-pressure Liquid Chromatography. J. Chromatogr. 122, 85 (1976)

11. Tables B. 1–5

Table B. 1. Seized narcotic drugs in the world during the year 1978 according to the International Narcotics Control Board of the United Nations. The list is abridged and simplified

Type of narcotic drug	Tons	Tons (total)
Herbal cannabis		5718.3
USA	2920.3	
Colombia	2062.0	
Brazil	276.7	
South African Union	238.0	
Bahamas	188.5	
Other countries	32.8	
Cannabis resin		30.3
USA	11.8	
India	3.3	
United Kingdom	2.7	
France	1.3	
Other countries	11.2	
Heroin		2.0
Thailand	0.58	
USA	0.37	
Hong Kong	0.33	
Federal Republic of Germany	0.16	
Other countries	0.56	
Morphine		1.0
Opium		17.8
Burma	6.0	
India	5.8	
Iran	3.3	
Other countries	2.7	
Coca leaves		27.7
Bolivia	14.9	
Argentina	11.5	
Colombia	1.1	
Other countries	0.2	
Cocaine		6.9
USA	5.1	
Colombia	0.6	
Bahamas	0.4	
Other countries	0.8	

Table B. 2. Tests used (alone or in combination with others) by 200 American forensic laboratories in 1977/78 for identifying controlled substances according to Ref. [13], listed in decreasing frequency

Test or method	Used by % of laboratories
Color tests	83
UV-spectrophotometry	61
IR-spectrophotometry	50
Crystalline tests	49
Gas chromatography (GC)	40
Thin-layer chromatography (TLC)	25
Melting point (MP)	7
GC/Mass spectrometry (MS)	4
Nuclear magnetic resonance (NMR)	2
MS (direct probe)	2
Permanganate oxidation	1
Flame test	1
Commercial kit	0.5

Table B. 3. Tests and methods used (alone or in combination with others) by 179 American forensic laboratories (test no. 1) and 146 laboratories (test no. 2) in 1977/78 for analyzing street drugs according to Ref. [13]

Test or method[a]	Used by % of laboratories	
	Test no. 1	Test no. 2
Color tests	86	87
TLC	67	66
GC	66	71
UV-spectrometry	66	79
Microcrystalline tests	54	45
IR-spectrophotometry	37	42
GC/MS	16	23
Extraction procedures	15	11
Column chromatography	10	–
MP	3	1
Precipitation reactions	2	–
X-ray diffraction	1	7
Other methods	10	21

[a] for abbreviations cf. Table B. 2.

Table B. 4. Examples of publications about methods for the analysis of illicit drug samples covering the period 1970–79 and arranged in alphabetical order

1. General papers

Anonymous: Bibliography of References on the Analysis for Drugs of Abuse. J. Chromatogr. Sci. *10*, 352 (1972). With 466 references

J. K. Brown, M. H. Malone: Qualitative Analytical Results of a Street Drug Monitoring Program – A New Aspect of Toxicology. Proc. West. Pharmacol. Soc. *16*, 134 (1973)

N. H. Choulis: Separation and Quantitation of Mixtures of the Most Commonly Abused Drugs. J. Pharm. Sci. *62*, 112 (1973)

R. S. Frank, J. W. Gunn: Standardization of Forensic Drug Analyses. J. Forens. Sci. *19*, 163 (1974)

V. J. McLinden, A. M. Stenhouse: A Chromatography System for Drug Classification. Forens. Sci. Intern. *13*, 71 (1979)

D. S. Sanger, I. J. Humphreys, J. R. Joyce: A Review of Analytical Techniques for the Comparison and Characterization of Illicit Drugs. J. Forens. Sci. Soc. *19*, 65 (1979)

2. Physical identification of commercial solid dosage forms

L. L. Alber, M. W. Overton: Individual Tablet Assay Program for a Small Computer. J. Assoc. Off. Anal. Chem. *54*, 1449 (1971)

A. H. Tillson, D. W. Johnson: Identification of Drug Tablet and Capsule Evidence as to Source. J. Forens. Sci. *19*, 873 (1974)

3. Thin-layer chromatography (TLC)

N. H. Choulis: Separation and Quantitation of Mixtures of the Most Commonly Abused Drugs. J. Pharm. Sci. *62*, 112 (1973)

R. A. Egli, S. Tanner: Universelle halogenfreie DC-Fliessmittel für Arzneisubstanzen. Z. Anal. Chem. *295*, 398 (1979)

J. C. Hudson, W. P. Rice: Rapid Detection of Some Basic Drugs by Thin-layer Chromatography. J. Chromatogr. *117*, 449 (1976)

A. C. Moffat, K. W. Smalldon: Optimum Use of Paper, Thin-layer and Gas-liquid Chromatography for the Identification of Basic Drugs. II Paper and Thin-layer Chromatography. J. Chromat. *90*, 9 (1974)

A. C. Moffat, B. Clare: Choice of Paper and Thin-layer Chromatographic Systems for Analysis of Basic Drugs. J. Pharm. Pharmacol. *26*, 665 (1974)

A. C. Moffat: The Standardisation of Thin-layer Chromatographic Systems for the Identification of Basic Drugs. J. Chromat. *110*, 341 (1975)

E. Stahl, J. Brombeer: Rapid Identification of Narcotic Drugs by Thin-layer Chromatography in the Pharmaceutical Laboratory. Deutsche Apotheker-Zeitung *118*, 1527 (1978)

J. A. Vinson, J. E. Hooyman: A Universal Thin-layer Chromatographic Visualization Reagent for Drugs. J. Chromat. *105*, 415 (1975)

R. A. van Welsum: A Simplified Procedure for the Identification of Drugs from the Illicit Street Market by Thin-layer Chromatography. J. Chromat. *78*, 237 (1973)

W. J. Woodford: Rapid Screening Procedure for Some Street Drugs by Thin-layer Chromatography. J. Chromat. *115*, 678 (1975)

4. Thin-layer chromatography combined with mass spectrometry (TLC/MS)

G. J. Down, S. A. Gwyn: Investigation of Direct Thin-layer Chromatography-Mass Spectrometry as a Drug Analysis Technique. J. Chromat. *103*, 208 (1975)

5. Gas chromatography (GC)

a. *Conventional GC*

J. K. Baker: Identification and Chemical Classification of Drugs Based on the Relative Response of a Nitrogen Selective Detector and a Flame Ionization Detector in Gas Chromatographic Analysis. Anal. Chem. *49*, 906 (1977)

R. C. Gupta, S. H. Montgomery, G. D. Lundberg: Quantitative Determinations of Street Drugs in the Los Angeles Area. Clin. Toxicol. *7*, 241 (1974)

J. C. Kok, P. E. Kamp: Analysis of Substance Used as Drugs (Street drugs). Pharm. Weekbl. *108*, 990 (1973)

V. J. Mc Linden, A. M. Stenhouse: A Chromatography System for Drug Identification. Forens. Sci. Intern. *13*, 71 (1979)

A. C. Moffat, A. H. Stead, K. W. Smalldon: Comparison of Stationary Phases for the Gas-liquid Chromatography of Basic Drugs. J. Pharm. Pharmacol. *25*, 155 (1973)

A. C. Moffat, A. H. Stead, K. W. Smalldown: Optimum Use of Paper, Thin-layer and Gas-liquid Chromatography for the Identification of Basic Drugs. III Gas-liquid Chromatography. J. Chromat. *90*, 19 (1974)

J. D. Nicholson: Derivative Formation in the Quantitative Gas-Chromatographic Analysis of Pharmaceuticals. Part I. The Analyst *103*, 1 (1978)

I. M. Riabtseva, M. I. Kuleshova, B. A. Rudenko: Quantitative Analysis of Drug Mixtures by the Gas-liquid Chromatography Method. Farmatsiia *20*, 23 (1971) (In Russian)

L. Zoccolillo: Use of Glass Beads in Gas Chromatographic Analysis of Drugs. J. Chromat. *178*, 311 (1979)

b. *Capillary GC*

A. S. Christophersen, K. E. Rasmussen: Glass Capillary Column Gas Chromatography of Narcotic Drugs after Flash-heater Trimethylsilylation. J. Chromat. *174*, 454 (1979)

J. B. F. Lloyd, K. Hadley: Some Applications of a Capillary Column Gas Chromatography System in Forensic Science. SCAN (3), 13 (1973) (Pye-Unicam)

M. Novotný, R. Farlow: A Simple Method for Concentrating Dilute High-boiling Samples for Capillary Gas Chromatography. J. Chromat. *103*, 1 (1975)

c. *Pyrolysis GC*

C. Dimitrov, N. Petsev, R. Daskalov: Identification of Narcotics and Psychotropics through Pyrolytic Gas Chromatography. Dokl. Bolg. Akad. Nauk. *31*, 1027 (1978)

6. High pressure liquid chromatography (HPLC)

J. K. Baker, R. E. Skelton, Ch-Yu Ma: Identification of Drugs by High-Pressure Liquid Chromatography with Dual Wavelength Ultraviolet Detection. J. Chromatogr. *168*, 417 (1979)

M. L. Chan, C. Whetsell, J. D. Mc Chesney: Use of High Pressure Liquid Chromatography for the Separation of Drugs of Abuse. J. Chromatogr. Sci. *12*, 512 (1974)

C. N. Hodnett, R. D. Eberhardt: Optimization of the Determination of Drugs by High-Pressure Liquid Chromatography (HPLC). J. Anal. Toxicol. *3*, 187 (1979)

I. Jane: The Separation of a Wide Range of Drugs of Abuse by High-pressure Liquid Chromatography. J. Chromatogr. *111*, 227 (1975)

I. M. Johansson, K-G. Wahlund, G. Schill: Reversed-Phase Ion-Pair Chromatography of Drugs and Related Organic Compounds. J. Chromatogr. *149*, 281 (1978)

P. Larsson et al.: Liquid Chromatography of Analgesic Drugs on Ion-Exchange Resins. Anal. Chem. *45*, 2306 (1973)

I. Lurie: Application of Reverse Phase Ion-Pair Partition Chromatography to Drugs of Forensic Interest. J. Assoc. Off. Anal. Chem. *60*, 1035 (1977)

K. Macek, V. Řehák: High-Resolution Liquid Chromatography of Some Psychotropic Drugs. J. Chromatogr. *105*, 182 (1975)
W. W. Mc Gee: Use Of Gel Permeation Chromatography in the Crime Laboratory. J. Liq. Chromatogr. *2*, 287 (1979)
J. S. Mayell, C. F. Hiskey, L. Lachman: Quantitative Analysis of a Multicomponent Drug Product Using Liquid Chromatography. Anal. Chem. *46*, 449 (1974)
P. J. Twitchett, A. C. Moffat: High-Pressure Liquid Chromatography of Drugs. An Evaluation of an Octadecylsilane Stationary Phase. J. Chromatogr. *111*, 149 (1975)
B. B. Wheals: Forensic Applications of Liquid Chromatography. Proc. Anal. Div. Chem. Soc. *13*, 164 (1976)
B. B. Wheals: Forensic Aspects of High-Pressure Liquid Chromatography. J. Chromatogr. *122*, 85 (1976)
B. B. Wheals, I. Jane: Analysis of Drugs and Their Metabolites by High-Performance Liquid Chromatography. A Review. Analyst *102*, 625 (1977)

7. Ultraviolet spectrophotometry

N. H. Choulis: Separation and Quantitation of Mixtures of the Most Commonly Abused Drugs. J. Pharm. Sci. *62*, 112 (1973)
T. D. Doyle, F. R. Fazzari: Determination of Drugs in Dosage Forms by Difference Spectrophotometry. J. Pharm. Sci. *63*, 1921 (1974)
M. Kuhnert-Brandstätter, G. Kramer, P. D. Lark: Thermomicroscopic and Ultraviolet Spectrophotometric Identification of Psychotherapeutic Drugs. Microchem. J. *17*, 739 (1972)
A. E. Mc Dowell, H. L. Pardue: Simultaneous Multicomponent Drug Determination with a Vidicon Spectrometer. J. Pharm. Sci. *67*, 822 (1978)
T. J. Siek: Identification of Drugs and Other Toxic Compounds from Their Ultraviolet Spectra. Part I: Ultraviolet Absorption Properties of Sixteen Structural Groups. J. Forens. Sci. *19*, 193 (1974)
T. J. Siek, R. J. Osiewicz: Identification of Drugs and Other Toxic Compounds from Their Ultraviolet Spectra. Part II: Ultraviolet Absorption Properties of Thirteen Structural Groups. J. Forens. Sci. *20*, 18 (1975)
T. J. Siek, R. J. Osiewicz, R. J. Bath: Identification of Drugs and Other Toxic Compounds from Their Ultraviolet Spectra. Part III: UV-Absorption Properties of 22 Structural Groups. J. Forens. Sci. *21*, 525 (1976)
K. W. Smalldon, L. A. King: Micro-Sampling Techniques in Visible, and Ultra-Violet Spectroscopy and Related Problems in Forensic Science. UV Spectrom. Grp. Bull. (2), 13 (1974)

8. Infrared spectrophotometry (IR)

W. L. Brannon: Comparative Study of Infrared Microtechniques (for drugs). J. Assoc. Off. Anal. Chem. *53*, 599 (1970)
N. H. Choulis, J. M. Krall, N. G. Clarke: Computers in Drug Analysis. I. Identification of Drugs of Abuse via Infrared Spectroscopy. Pharmazie *34*, 647 (1979)
S. Goenechea: IR-Untersuchungen sauer und neutral reagierender Arzneimittel. II. Bromureide und Carbamate. Mikrochim. Acta (3), 266 (1972)
S. Goenechea: IR-Untersuchungen sauer und neutral reagierender Arzneimittel. III. Hydantoine und Purinderivate. Mikrochim. Acta (3), 276 (1972)
S. C. Mutha, W. B. Ludemann: Solid-State Anomalies in IR Spectra of Compounds of Pharmaceutical Interest. J. Pharm. Sci. *65*, 1400 (1976)
C. S. Rann: Automatic Sorting of Infrared Spectra. Anal. Chem. *44*, 1669 (1972)

9. GC – IR

J. Bus, T. J. Liefkens: Collection of Microgram GLC Fractions for Infrared Spectroscopy. Z. Anal. Chem. *250*, 294 (1970)

10. Mass spectrometry (MS)

a. General papers on MS and GC-MS

D-P. Beggs, A. G. Day III: Chemical Ionization Mass Spectrometry: A Rapid Technique for Forensic Analyses. J. Forens. Sci. *19*, 891 (1974)

A. Callieux, P. Allain: Superiority of Chemical Ionization on Electron Impact for Identification of Drugs by Gas-Chromatograph-Mass Spectrometer. J. Anal. Toxicol. *3*, 39 (1979)

D. Frangi-Schnyder, H. Brandenberger: Anion-CI-Mass Spectrometry of the Main Classes of Hypnotics. Z. Anal. Chem. *290*, 153 (1978)

B. Holmstedt, J. E. Lindgren: Nachweis von Suchtmitteln und Halluzinogenen mit Hilfe der Kombination Gas-Chromatographie/Massenspektrometrie. Z. Anal. Chem. *261*, 291 (1972)

D. H. Hunnemann: Massenspektrometrische Charakterisierung populärer Rauschmittel. Chemiker-Ztg. *98*, 28 (1974)

L. Kazyak: Dual Column Operation for Gas Chromatograph – Mass Spectrometer. Anal. Chem. *48*, 1826 (1976)

G. W. A. Milne, H. M. Fales, T. Axenrod: Identification of Dangerous Drugs by Isobutane Chemical Ionization Mass Spectrometry. Anal. Chem. *43*, 1815 (1971)

R. Saferstein, J-M. Chao: Identification of Drugs by Chemical Ionization Mass Spectroscopy. J. Assoc. Off. Anal. Chem. *56*, 1234 (1973)

R. C. Sullivan et al.: Evaluation and Selection of Gas Chromatography/Mass Spectrometry Systems for the Identification of Dangerous Drugs. J. Police Sci. Adm. *2*, 185 (1974)

A. Tatematsu et al.: Mass Spectrometry of Medicines. Quantitative Determination by Direct Probe Inlet System Mass Chromatography. Biomed. Mass. Spectrom. *5*, 192 (1978)

b. Computer programs for MS

E. P. Abramson: Automated Identification of Mass Spectra by the Reverse Search. Anal. Chem. *47*, 45 (1975)

R. G. Dromey: Simple Index for Classifying Mass Spectra with Application to Fast Library Searching. Anal. Chem. *48*, 1464 (1976)

R. G. Dromey: Mass Spectrum Dictionary for Library Searching. Anal. Chem. *51*, 229 (1979)

B. S. Finkle, D. M. Taylor: A GC/MS Reference Data System for the Identification of Drugs of Abuse. J. Chromatogr. Sci. *10*, 312 (1972)

F. W. Mc Lafferty, R. Venkataraghavan: Computer Techniques for Mass Spectral Identification. J. Chromatogr. Sci. *17*, 24 (1979)

H. Nau, K. Biemann: Utilization of Automatically Assigned Retention Indices for Computer Identification of Mass Spectra of Drugs and Metabolites in Biological Fluids. Anal. Lett. *6*, 1071 (1973)

K. Wrabetz: Ein einfaches Verfahren zur programmierten MS-Bibliotheksuche. Z. Anal. Chem. *282*, 135 (1976)

J. Yinon, S. Zitrin: Processing and Interpreting Mass Spectral Data in Forensic Identification of Drugs and Explosives. J. Forens. Sci. *22*, 742 (1977)

c. Combination HPLC-MS

J. D. Henion: Drug Analysis by Continuously Monitored Liquid Chromatography/Mass Spectrometry with a Quadrupole Mass Spectrometer. Anal. Chem. *50*, 1687 (1978)

R. E. Lovins et al.: Liquid Chromatography – Mass Spectrometry – Coupling of a Liquid Chromatograph to a Mass Spectrometer. Anal. Chem. *45*, 1553 (1973)

T. Takeuchi, Y. Hirata, Y. Okumura: On-line Coupling of a Micro Liquid Chromatograph and Mass Spectrometer through a Jet Separator. Anal. Chem. *50*, 659 (1978)

S. Tsuge, Y. Hirata, T. Takeuchi: Vacuum Nebulizing Interface for Direct Coupling of Micro Liquid Chromatograph and Mass Spectrometer. Anal. Chem. *51*, 166 (1979)

11. Fluorescence

M. Hoshino, A. Tsuji: A Fluorometric Determination of Basic Drugs by Using a Fluorescent Dye Salt Procedure; Tetrabromofluorescein Ethyl Ester Method. Japan Analyst *21*, 1523 (1972). Cit. fr. Z. Anal. Chem. *268*, 402 (1974)

J. A. F. de Silva, N. Strojny, K. Stika: Luminescence Determination of Pharmaceuticals of the Tetrahydrocarbazole, Carbazole, and 1,4-Benzodiazepine Class. Anal. Chem. *48*, 144 (1976)

P. De Zan, R. F. Canaff, R. Bianchi: Fluorimetric Characteristics of Some Narcotic and Dangerous Drugs. J. Assoc. Off. Anal. Chem. *54*, 925 (1971)

12. Nuclear magnetic resonance (NMR)

T. C. Kram: Analysis of Illicit Drug Exhibits by Hydrogen-1 Nuclear Magnetic Resonance Spectroscopy. J. Forens. Sci. *23*, 456 (1978)

Table B. 5. Examples of publications on analytical methods for some classes of illicit drugs from 1970–1979 (in alphabetical order)

1. Opium alkaloids

a. Several methods in parallel or in combination

C. C. Clark: A Study of Procedures for the Identification of Heroin. J. Forens. Sci. *22*, 418 (1977)

C. C. Clark: Brown Heroin Comparison Parameters. Microgram *11*, 5 (1978)

R. Fuelster: Extraction and Identification of Morphine and Codeine. Microgram *8*, 135 (1975)

S. E. Hays, L. T. Grady, A. V. Kruegel. Purity Profiles for Heroin, Morphine, and Morphine Hydrochloride. J. Pharm. Sci. *62*, 1509 (1973)

D. J. Reuland, W. A. Trinler: An Unequivocal Determination of Heroin in Simulated Street Drugs buy a Combination of High-Performance Liquid Chromatography and Infrared Spectrophotometry Using Micro-Sampling Techniques. Forens. Sci. *11*, 195 (1978)

R. C. Shaler, J. H. Jerpe: Identification and Determination of Heroin in Illicit Seizures by Combined Gas Chromatography – Infrared Spectrophotometry. J. Forens. Sci. *17*, 668 (1972)

D. Sohn et al.: Screening for Heroin – a Comparison of Current Methods. Anal. Chem. *45*, 1498 (1973)

b. Thin-layer chromatography (TLC)

D. J. Doedens, R. B. Forney: Confirmation of Morphine on Thin-Layer Plates by Fluorometry. J. Chromatogr. *100*, 225 (1974)

A. Gyeresi, G. Racz: New Solvent for Separation of the Main Alkaloids of Opium (by Thin-Layer Chromatography). Pharmazie *28*, 271 (1973)

E. Klug: Über einen spezifischen dünnschichtchromatographischen Nachweis von Morphin. Z. Anal. Chem. *260*, 31 (1972)

J. Paul, F. Conine: Rapid Thin-Layer Chromatographic Separation of Cocaine from Codeine, Heroin (Diamorphine), 6-Acetylnurphine, Morphine and Quinine in Microscope Slides. Microchem. J. *18*, 142 (1973)

K. Roder, E. Eich, E. Mutschler: Direct Quantitative Evaluation of Thin-Layer Chromatography with Remission and Fluorescence Determinations. 3. Determination of Morphine, Codeine, Thebaine, and so on. Arch. Pharm. *304*, 297 (1971)

S. N. Tewari, D. N. Sharma: Detection and Identification of Morphine in Adulterated Opium Samples by Thin-Layer Chromatography. Z. Anal. Chem. *281*, 381 (1976)

c. Gas chromatography (GC)

A. Bechtel: Gas-chromatographische Identifizierung und quantitative Bestimmung von Morphin, Codein, Thebain, Papaverin und Narkotin in Opiumextrakt. Chromatographia *5*, 404 (1972)

P. De Zan, J. Fasanello: The Quantitative Determination of Heroin in Illicit Preparations by Gas Chromatography. J. Chromatogr. Sci. *10*, 333 (1972)

D. Furmanec: Quantitative Gas Chromatographic Determination of the Major Alkaloids in Gum Opium. J. Chromatogr. *89*, 76 (1974)

H-Y. Lim, S-T. Chow: Heroin Abuse and a Gas Chromatographic Method for Determining Illicit Heroin Samples in Singapore. J. Forens. Sci. *23*, 319 (1978)

J. M. Moore, F. E. Bena: Rapid Gas Chromatographic Assay for Heroin in Illicit Preparations. Anal. Chem. *44*, 385 (1972)

J. M. Moore: Rapid and Sensitive Gas Chromatographic Quantitation of Morphine, Codeine and O^6-Acetylmorphine in Illicit Heroin Using an Electron Capture Detector. J. Chromatogr. *147*, 327 (1978)

K. D. Parker et al.: Preliminary Report on the Detection and Quantitation of Opiates and certain other Drugs of Abuse as Trimethylsilyl Derivatives by Gas-Liquid Chromatography. Forens. Sci. Soc. J. *10*, 17 (1970)

M. J. Prager, S. M. Harrington, T. F. Governo: Gas-Liquid Chromatographic Determination of Morphine, Heroin, and Cocaine: J. Assoc. Off. Anal. Chem. *62*, 304 (1979)

E. P. J. van der Slooten, H. J. van der Helm: Analysis of Heroin in Relation to Illicit Drug Traffic. Forens. Sci. *6*, 83 (1975)

A. Vialo et al.: Gas-Chromatographische Identifizierung von Heroin und Morphin in Verdacht erregenden Produkten. J. European Toxicol. *4*, 375 (1971)

d. High pressure liquid chromatography (HPLC)

J. Albanbauer et al.: Quantitative Hochdruck-Flüssigkeitschromatographie von Rauschgiften. 1. Mitteilung: Heroin und seine Zubereitungen. Arch. Kriminol. *162*, 103 (1978)

P. J. Cashman, J. I. Thornton: High Speed Liquid Adsorption Chromatography in Criminalistics. II The Separation of Heroin, O^6-Monoacetylmorphine and Morphine. J. Forens. Sci. Soc. *12*, 417 (1972)

I. Jane, J. F. Taylor: Characterisation and Quantitation of Morphine in Urine Using High-Pressure Liquid Chromatography with Fluorescent Detection. J. Chromatogr. *109*, 37 (1975)

J. H. Knox, J. Jurnad: Application of High-Speed Liquid Chromatography to the Analysis of Morphine, Heroin, 6 (O-Acetyl) Morphine and Methadone. J. Chromatogr. *87*, 95 (1973)

C. Olieman et al.: Chemistry of Opium Alkaloids and Related Compounds by Ion-Pair High-Performance Liquid Chromatography. J. Chromatogr. *133*, 382 (1977)

G. K. Poochikian, J. C. Cradock: Simple High-Performance Liquid Chromatographic Method for the Separation of 3,6-Diacetylmorphine Hydrochloride (Heroin) and Hydrolysis Products. J. Chromatogr. *171*, 371 (1979)

P. J. Twitchett: Analysis of Illicit Diamorphine Preparations by High-Pressure Liquid Chromatography. J. Chromatogr. *104*, 205 (1975)

P. G. Vincent, B. F. Engelke: High Pressure Liquid Chromatographic Determination of the Five Major Alkaloids in Papaver Somniferum L. and Thebaine in Papaver Bracetatum Lindl. Capsular Tissue. J. Assoc. Off. Anal. Chem. *62*, 310 (1979)

H. W. Ziegler, T. H. Beasley Sr, D. W. Smith: Simultaneous Assay for Six Alkaloids in Opium, Using High-Performance Liquid Chromatography. J. Assoc. Off. Anal. Chem. *58*, 888 (1975)

e. Ultraviolet spectrophotometry (UV)

E. Lang: Photometrische Identifizierung und Bestimmung von Morphin und Morphinderivaten in Arzneimitteln: Pharmazie *26*, 245 (1971)

R. V. Rondina, A. L. Bandoni, J. D. Coussio: Quantitative Determination of Morphine in Poppy Capsules by Differential Spectrophotometry. J. Pharm. Sci. *62*, 502 (1973)

f. Infrared spectrophotometry (IR)

L. Borka: The Polymorphism of Heroin and its Forensic Aspects. Acta Pharm. Suec. *14*, 210 (1977)

g. Gas Chromatography – mass spectrometry (GC-MS)

J. M. Moore, M. Klein: Identification of O^3-Monoacetyl-Morphine in Illicit Heroin by Using Gas Chromatography-Electron-Capture Detection and Mass Spectrometry. J. Chromatogr. *154*, 76 (1978)

G. R. Nakamura et al.: Forensic Identification of Heroin in Illicit Preparations Using Integrated Gas Chromatography and Mass Spectrometry. Anal. Chem. *44*, 408 (1972)

R. Saferstein, J. Manura, T. A. Brettell: Chemical lonization Mass Spectrometry of Morphine Derivatives J. Forens. Sci. *24*, 312 (1979)

R. M. Smith: Forensic Identification of Opium by Computerized Gas Chromatography/Mass Spectrometry. J. Forens. Sci. *18*, 327 (1973)

2. Cocaine (all methods)

J. K. Brown et al.: A Rapid Screening Procedure for Some "Street-Drugs" by Thin-Layer Chromatography. II. Cocaine, Heroin, Local Anesthetics and Mixtures. J. Chromatogr. *87*, 211 (1973)

C. C. Clark: Gas-Liquid Chromatographic Quantitation of Cocaine. HCl in Powders and Tablets: Collaborative Study. J. Assoc. Off. Anal. Chem. *61*, 683 (1978)

R. M. Dews: Separation of Mixtures of Procaine, Cocaine, and Phencyclidine Using Gas Chromatography. Microgram *6*, 60 (1971)

D. Eskes: Thin-Layer Chromatographic Procedure for the Differentiation of the Optical Isomers of Cocaine. J. Chromatogr. *152*, 589 (1978)

J. M. Moore: Identification of Cis- and Trans-cinnamoylcocaine in Illicit Cocaine Seizures. J. Assoc. Off. Anal. Chem. *56*, 1199 (1973)

3. Central nervous system stimulating amines

a. Several methods in parallel or in combination

K. Bailey, D. Legault, D. Verner: Spectroscopic and Chromatographic Identification of Dimethoxyamphetamines. J. Assoc. Off. Anal. Chem. 57, 70 (1974)

K. Bailey, D. R. Gagné, R. K. Pike: Investigation and Identification of the Bromination Products of Dimethoxyamphetamines. J. Assoc. Off. Anal. Chem. *59*, 1162 (1976)

R. C. Briner: Procedure for Distinguishing Between Phenmetrazine and Phendimetrazine. Microgram *5*, 142 (1972)

B. M. Courtney: Possible Erroneous Identification of Phenelzine as Amphetamine or Metamphetamine in Spectrophotometric Determinations. Microgram 5, 108 (1972)

T. R. Gaston, G. T. Rasmussen: Identification of 3,4-Methylenedioxymethamphetamine. Microgram *5*, 60 (1972)

T. Inoue et al.: Analysis of Stimulants and Their Analogous Drugs. Rep. Nat. Res. Inst. Police Sci. (Tokyo) *31*, 261 (1978)

T. C. Kram: Identification of an Impurity in Illicit Amphetamine Tablets. J. Pharm. Sci. *66*, 443 (1977)

S. J. Mulé: Methods for the Analysis of Narcotic Analgesics and Amphetamines. J. Chromatogr. Sci. *10*, 275 (1972)

M. Steinigen: "Practical Notes on the Analysis of Sympathomimetics of the Phenylalkyl- and Phenylalkanol- Aminetype". Apothekerzeitung *113*, 81 (1973)

R. J. Warren, P. P. Begosh, J. E. Zarembo: Identification of Amphetamines and Related Sympathomimetic Amines. J. Assoc. Off. Anal. Chem. *54*, 1179 (1971)

b. Thin-layer chromatography

D. Eskes: A Procedure for the Differentiation of the Optical Isomers of Amphetamine and Methamphetamine by Thin-Layer Chromatography. J. Chromatogr. *117*, 442 (1976)

Y. Hashimoto et al.: Quantitative Determination of (−)−and (+)−Pseudo-Ephedrine in Ephedra spp by Thin-Layer Chromatography. Yakugaku Zasshi *97*, 594 (1977)

Y. Hashimoto et al.: Quantitative Analysis of Phenethylamine Derivatives by Thin-Layer Chromatography. Determination of Psychotropic Drugs and Ephedra Bases. Mikorchim. Acta II, 493 1978

E. Röhm, H. V. Schmid: Empfindliche Nachweismethoden für das Stimulans Khat. Arch. Kriminol. *155*, 155 (1975)

M. A. Shaw, H. W. Peel: Thin-Layer Chromatography of 3,4-Methylenedioxyamphetamine, 3,4-Methylenedioxymethamphetamine and Other Phenethylamine Derivatives. J. Chromatogr. *104*, 201 (1975)

R. A. van Welsum: Simplified Procedure for the Identification of Drugs from the Illicit Street Market by Thin-Layer Chromatography. J. Chromatogr. *78*, 237 (1973)

c. Gas chromatography

J. L. Cashaw et al.: Simultaneous Determination of Biogenic Amines and Narcotic Alkaloids by GLC. J. Chromatogr. Sci. *9*, 98 (1971)

M. Donike, D. Straman: Gas-chromatographische Identifizierung von Stimulantien der Phenyläthylaminreihe mit Hilfe der Retentions-Indices. Z. Anal. Chem. *279*, 129 (1976)

L. Mc Cullagh: Rapid Qualitative GLC Identification of Amphetamine and Various Hallucinogens. Microgram *4*, 135 (1971)

L. T. Sennello: Gas-Liquid Chromatographic Determination of Methylamphetaminehydrochloride in Methyl Acrylatemethyl Methacrylate Sustained-Release Tablets. J. Pharm. Sci. *60*, 595 (1971)

R. W. Souter: Gas Chromatographic Resolution of Enantiomeric Amphetamine and Related Amines. II Effect of Cyclic Structures on Diastereomer and Enantiomer Resolution. J. Chromatogr. *114*, 307 (1975)

L. Strömberg: Comparative Gas Chromatographic Analysis of Narcotics. II Amphetamine Sulphate. J. Chromatogr. *106*, 335 (1975)

L. Strömberg, A. C. Maehly: Comparative Gas Chromatographic Analysis of Narcotics. III Phenmetrazine Hydrochloride. J. Chromatogr. *109*, 67 (1975)

C. E. Wells: GLC Determination of Optical Isomers of Amphetamine. J. Assoc. Off. Anal. Chem. *53*, 113 (1970)

C. E. Wells: Collaborative Study of the Gas-Liquid Chromatographic Method for Determination of Stereochemical Composition of Amphetamine. J. Assoc. Off. Anal. Chem. *55*, 146 (1972)

d. High-pressure liquid chromatography (HPLC)

P. J. Cashman, J. I. Thornton, D. L. Shelman: High-Pressure Liquid Chromatographic Separation of Phenethylamines of Forensic Interest. J. Chromatogr. Sci. *11*, 7 (1973)

e. Infrared spectrophotometry (IR)

V. C. Bass: The Identification of Sympathomimetic Amines by Raman Spectroscopy. Forens. Sci. *11*, 57 (1978)

J. A. Heagy: Infra-Red Method for Distinguishing Optical Isomers of Amphetamine. Anal. Cheml *42*, 1459 (1970)

R. J. Mesley, W. H. Evans: Infra-Red Identification of Some Hallucinogenic Derivatives of Tryptamine and Amphetamine. J. Pharm. Pharmacol. *22*, 321 (1970)

f. Spectrophotofluorimetry (F)

A. Deleenheer, J. E. Sinsheimer, J. H. Burckhalter: Fluorimetric Determination of Primary and Secondary Aliphatic Amines by Reaction with 9-Isothiocyanatoacridine. J. Pharm. Sci. *62*, 1370 (1973)

A. C. Mehta, S. G. Schulman: Comparison of Fluorometric Procedures for Assay of Amphetamine. J. Pharm. Sci. *63*, 1150 (1974)

J. M. Meola, H. H. Brown: Fluorescamine as a Fluorogenic Reagent for D-Amphetamine Identification by TLC. Clin. Chem. *20*, 88 (1974)

C. I. Miles, G. H. Schenk: Fluorescence and Phosphorescence of Phenethylamines and Barbiturates: Analysis of Amphetamine and Barbiturate Preparations. Anal. Chem. *45*, 130 (1973)

C. R. Nix, A. S. Hume: A Spectrophotofluorometric Method for the Determination of Amphetamine. J. Forens. Sci. *15*, 595 (1970)

J. T. Stewart, D. M. Lotti: Fluorometric Determination of Amphetamines with 3-Carboxy-7-hydroxycoumarin. J. Pharm. Sci. *60*, 461 (1971)

g. Gas chromatography – mass spectrometry (GC-MS)

H. Brandenberger. D. Schnyder: Toxikologischer Spurennachweis von biologisch wirksamen Aminen (Amphetamine, Catecholamine, Halluzinogene) durch Gas-Chromatographie mit massenspezifischer Detektion (GC-MD). Z. Anal. Chem. *261*, 297 (1972)

A. K. Cho et al.: Deuterium Substituted Amphetamine as an Internal Standard in a Gas Chromatographic/Mass Spectrometric (GC/MS) Assay for Amphetamine. Anal. Chem. *45*, 570 (1973)

B. L. Flamm, J. Gal: Thermal Decomposition of Methylphenidate in the Gas Chromatograph-Mass Spectrometer. Biomed. Mass. Spectrom. *2*, 281 (1975)

M. T. Gilbert, C. J. W. Brooks: Characterisation of Diastereomeric and Enantiomeric Ephedrines by Gas Chromatography Combined with Electron-Impact Mass Spectrometry and Isobutane Chemical-Ionisation Mass Spectrometry. Biomed. Mass Spectrom. *4*, 226 (1977)

L. Strömberg, I. Wistedt: Gas Chromatography – Mass Spectrometry. Rapid Analysis of Methamphetamine, Phentermine, N,N-Dimethylamphetamine, N-Ethylamphetamine and Mephentermine in the Presence of Each Other. Microgram *10*, 59 (1977)

h. Melting point (m. p.)

G. J. Lowell: Determination of Isomeric Composition of Amphetamine Mixtures from Melting Points of Monohydrogen Succinate Salts. J. Pharm. Sci. *61*, 1976 (1972)

W. M. Ment, V. S. Marino: Stereochemical Composition of d- and l-Amphetamine Mixtures by Thermal Analysis of the Benzoyl Derivatives. J. Assoc. Off. Anal. Chem. *53*, 1087 (1970)

4. Lysergic acid diethylamine (LSD)

a. Several methods in parallel or in combination

R. E. Ardrey, A. C. Moffat: A Compilation of Analytical Data for the Identification of Lysergide and its Analogues in Illicit Preparations: J. Forens. Sci. Soc. *19*, 253 (1979)

K. Bailey, D. Verner, D. Legault: Distinction of Some Dialkyl Amides of Lysergic and Isolysergic Acids from LSD. J. Assoc. Off. Anal. Chem. *56*, 88 (1973)

S. W. Bellman, J. W. Turczan, T. C. Kram: Spectrometric Forensic Chemistry of Hallucinogenic Drugs. J. Forens. Sci. *15*, 261 (1970)

C. G. A. Bos, J. G. J. Frijns: Quantitative Analysis of Dihydroergotamine in Pharmaceuticals Using Fluorescent Spectroscopy and Thin-Layer Chromatography. Pharm. Weekbl. *107*, 111 (1972)

A. B. Clark: Lysergic Acid Diethylamide and Lysergic Acid Dimethylamide. Microgram, *6*, 37 (1973)

M. D. Cunningham: Lysergic Acid Diethylamide. Mass Spectroscopy. Microgram *6*, 19 (1973)

I. Jane, B. B. Wheals: The Characterization of LSD in Illicit Preparations by Pressure-Assisted Liquid Chromatography and Gas-Chromatography. J. Chromatogr. *84*, 181 (1973)

A. R. Sperling: Analysis of Hallucinogenic Drugs. J. Chromatogr. Sci. *10*, 268 (1972)

A. R. Sperling: Thin-Layer and Gas Chromatographic Identification of LSD. J. Chromatogr. Sci. *12*, 265 (1974)

M. Steinigen: Nachweis und quantitative Bestimmung von Lysergsäurediäthylamid (LSD). Deut. Apotheker-Ztg. *112*, 51 (1972)

b. Thin-layer chromatography (TLC)

G. V. Alliston, M. J. De Fauber-Maunder, G. F. Phillips: A Novel Thin-Layer Chromatographic System for Lysergide (LSD). J. Pharm. Pharmacol. *23*, 555 (1971)
E. Eich, W. Schunack: Die direkte Auswertung von Dünnschichtchromatogrammen durch Remissions- und Fluorozenzmessungen. IV. Ergotalkaloide. Planta medica *27*, 58 (1975)
R. Fowler, P. J. Gomm, D. A. Patterson: Thin-Layer Chromatography of Lysergide and Other Ergot Alkaloids. J. Chromatogr. *72*, 351 (1972)
B. Newman, V. T. Sullivan, A. Dihrberg: Thin-Layer Chromatography of Lysergic Acid Diethylamide (LSD), N,N-Dimethyl Tryptamine (DMT), Methylene Dioxyphenyl Isopropylamine (STP), and Ibogaine. J. Crim. Law Criminol. *61*, 112 (1970)
T. Niwaguchi, T. Inoue: Studies on Quantitative in Situ Fluorometry of Lysergic Acid Diethylamide (LSD) on Thin-Layer Chromatograms. J. Chromatogr. *59*, 127 (1971)
T. Niwaguchi, T. Inoue: Direct Quantitative Analysis of Lysergic Acid Diethylamide (LSD) and 2,5-Dimethoxy-4-Methyl-Amphetamine (STP) on Thin-Layer Chromatograms. J. Chromatogr. *121*, 165 (1976)
J. Reichelt, S. Kudrnác: Analytical Studies on Ergot Alkaloids and TheirDerivatives. I Separation of Ergot Alkaloids of the Ergotoxine and Ergotamine Groups by Thin-Layer Chromatography. J. Chromatogr. *87*, 433 (1973)
D. A. Stoney, J. I. Thornton: Solvent-Dependent Photolysis for Identification of Lysergic Acid Diethylamide and Other Indolamines. Anal. Chem. *51*, 1341 (1979)

c. Gas chromatography (GC)

S. Agurell, A. Ohlsson: Gas Chromatography of Ergot Alkaloids. J. Chromatogr. *61*, 339 (1971)

d. High pressure liquid chromatography (HPLC)

R. A. Heacock et al.: A Preliminary Investigation of the High-Speed Liquid Chromatography of Some Ergot Alkaloids. J. Chromatogr. *77*, 425 (1973)
J. D. Wittwer, J. H. Kluckholm: Liquid Chromatographic Analysis of LSD. J. Chromatogr. Sci. *11*, 1 (1973)

e. Ultraviolet spectrophotometry (UV)

K. Khashalov: Ultraviolet Spectroscopic Detection and Determination of Lysergic Acid Diethylamide in the Presence of Morphine and Hashish. Farmatsiya *24*, 9 (1974)

f. Spectrophotofluorimetry (F)

R. F. Canaff, P. De Zan: Determination of LSD in Illicit Preparations by Fluorescence Spectroscopy. Microgram *3*, 194 (1970)
D. M. Fabrick, J. D. Winefordner: Quantitative Measurement of Mixtures of Hallucinogens by Fluorimetry and Phosphorimetry. Talanta *20*, 1220 (1973)
V. Gillis, T. A. Kubic: Spectrofluorometric Analysis of Illicit Drug Samples, Employing a Corrected Excitation Spectrofluorometer. Adv. Instrum. *29*, 750 (1974)

g. Mass spectrometry (MS)

K. W. Crawford: The Identification of Lysergic Acid Amide in Baby Hawaiian Woodrose by Mass Spectrometry. J. Forens. Sci. *15*, 588 (1970)
R. W. Urich et al.: Mass Spectral Studies of Ultraviolet Irradiated and Non-Irradiated Lysergic Acid Diethylamide Extracts from Illicit Preparations. Anal. Chem. *47*, 581 (1975)
D. Voigt, S. Johne, D. Groeger: Mass Spectrometry of Ergot Alkaloids. Pharmazie *29*, 697 (1974)

h. Column chromatography

A. R. Sperling: Analysis of LSD-STP Mixtures. J. Forens. Sci. *15*, 86 (1970)

i. Thin-layer electrophoresis

A. Cavallaro, G. Elli: Thin-Layer Electrophoresis of Hallucinogenic Substances. Boll. Lab. Chim. Prov. *23*, 27 (1972)

j. Radio-immunoassay (RIA)

L. J. Loeffler, J. V. Pierce: Radioimmunoassay of Lysergide (LSD) in Illicit Drugs and Biological Fluids. J. Pharm. Sci. *62*, 1817 (1973)

k. Nuclear magnetic resonance (NMR)

N. J. Bach et al.: Carbon-13 Nuclear Magnetic Resonance Spectroscopy of Naturally Occurring Substances. XXI. Nuclear Magnetic Resonance Spectral Analysis of the Ergot Alkaloids. J. Org. Chem. *39*, 1272 (1974)

5. Hallucinogenic drugs in general except for LSD, PCP and cannabis

J. J. Aaron, L. B. Sanders, J. D. Winefordner: Analytical Study of Some Important Hallucinogens by a Combined Fluorimetric and Phosphorimetric Method. Clin. Chim. Acta *45*, 375 (1973)
K. Bailey: The Mass Spectra of Dimethoxyamphetamine Hydrochlorides. Anal. Chim. Acta *60*, 287 (1972)
D. S. Miller: Identification of Psilocybin in Mushrooms. Microgram *5*, 5 (1972)
N. Narasimhachari, J. Spaide, B. Heller: Gas-Liquid Chromatographic and Masspectrometric Studies on Trimethylsilyl Derivatives of N-Methyl and N, N-Dimethyltryptamines. J. Chromatogr. Sci. *9*, 502 (1971)
D. B. Repke et al.: GLC-Mass Spectral Analysis of Psilocin and Psilocybin. J. Pharm. Sci. *66*, 743 (1977)
P. C. White: Analysis of Extracts from *Psilocybe Semilanceata* Mushrooms by High-Pressure Liquid Chromatography. J. Chromatogr. *169*, 453 (1979)
P. De Zan, R. F. Canaff, R. Bianchi: Fluorimetric Characteristics of Some Narcotic and Dangerous Drugs. J. Assoc. Off. Anal. Chem. *54*, 925 (1971)

6. Cannabis sativa (1975–1979)

a. Several methods in parallel or in combination

R. T. Coutts, G. R. Jones: A Comparative Analysis of Cannabis Material. J. Forens. Sci. *24*, 291 (1979)
K. Fonseka, M. Widman, S. Agurell: Chromatographic Separation of Cannabinoids and Their Monooxygenated Derivatives. J. Chromatogr. *120*, 343 (1976)
E. E. Knaus, R. T. Coutts, C. W. Kazakoff: The Separation, Identification, and Quantitation of Cannabinoids and Their t-Butyldimethylsilyl, Trimethylsilylacetate, and Diethylphosphate. J. Chromatogr. Sci. *14*, 525 (1976)
F. J. E. M. Küppers et al.: Cannabis XIV. Pyrolysis of Cannabidiol – Analysis of the Volatile Constituents. J. Chromatogr. *108*, 375 (1975)
Z. Mobarak, D. Bieniek, F. Korte: Studies on Non-Cannabinoids of Hashish Isolation and Identification of Some Hydrocarbons. Chemosphere *3*, 5 (1974)
R. M. Smith, K. D. Kempfert: Identification of Δ^1-3,4-*cis*-Tetrahydrocannabinol in Marijuana. Microgram *10*, 63 (1977)

D. Tiess: Beitrag zur systematischen chemisch-analytischen Bestimmung und Differenzierung von Cannabis-Präparaten mittels Dünnschicht- und Gaschromatographie. Kriminal. Forens. Wiss. *25*, 93 (1975)

B. B. Wheals, R. N. Smith: Comparative Cannabis Analysis. A Comparison of High-Pressure Liquid Chromatography with Other Chromatographic Techniques. J. Chromatogr. *105*, 396 (1975)

b. Thin-layer chromatography (TLC)

M. J. De Faubert Maunder: The Forensic Significance of the Age and Origin of Cannabis. Med. Sci. Law *16*, 78 (1976)

R. Fowler, R. A. Gilhooley, P. B. Baker: Thin-Layer Chromatography of Cannabinoids. J. Chromatogr. *171*, 509 (1979)

R. B. Hughes, R. R. Kessler: Increased Safety and Specificity in the Thin-Layer Chromatographic Identification of Marihuana. J. Forens. Sci. *24*, 842 (1979)

A. B. Segelman, F. P. Segelman: *Cannabis Sativa L.* (Marijuana). VII The Relative Specificity of the RIM test. J. Chromatogr. *123*, 79 (1976)

J. A. Vinson, J. E. Hooyman: Studies in Laboratory-Use Reagents. III Simple Thin-Layer Chromatographic System for the Separation of Cannabinoids. J. Chromatogr. *106*, 196 (1975)

c. Gas chromatography (GC)

K. Bailey, D. Gagne: Distinction of Synthetic Cannabidiol, Cannabichromene and Cannabivarin by Gas-Liquid Chromatography Using On-Column Methylation. J. Pharm. Sci. *64*, 1719 (1975)

K. D. Brunnemann, D. Hoffmann: Chemical Studies on Tobacco Smoke. XXIV. Gas-Chromatographic Determination of Ammonia in Cigaret and Cigar Smoke. J. Chromatogr. Sci. *13*, 159 (1975)

L. Crombie, W. M. L. Crombie: Cannabinoid Bishomologs. Miniaturized Synthesis and GLC (Gas-Liquid Chromatography) Study. Phytochemistry *14*, 213 (1975)

K. Fairlie, B. L. Fox: Rapid, Quantitative Determination of Tetrahydrocannabinol in Marihuana by Gas Chromatography. J. Chromatogr. Sci. *14*, 334 (1976)

L. V. S. Hood, G. T. Barry: Headspace Volatiles of Marihuana and Hashish: Gas Chromatographic Analysis of Samples of Different Geographic Origin. J. Chromatogr. *166*, 499 (1978)

N. K. Mc Callum, E. R. Cairns: Simple Device for GLC Separations of Cannabinoids Using a Surface-Coated Open Tube Column without Stream Splitting. J. Pharm. Sci. *66*, 114 (1977)

Z. Mobarak, D. Bieniek, F. Korte: Some Chromatographic Aspects of Hashish Analysis. II. Forens. Sci. *11*, 189 (1978)

G. M. Patwardhan, M. D. Pundlik, S. K. Meghal: Gas-chromatographic Detection of Resins in Cannabis Seeds. Indian J. Pharm. Sci. *40*, 166 (1978)

K. E. Rasmussen: Analysis of Cannabinoids in Cannabis by Means of Gas-Liquid Chromatography and Solid Injection. Improvements to the Method. J. Chromatogr. *109*, 175 (1975)

K. E. Rasmussen: On-Column Silylation of Cannabinoids after Injection of Solid Plant Material and Cold Trapping. J. Chromatogr. *114*, 250 (1975)

J. C. Schoolar, B. T. Ho, V. S. Estevez: Comparison of various Solvent Extractions for the Chromatographic Analysis of Delta-9-THC and its Metabolites. Marihuana: Chem., Biochem., Cell. Eff., Proc. Satell. Sympos., Houston, Texas, pp. 63–69, 1976

C. E. Turner et al.: Constituents of Cannabis sativa. VIII. Possible Biological Application of a new Method to Separate Cannabidiol and Cannabichromene. J. Pharm. Sci. *64*, 810 (1964)

d. High Pressure liquid chromatography (HPLC)

S. R. Abbott et al.: High Pressure Liquid Chromatography of Cannabinoids as their Fluorescent Dansyl Derivatives. Res. Commun. Chem. Pathol. Pharmacol. *10*, 9 (1975)

S. L. Kanter, M. R. Musumeci, L. E. Hollister: Quantitative Determination of Δ^9-Tetrahydrocannabinol and Δ^9-Tetrahydrocannabinolic Acid in Marihuana by High-Pressure Liquid Chromatography. J. Chromatogr *171*, 504 (1979)

R. N. Smith: High Pressure Liquid Chromatography of Cannabis. Identification of Separated Constituents. J. Chromatogr. *115*, 101 (1975)

P. J. Twitchett, P. L. Williams, A. C. Moffat: Photochemical Detection in High-Performance Liquid Chromatography and its Application to Cannabinoid Analysis. J. Chromatogr. *149*, 683 (1978)

e. Mass spectrometry (MS)

J. F. Fiechtl, G. Spiteller: Neue Cannabinoide -1. Tetrahedron *31*, 479 (1975)
D. J. Harvey, W. D. M. Paton: Use of Trimethylsilyl and Other Homologous Trialkylsilyl Derivatives for the Separation and Characterization of Mono- and Dihydroxycannabinoids by Combined Gas Chromatography and Mass Spectrometry. J. Chromatogr. *109*, 73 (1975)
D. J. Harvey: Characterisation of Butyl Homologues of Δ^1-Tetrahydrocannabinol, Cannabinol and Cannabidiol in Samples of Cannabis by Combined Gas Chromatography and Mass Spectrometry. J. Pharm. Pharmac. *28*, 280 (1976)
M. L. Lee, M. Novotny, K. D. Bartle: Gas Chromatography/Mass Spectrometric and Nuclear Magnetic Resonance Spectrometric Studies of Carcinogenic Polynuclear Aromatic Hydrocarbons in Tobacco and Marijuana Smoke Condensates. Anal. Chem. *48*, 405 (1976)
J. H. Liu, M. P. Fitzgerald, G. V. Smith: Mass Spectrometric Characterization of Cannabinoids in Raw *Cannabis sativa L.* Samples. Anal. Chem. *51*, 1875 (1979)
M. Novotny et al.: Analysis of Marijuana Samples from Different Origins by High-Resolution Gas-Liquid Chromatography for Forensic Application. Anal. Chem. *48*, 24 (1976)
T. B. Vree: Mass Spectrometry of Cannabinoids. J. Pharm. Sci. *66*, 1444 (1977)

7. Phencyclidine and its analogs

K. Bailey, D. R. Gagné, R. K. Pike: Identification of Some Analogs of the Hallucinogen Phencyclidine. J. Assoc. Off. Anal. Chem. *59*, 81 (1976)
K. Bailey, D. Legault: Identification of Analogs of the Hallucinogen Cyclohexamine. J. Assoc. Off. Anal. Chem. *62*, 1124 (1979)
C. C. Clark: Gas-Liquid Chromatographic Quantitation of Phencyclidine Hydrochloride in Powders: Collaborative Study. J. Assoc. Off. Anal. Chem. *62*, 560 (1979)
E. J. Cone et al.: Separation and Identification of Phencyclidine Precursors, Metabolites and Analogs by Gas and Thin-Layer Chromatography and Chemical Ionization Mass Spectrometry. J. Chromatogr. *177*, 149 (1979)
B. Goldston, M. D. Miller: Separation of LSD and PCP by Micro Alumina Column Chromatography. Microgram *4*, 64 (1971)
J. K. Raney, G. T. Skowronski, R. J. Wagenhofer: Gas Liquid Chromatographic Screening Procedure for Components Found in Clandestine Phencyclidine Reaction Mixtures. Microgram *11*, 139 (1978)
K. G. Rao, S. K. Soni, M. Mullen: A Spectroscopic and Chromatographic Study of Phencyclidine (PCP) and its Analogs. Microgram *13*, 52 (1980)
J. E. Smialek et al.: Cyclohexamine ("Rocket Fuel")-Phencyclidine's Potent Analog. J. Anal. Toxicol. *3*, 209 (1979)

II. Explosives

1. Introduction

Explosives are solid, semisolid, or liquid compounds or mixtures of compounds which are prepared and sold for blasting, propulsion, ignition or pyrotechnical effects.

Explosions can be brought about by a large number of other substances, including dust, aerosols and gases, but these substances are not considered to be explosives in the narrower sense described above, nor are those compounds which cannot be handled safely (hydroazoic acid, nitrogen iodide etc).

Explosives can burn with relatively low speed, deflagrate, or detonate. In deflagration, the reaction may not be complete so that intermediate products or even unreacted starting materials can be found afterwards. The explosion itself is a consequence of detonation.

As a rule, commercial explosives require a detonating device for initiating its detonation. The detonator contains a primary explosive (primer) which, in turn, is much more sensitive to initiation by heat, electricity or mechanical shock (percussion). For blasting operations, the detonators are stored separately from the explosive proper and are only adapted shortly before use. In firearm ammunition, the detonator is, as a rule, built into the cartridge case.

2. Classification of Explosives

a. Basic Substances

While there are many ways for classifying this group of substances, one established system, combining chemical and physical properties for subdividing the various groups of explosives has been compiled by Kühne [F]. With slight changes it looks as follows:

I. Monocomponent explosives
 a. Nitric acid esters
 a. Other nitrocompounds
 c. Nitramines
 d. Salts

II. Multicomponent explosives
 a. Black powder
 b. Mixtures of nitro products

III. Primary explosives (primers)
 a. Mercury fulminate
 b. Lead azide
 c. Lead salts of nitrophenols
 d. Diazodinitrophenol
 e. Tetrazene

IV. Propellant powders
 a. Single-Base powders
 b. Double-base powders
 c. Triple-base powders

V. Rocket propellants

VI. Pyrotechnical mixtures

Some of the properties of the explosives listed above are assembled in Table B.6. (p. 76).

b. Commercial Explosives

Commercial explosives generally contain combinations of the basic components described in B.II.2a. The main groups are:

1. Explosives in powder form (d 0.9–1.1)
2. Plastic explosives (d 1.5–1.6)

To group 1 belong black powder, ammonium nitrate, chlorate mixtures (such as chloratite), and – only of historical interest – dynamite on diatomean earth base.

Group 2 comprises mixtures containing nitroglycol, and especially nitroglycerine. The other components of these mixtures can be nitrocellulose, collodium wool, aromatic nitro compounds, sodium nitrate, ammonium nitrate, sulfur, sawdust etc.

The composition of these explosives varies from manufacturer to manufacturer and from country to country. The forensic chemist must therefore establish his own reference collection, based on the products of the national manufactures.

One typical composition for commercial dynamite (ammonia gelatin, US Manufacture) is:

26.2% nitroglycerine
49,6% sodium nitrate
8.0% ammonium nitrate
5.6% sulfur
0.8% calcium carbonate
0.4% nitrocellulose
8.0% filler

Typical dynamites on the Scandinavian market contain:

A. 55% ammonium nitrate 25–30% nitroglycerine + nitroglycol	15–20% dinitrotoluene + nitrocellulose + sawdust + chalk + rhodamine B (dye)
B. 50–53% nitroglycerine 25–28% sodium nitrate 9–12% barium sulfate 2–3% nitrocellulose	0.2–0.7% diammonium phosphate 6–8% sawdust 2–4% wheat flour

3. Purpose of the Analysis

From the forensic point of view, there are two situations in which an analysis for explosives is needed. One is the detection and identification of non-detonated explosives, the other the isolation and identification of explosive residues (cf. sect. C.I.3). If explosives are found prior to their detonation, or after incomplete explosion, rather large amounts can be secured, and an analysis can be carried out by relatively simple techniques. This situation is, however, relatively rare, and the analytical methods must be able to cope with milligram or microgram amounts of explosives or their residues.

4. The Analysis

a. Introduction

The analyst's tasks can be divided into three levels of ambition, i. e.

detection,
identification,
quantitative analysis.

Many articles on systematic approaches to the analysis of explosives, and especially of explosive residues, have recently been published [2, 3, 5, 6, 8, 9, 10, 15, 21, 24, 25, 27–29, 34–36].

b. Detection

Detection of explosives plays a role in the search for explosive devices and for the screening of individuals suspected of having handled explosives. In regions with a high incidence of terrorism, these analyses have become very necessary. But also, the cracking of safes, gasoline automats, night service bank boxes etc. has increased in many countries, requiring investigation.

One relatively sophisticated approach for this purpose is the detection of traces of vapor of the more volatile explosives, notably nitroglycerin and the aromatic nitro compounds (TNT, RDX). The vapor is analyzed in specially constructed gas chromatographs using flame ionization or electron capture detectors. Positive responses are indicated acoustically or on a pen recorder [15, 35]. At least two commercial versions of such instruments are at present available (from Great Britain and Israel). In some cases, sampling can be done "in the field", and analysis of the sample in the laboratory.

Vapors from explosives can be detected at the scene of the explosion in containers or spaces where explosives had been stored, in containers with explosion debris brought to the laboratory, or even from the hands and the clothing of suspects.

The detection of solid residues from explosives is more complicated and time-consuming but it is a necessary prerequisite for the subsequent identification of the explosive. If it is possible to find particles of explosives in the debris of an explosion, e. g. the clothing of a suspect, the vacuum sweepings of his car etc., then the investigator has a great advantage. Not only does he avoid the drawbacks of an extraction step (dilution of the residues, introduction of large amounts of impurities and false positive results from non-explosive chemicals) but he can observe the physical properties of these particles (color, shape, hardness etc.). Debris from explosions should therefore be carefully inspected under a low-power zoom macroscope. Reference samples of explosive products, and appropriate tools (dissecting needle, fine tweezers, petri dishes) shall be within reach of the investigator [15, 20, 34].

As a last resort, extraction of the residues from debris or vacuum sweepings (clothes, car, apartment etc. of suspect) may be used [8, 15, 21, 27]. The extraction can first be carried out with ether which extracts most of nitroglycerine, ethylene glycol dinitrate as well as di- and trinitrotoluene. This is followed by acetone which will dissolve the rest of the nitrated aromatics and most of the nitrated aliphatic compounds. A third extraction, using water, will recover ammonium salts and nitrates. In many cases, the ether extraction can be left out without great losses. The "extraction" of a suspect's hands presents a special case [15, 22]. Cotton swabs, moistened with acetone, are recommended for this purpose.

c. Identification

The isolation of the explosives by mechanical means or by extraction and evaporation of the extracts is followed by chemical analysis. If the particles have been isolated as such, valuable hints as to the type of explosives used have been gathered, and the analytical methods can be selected accordingly.

The choice of the methods used for identification depends on the level of ambition of the investigator or the forensic science laboratory. It may further depend on the seriousness or importance of the crime investigated.

The reliability of the identification but also the amount of work and the complexity of instrumentation increase when the following methods are used:

spot test and color reactions,
thin-layer chromatography (TLC),
gas liquid chromatography (GLC) (limited applicability),
high pressure liquid chromatography (HPLC),
mass spectrometry (MS).

The latter four methods will be discussed in some detail.

In addition, elemental analysis (by chemical means or by X-ray spectrometry), X-ray diffraction [3, 4, 36], infrared spectrophotometry (IR) [3, 5, 21, 24, 25, 29, 30, 36], ultraviolet spectrophotometry (UV) [16, 18, 21], mass spectrometry [37–39], nuclear magnetic resonance (NMR) measurements [13], and mixed melting point determinations can be useful.

Thin-layer Chromatography

Spot tests and color reactions shall not be discussed here separately, since their specificity is quite low, and since the cheap and rapid method of thin-layer chromatography (TLC) is now readily available. In TLC, both R_f values and color reactions are usually combined, giving a much better selectivity.

The first applications of TLC to the analysis of explosives were published in the 1960's, and it would be unwise to list all papers on this subject (more than 20 up to 1978). Before discussing various TLC systems for laboratory use, it should be mentioned that Fisco [10] has described a portable explosion identification kit based on thin-layer chromatography and color reactions.

Midkiff and Washington [25] have evaluated many of the TLC procedures published in this field and cite 12 references. They recommend silica gel plates, and as solvents for

(1) differentiating TNT from DNT: carbon tetrachloride-1,2-dichloroethane (4+1). R_f for TNT 0.40, for DNT 0.50.
(2) separating RDX and HMX: chloroform-acetone (2+1). R_f for RDX 0.80, for HMX 0.55.
(3) Separating PETN and NG: 1,2–dichloroethane-petroleum ether (1+2). R_f for PETN 0.43, for NG 0.53. The separation is not always easy to carry out.

In general, it was found by most authors that a single TLC system cannot be used for separating all explosives. There could be a general screening system, which then is followed by other systems designed for specific groups of explosives. This approach has been suggested by Kaplan and Zitrin [21]. They report the R_f values collected in Table B.7 (p. 77). Parihar et al. [26] have chosen to chromatograph π-complexes of aromatic nitro compounds with aromatic amines with good results. Élie-Calmet and Forestier [8] have written an extensive report on the TLC examination of traces of explosives after detonation. They also used 2-dimensional thin-layer systems.

Gas Chromatography

Early attempts to use gas chromatographic methods in the analysis of explosives were disappointing, due to the low volatility of some of the nitric acid esters (NG,

EDGN), the high volatility of nitroamines (esp. HMX), and the instability of many explosive compounds. In recent years many of the difficulties in GC-analysis of explosives have been overcome. It is important to use all-glass systems, on-column injection, short columns, relatively low concentrations of liquid phase, to take care in preparing column fillings of high standard, and to optimize the temperatures for the injection region and the column proper. Some examples of GC analyses of explosives from published papers are collected in Table B.8 (p. 77 and 78).

High Pressure Liquid Chromatography

The development of high pressure instrumentation for liquid chromatography (HPLC) has made it possible to separate rather complex mixtures into their components. Some of the advantages of HPLC over other chromatographic methods, notably thin-layer chromatography (TLC) and gas liquid chromatography (GLC) are:

1. The chromatogram can be run at room temperature, avoiding decomposition of labile compounds (such as explosives) as in GLC.
2. The speed of the analysis is comparable to TLC and GLC.
3. Quantitative analysis is possible and more exact than in TLC.
4. The effluent fractions can be conveniently collected for subsequent analysis by e.g. infrared spectrophotometry (IR) or mass spectrometry (MS).

Among the disadvantages of HPLC can be mentioned the somewhat lower sensitivity than in GLC and the high price of equipment and columns.

The HPLC techniques published so far in the field of explosives are listed in Table B.9 (p. 79). It can be seen that only a few explosives have hitherto been investigated. It should be mentioned that, whereas acetone followed by water are the solvents of choice for the extraction of explosives from debris (cf. above), acetone is not suitable as a mobile phase due to its strong absorption in the ultraviolet part of the spectrum.

The fractions obtained by HPLC can be collected and used for identification by IR or MS techniques. Whenever the amounts are sufficient for IR spectrophotometry, this method should be used for its high discriminating power. The compilation of IR spectra for 68 explosive compounds published by Pristera et al. 1960 [30] is an excellent reference.

Mass Spectrometry

Whenever the amount of purified explosive compounds is insufficient for infrared spectrophotometry it is desirable to identify these substances by another discriminating method. Mass spectrometry (MS) has now become the method of choice for most discriminating organic chemical analyses. In the case of explosives, however, there are inherent difficulties in the application of MS. As we have seen above in the discussion of gas chromatography, many explosive compounds are either unstable or quite volatile, or both already at moderate temperatures (above, say, 100–150° C). In addition, the fragmentation patterns of some compounds within the same group are frequently very similar.

Published mass spectra of the major explosive substances are collected in Table B.10 (p.). Until 1972, electron impact mass spectrometry (EI-MS) has been used [11, 23, 32]. Due to the instability of many explosives, the molecular peaks were absent (NG, DEGDN) or very small (RDX, HMX, tetryl, TNT). Only nitroguanidine and TNB showed sizable molecular peaks.

Since 1974, several papers on chemical ionization MS (CI-MS) of explosives have appeared. With this technique MH^+ or MNH_4^+ peaks of high intensity for RDX, HMX, PETN, TNT, 2,4-DNT, 4-nitrotoluene, tetryl, PA, and even NG have been obtained [14, 31, 37–39].

The combination gas chromatography – mass spectrometry which is otherwise very useful in analytical chemistry cannot be used with most explosives again due to their instability.

An off-line combination of liquid chromatography (cf. above), collection of fractions and insertion of the residues into the direct inlet probe of the mass spectrometer is at present the method of choice.

Since only few forensic science laboratories have easy access to CI-MS at the present time, mass spectrometry is of limited value in the field of explosives.

With the development of less expensive instrumentation, however, the data of Table B.10 (p. 80 and 81) may become more interesting.

5. Practical Examples

a. Evidence in the Trial of a Dynamiting Gang

In 1977 police arrested four persons suspected of having blown up several night-service bank boxes and two gasoline automats. The laboratory received 50 different exhibits, including pieces of suspected explosive, glass, plastic, metal, cellotape, matches, wire, paint, paint flakes, and sand for comparison purposes. Also sent in were blasting caps and ammunition (for identification) as well as a pair of pliers and several items with toolmarks for comparison purposes. The exhibits had been secured at [1] the crime scenes and [2] in the apartments and automobiles of the four suspects.

Explosives and suspected explosive residues on 23 items were analyzed, and in 14 exhibits positive identification was accomplished.

Two kinds of dynamites were found, which here will be called A and B respectively. A contains about 60% nitroglycerine and nitroglycol, 5% $BaSO_4$, 10% $NaNO_3$ and plasticizers. B contains about 60% NH_4NO_3 and $NaNO_3$, about 25% nitroglycerine and nitroglycol, small amounts of 2,4-DNT and various additives. The results are collected in Table B.11 (p. 81 and 82).

The analyses showed that the same type of explosive [A] was found at the scence of crime no. 1 and on various materials belonging to, or rented by several suspects. The laboratory report ends with the following statement.

"Based on the results of our investigation the following facts can be stated:

1. Materials 1 and I consist of dynamite, type A.
2. Material XI consists of dynamite, type B.
3. There is strong support for the assumption that residues, resembling explosives and isolated from materials II through X, stem from dynamite, type A.
4. There is support for the assumption that also the residue, resembling an explosive, isolated from material 2 stems from dynamite, type A."

b. Dynamite Trace on a Pencil

Burglars had blown open a safe in a rural post-office and disappeared with the equivalent of $ 12.000. No traces of the explosive could be found but the burglars had left a pencil at the scene of the crime. The police suspected that the pencil had been used for pushing the charge into the keyhole of the safe and sent it to the laboratory.

The pencil had traces of an organic material and of white crystals. The crystals gave a positive test for nitrate ions, and their melting point (169° C) corresponded to that of ammonium nitrate. The remainder of the trace material was scraped off the pen and extracted with acetone. Gas chromatography gave the same pattern of nitrated toluenes as for a reference sample of a Swedish brand of dynamite, Dynamex B, which also contains ammonium nitrate.

Conclusion: There were traces of Dynamex B on the pencil sent in by the police.

c. Attempted Linking of a Gang of "Dynamiters" to Two Scenes of Crime

Self-service gas stations in various parts of Sweden were robbed at night, and it was suspected that one and the same gang was responsible for these deeds. The standard self-service pumps in Sweden accept 10 kronor bills, each giving (at present) about 3 liters of gasoline. The robbers had blown up these automatic registers and successfully removed the accumulated bills.

The police from three districts in different parts of Middle Sweden had cooperated and sent in a number of pertinent exhibits from two crime scenes and from a suspected person's home. Eight of these exhibits were analyzed in the laboratory. The nature of these materials and the analytical results obtained are listed in Table B.12 (p. 82 and 83).

Conclusions:

1. Exhibit A 1 corresponds to the "pitch" used in as far as the results show.
2. Exhibit A 2 consists of trinitrotoluene (trotyl).
3. Exhibits A 1 and B 4 correspond to exhibit C 3 (Bostik) as far as the test results show.
4. Exhibits B 1, B 2 and C4 consist of Dynamex (a Swedish brand of dynamite).

Comment: In this case, the linking of the suspect to the two scenes of crime was indirect. In all three places, Bostik was found; but this is a widely used material. The explosive Dynamex B was found in crime scene nr 2 and in the suspect's home. This again is circumstancial evidence. The finding of trinitrotoluene in the suspect's home has no connection with the actual scenes of crime. – The example is, nevertheless, a nice laboratory exercise.

6. General References (In Chronological Order)

A) T. L. Davis: The Chemistry of Powder and Explosives, New York: John Wiley & Sons 1943
B) M. A. Cook: The Science of High Explosives. New York: Reinhold 1958
C) R. McAdam, R. Westwater: Mining Explosives, London: Oliver & Boyd 1958
D) B. T. Federoff, O. E. Sheffield: Encyclopedia of Explosives and Related Items. Dover N. J.: Picatinni Arsenal. First volume 1960; new volumes are being published
E) R. E. Kirk, D. F. Othmer: Encyclopedia of Chemical Technology, vol. 8. New York: Interscience 1965
F) S. Kühne et al. Sprengstoffe. In: Ullmans Encyklopädie der technischen Chemie, 3rd edit., vol, 16. Foerst, W. (ed.). Berlin: Urban&Schwarzenbach 1965, pp. 56–108
G) T. Urbanski: Chemistry and Technology of Explosives, New York: Pergamon Press 1967
H) R. Meyer: Explosivstoffe, 5th edit. Essen: Verlag Chemie 1979
I) H. J. Yallop: Explosion Investigation, Harrogate, England: The Forensic Science Society 1980

7. Specific References

1) B. J. Alley, H. W. H. Dykes: Gas-Liquid Chromatographic Determination of Nitrate Esters Stabilizers and Plasticizers in Nitrocellulose-Base Propellants. J. Chromatogr. *71*, 23 (1972)
2) A. Basch, S. Kraus: Analysis and Characterization of Military-Grade Trinitrotoluene by Gas Chromatography. J. Forens. Sci. *24*, 870 (1979)
3) A. D. Beveridge, et al. Systematic Analysis of Explosive Residues. J. Forens. Sci. *20*, 431 (1975)
4) D. V. Canfield, P. R. De Forest: The Use of the Gandolfi Camera as a Screening and Confirmation Tool in the Analysis of Explosive Residues: J. Forens. Sci. *22*, 337 (1977)
5) D. E. Chasan, G. Norwitz: Qualitative Analysis of Primers, Tracers, Igniters, Incendiaries, Boosters, and Delay Compositions on a Microscale by Use of Infrared Spectroscopy. Microchem. J. *17*, 31 (1972)
6) F. J. Conrad, T. A. Burrows, W. D. Williams: Selection of a Gas Chromatographic Material for Use in Explosives Vapor Preconcentration: J. Chromatogr. *176*, 37 (1979)
7) I. W. F. Davidson, F. J. Dicarlo, E. I. Szabo: Gaschromatographic Separation and Detection of Pentaerythrol Nitrates and other Organic Nitrate Esters. J. Chromatogr. *57*, 345 (1971)
8) J. Élie-Calmet, H. Forestier: Characterisation of Explosives Traces After an Explosion. Intern. Crim. Police Rev. Part I: *28*, 99 (1974), Part II: *34*, 38 (1979), Part III: *34*, 62 (1979)
9) M. Fariwar-Mohsenie, E. Ripper, K. H. Habermann: Analyse von Explosivstoffen durch Hochdruck-Flüssig-Chromatographie: Fresenius Z. Anal. Chem. *296*, 152 (1979)
10) W. Fisco: A Portable Explosives Identification Kit for Field Use. J. Forens. Sci. *20*, 141 (1975)
11) R. T. M. Fraser, N. C. Paul: The Mass Spectrometry of Nitrate Esters and Related Compounds. Part 1: J. Chem. Soc. *B* 1968, 659
12) D. H. Freeman, R. M. Angeles, I. C. Poinescu: High-Performance Liquid Chromatographic Separation of RDX and HMX Explosives on Adsorptive Polymers: J. Chromatogr. *118*, 157 (1976)
13) D. G. Gehring, G. S. Reddy: Nuclear Magnetic Resonance Examination and Determination of the Di- and Trinitrotoluene Isomers in 2, 4, 6-Trinitrotoluene. Anal. Chem. *40*, 792 (1968)
14) R. G. Gillis, M. J. Lacey, J. S. Shannon: Chemical Ionisation Mass Spectra of Explosives. Organ. Mass Spectrom. *9*, 359 (1974)
15) C. M. Hoffman, E. B. Byall: Identification of Explosive Residues in Bomb Scene Investigations. J. Forens. Sci. *19*, 54 (1974)

16) J. Hoffsommer, J. McCullough: Quantitative Analysis of Polynitroaromatic Compounds in Complex Mixtures by Combination Thin-Layer Chromatography and Visible Spectrometry. J. Chromatogr. *38*, 508 (1968)
17) J. Hoffsommer: Quantitative Analysis of Nitro Compounds in the Micro Picogram Range by a Combination of Thin-Layer and Vapor Phase Chromatography with the Nickel -63 Electron Capture Detector: J. Chromatogr. *51*, 243 (1970)
18) J. Hoffsommer, D. Glover: Capillary Decomposition of Explosives and Analysis by Either Vapor-Phase Chromatography or the Combination of Thin-Layer Chromatography and Visible Spectrometry. J. Chromatogr. *62*, 417 (1971)
19) J. C. Hoffsommer, J. M. Rosen: Analysis of Explosives in Water. Bull. Environm. Contamin & Toxicol. *7*, 177 (1972)
20) J. E. House Jr., P. J. Zack: Thermal Decomposition of Nitrocellulose Propellants. J. Forens. Sci. *22*, 332 (1977)
21) M. A. Kaplan, S. Zitrin: Identification of Post-Explosion Residues. J. Assoc. Off. Anal. Chem. *60*, 619 (1977)
22) C. R. Kempe, W. K. Tannert: Detection of Dynamite Residues on the Hands of Bombing Suspects. J. Forens. Sci. *17*, 323 (1972)
23) J. T. Larkin, F. E. Saalfeld, L. Kaplan: Electron-Impact Fragmentation of Alipathic Polynitro Compounds. Org. Mass Spectrom. *2*, 213 (1969)
24) C. R. Midkiff Jr., W. D. Washington: Systematic Approach to the Detection of Explosive Residues. III Commercial Dynamite. J. Ass. Off. Anal. Chem. *57*, 1092 (1974)
25) C. R. Midkiff Jr., W. D. Washington: Systematic Approach to the Detection of Explosive Residues. IV Military Explosives. J. Ass. Off. Anal. Chem. *59*, 1357 (1976)
26) D. B. Parihar, S. P. Sharma, K. K. Verma: Trace Analysis of Explosives as Pi Complexes. J. Forens. Sci. *13*, 246 (1968)
27) R. G. Parker et al.: Analysis of Explosives and Explosive Residues. Part 1: Chemical Tests. J. Forens. Sci. *20*, 133 (1975)
28) R. G. Parker, J. M. McOwen, J. A. Cherolis: Analysis of Explosives and Explosive Residues. Part 2: Thin-Layer Chromatography. J. Forens. Sci. *20*, 254 (1975)
29) R. G. Parker: Analysis of Explosives and Explosive Residues. Part 3: Monomethylamine Nitrate. J. Forens. Sci. *20*, 257 (1975)
30) F. Pristera et al.: Analysis of Explosives Using Infrared Spectroscopy. Anal. Chem. *32*, 495 (1960)
31) R. Saferstein, J. M. Chao, J. Manura: Isobutane Chemical Ionization Mass Spectrographic Examination of Explosives. J. Ass. Off. Anal. Chem. *58*, 734 (1975)
32) F. Volk, H. Schubert: Massenspektrometische Untersuchung von Explosivstoffen. Explosivstoffe *16*, 2 (1968)
33) P. Vouros et al.: Analysis of Explosives by High Performance Liquid Chromatography and Chemical Ionization Mass Spectroscopy. Anal. Chem. *49*, 1039 (1977)
34) W. D. Washington, C. R. Midkiff Jr.: Systematic Approach to the Detection of Explosive Residues. I Basic Techniques. J. Ass. Off. Anal. Chem. *55*, 811 (1972)
35) W. D. Washington, C. R. Midkiff Jr.: Systematic Approach to the Detection of Explosive Residues. II Trace Vapor Analysis. J. Ass. Off. Anal. Chem. *56:* 1239 (1973)
36) W. D. Washington, R. J. Kopec, C. R. Midkiff Jr.: Systematic Approach to the Detection of Explosive Residues. V Black Powders. J. Ass. Off. Anal. Chem. *60*, 1331 (1977)
37) J. Yinon, H. G. Boettger, W. P. Weber: Negative Ion Mass Spectrometry – A New Analytical Method for Detection of Trinitrotoluene. Anal. Chem. *44*, 2235 (1972)
38) J. Yinon: Identification of Explosives by Chemical Ionization Mass Spectrometry using Water as Reagent. Biomed. Mass Spectrom. *1*, 393 (1974)
39) S. Zitrin, J. Yinon: Chemical Ionization Mass Spectra of 2, 4, 6-Trinitroaromatic Compounds. Org. Mass Spectrom. *11*, 388 (1976)

8. Tables B.6–12

Table B. 6. Major types compounds used in explosives

Type (cf. text) and abbrev. used in this chapter	Names other than abbreviation	Molecular weight	mp (°C)	bp 760 mm (°C)	D_4^{20}
Ia EGDN	Nitroglycol, ethylene glycol dinitrate	$C_2H_4N_2O_6$ =152.1	−2	197.5	1.49
Ia NG	nitroglycerine, glycerine trinitrate, trinitrine	$C_3H_5N_3O_9$ =227.1	13.5	~256 expl.	1.60
Ia DEGDN	Diethylene glycol dinitrate	$C_4H_8N_2O$ =196.1	3	–	1.39
Ia PETN	Pentaerythrol tetranitrate, nitropenta, PTN, pentyl	$C_5H_8N_4O_2$ =316.2	142	150 dec.	1.77
Ia	Nitromannite, mannitol hexanitrate	$C_6H_8N_6O_{18}$ =452.2	113	~160	1.60
Ia	Nitrocellulose, cellulose nitrate	–	–	160 ign.	1.66
Ia	Nitrated starch	–	–	–	–
Ib TNB	1,3,5-trinitrobenzene	$C_6H_3N_3O_6$ =213.1	122	–	1.48
Ib MNT	p-nitrotoluene	$C_7H_7NO_2$ =137.1	52	238	1.30
Ib 2,4-DNT	2,4-dinitrotoluene	$C_7H_6N_2O_4$ =182.1	71	300 dec.	1.32
Ib 2,5-DNT	2,5-dinitrotoluene	$C_7H_6N_2O_4$ =182.1	53	–	1.28
Ib 2,6-DNT	2,6-dinitrotoluene	$C_7H_6N_2O_4$ =182.1	66	–	1.28
Ib	2,4,5-tri-nitrotoluene	$C_7H_5N_3O_6$ =227.1	104	240 expl.	–
Ib PA	Picric acid, 2,4,6-trinitro-phenol, lyddite, ecrasite	$C_6H_3N_3O_7$ =229.1	122	300 expl.	1.77
Ib TNT	2,4,6-trinitro-toluene, trotyl	$C_7H_5N_3O_6$ =227.1	81	240 expl.	1.65

Table B. 6 (continued)

Type (cf. text) and abbrev. used in this chapter	Names other than abbreviation	Molecular weight	mp (°C)	bp 760 mm (°C)	D_4^{20}
Ib	Hexanitro-diphenylamine, dipicrylamine, hexamine	$C_{12}H_5N_7O_{12}$ =439.2	204 dec.	–	1.65
Ic RDX	Cyclonite, hexogen, hexahydro-1,3,5-trinitro-s- triazine	$C_3H_6N_6O_6$ =222.1	204 dec.	–	1.82
Ic HMX	HDX, homocy-clonite, octogen, cyclotetramethy-lene-tetranitramine	$C_4H_8N_8O_8$ =296.1	280	–	1.96
Ic	Nitroguanidine	$CH_4N_4O_2$ =104.1	230 dec.	–	–
Ic Tetryl	Tetranitro-methylaniline, 2.4.6-trinitro-phenylmethyl-nitramine	$C_7H_5N_5O_8$ =287.2	130	187 expl.	1.57
Id	Ammonium nitrate	$H_4N_2O_3$ =80.0	170	210	1.72
Id	Ammonium picrate	$C_6H_6N_4O_7$ =246.1	280	300 expl.	1.72
IIa	Black powder	75% KNO_3 15% charcoal 10% sulfur	–	–	–
IIIa	Mercury fulminate	$Hg(CNO)_2$ =284.6	expl.	–	4.4
IIIb	Lead azide	$Pb(N_3)_2$ =291.2	–	350 expl.	–
IIIc	Lead trinitrores-orcinate, trizinate	$C_6HN_3PbO_8$ =370.2	276 dec.	–	3.00
IIId	Diazodinitro-phenol	$C_6H_2N_4O_5$ =210.1	157	–	1.63
IIIe	Tetrazene	$C_2H_8N_{10}O$ =188.2	140	–	1.65
IIIe	Methylamine nitrate	$CH_6N_2O_3$ =94.1	–	–	–
IVa	Nitrocellulose (cf. even Ia)				
IVb	Nitrocellulose and nitroglycerine, diglycol dinitrate etc.				
IVc	As for IVb and e. g. nitroguanidine				

Table B.7. R_f values and colors produced by explosives in 4 TLC systems[a]

	System			Spray reagent		
Explosive	1,2[b]	3[c]	4[d]	KOH	Griess	o-Tolidine[e]
2,4-DNT	0.58					Pale brown
TNT	0.59			Brown-red		Gray-violet
TNB	0.54			Orange-brown		Gray-violet
Tetryl	0.40			Orange-brown		Gray
EGDN	0.47	0.26			Rose	
NG	0.47	0.13			Rose	
PETN	0.55	0.08			Rose	
RDX	0.15		0.58		Rose	
HMX	0.05		0.25		Rose	

[a] cf. Ref. [21].
[b] System 1 and 2: Silica gel plates. Trichloroethylene-acetone (4+1).
[c] System 3: Aluminium oxide plates. Petroleum ether-1,2-dichloroethane (4+1).
[d] System 4: Aluminium oxide plates. 1,2-Dichloroethane-acetonitrile (9+1).
[e] For system no. 2.

Table B.8. Gas chromatographic analysis of explosives

Ref. and year	Detector	Column/temp.	Filling	Compound	Retention time (min)
[a] 1970	ECD (Ni-63)	4 ft. × 1/4 inch glass, temp. 155° C	2.53% apiezon on 60/80 mesh Diataport-S	RDX TNB 2,3,4-TNT 1,2-DNB	6.3 4.0 3.6 0.64
[b] 1971	FID and ECD (T)	67 inch × 1/4 inch glass, temp. program 65° C, then 10° C/min to 170° C	1% SE-30 on 100/120 mesh Sililated Chromosorb P	PETN*) Erythrol tetranitrate*) Mannitol hexanitrate*)	7.8 6.3 10.1
		72 inch × 1/4 inch glass, temp. program 65° C, then 10° C/min to 170° C	1% dexsil on 100/120 mesh Sililated Chromosorb W	PETN*) Erythrol tetranitrate*) Mannitol hexanitrate*)	9.6 8.2 11.6
[c] 1972	ECD (Ni-63)	4 ft. × 1/4 inch glass, temp. 165° C	2.92% dexsil 300 GC on 80/100 mesh chromosorb WAWDMCS	Dinitrobenzene TNT 2.31 RDX 6.09 Tetryl 14.5	1.00 (int. stand.)

Table B. 8 (continued)

Ref. and year	Detector	Column/temp.	Filling	Compound	Retention time (min)		
					No. 1	No. 2	No. 3
[d] 1972	FID	2 ft. × 1/8 inch stainless steel, temp. program 70° C, then 6° C/min to 220° C	Column no. 1 = 3.8% OV-101 on gas-chrom. Q Column no. 2 = 2.5% OV-210 on chromosorb W-HP Column no. 3 = 1.1% OV-225 on gas-chrom. Q	DEGDN	4.78	6.60	9.16
				NG	5.56	7.91	11.7
				2,6-DNT	6.07	7.42	10.3
				2,4-DNT	7.37	9.18	11.9
				3,4-DNT	8.77	11.0	14.3
				TNT	10.8	12.9	16.8
				2,5-DNT	16.2	18.4	23.0
[e] 1977	FID and ECD (Ni-63)	60–100 cm × 1/4 inch glass, temp. 160° C	30% OV-17 on 100 mesh gas-chrom. Q	2,6-DNT	0.70		
				2,4-DNT	0.91		
				TNT	2.18		
				TNB	2.41		
				PETN**)	2.50		
				RDX**)	7.71		
				tetryl	14.1		
					No. 1[g]	No. 2[g]	No. 3[g]
[f] 1979	FID and EC (Ni-63)	1.5 m × 1/4 inch glass with 60/80 mesh chromosorb W, acid washed, silanized	no. 1: 10% OV-225 at 220° C no. 2: 1.8% OV-225 + 10.2% OV-101 at 200° C no. 3: 0.6% OV-225 + 2.4% poly (phenyl ether) at 185° C	2,6-DNT	0.67	0.75	0.64
				2,5-DNT	0.82	0.86	0.86
				2,4-DNT	1.00	1.00	1.00
				2,3-DNT	1.18	1.08	0.94
				3,4-DNT	1.72	1.44	1.39
				2,4,6-TNT	2.65	2.19	3.14
				2,4,5-TNT	3.04	2.42	2.89
				trinitro benzene	3.27	2.50	3.95
				2,3,5-TNT	4.92	3.77	5.03
				2,3,4-TNT	5.52	4.17	4.58

*) as trifluoroacetates
**) possibly decomposition products
[a] cf. Ref. [17]
[b] cf. Ref. [7]
[c] cf. Ref. [19]
[d] cf. Ref. [1]
[e] cf. Ref. [21]
[f] cf. Ref. [2]
[g] relative retention times for trinitrobenzene = 1.00

Table B. 9. High Pressure Liquid chromatography of explosives

Ref. and year	Column	Detector(s)	Mobile phase	Retention time (min) for explosive compounds
[a] 1974	Corasil II (Waters Assoc.) 1 m glass	Refractive index and UV (254 nm)	60% hexane, 40% methylene chloride	Toluene (5.7), MNT (6.8), 2,4-DNT (8.9), TNT (11.2)
			90% cyclohexane, 10% dioxane	TNT (4.6), tetryl (15.2)
			70% cyclohexane, 30% dioxane	TNT (5.4), tetryl (6.4), RDX (20.1)
			65% cyclohexane, 35% dioxane	RDX (7.3), HMX (9.6)
			84% cyclohexane, 16% chloroform	NG (6.1)
			90% cyclohexane, 10% chloroform	NG (9.9)
			90% cyclohexane, 10% methylene chloride	NG (~8)
[b] 1976	5 different absorbants	Refractive index	Acetone	Separation of RDX and HMX under various conditions
[c] 1977	10-μm	UV (254 nm)	1,2-dichloroethane	TNT (~4), RDX (~5.5), HMX (~12)
	Partisil (Whatman) 25 cm × 3.1 mm		2% 2-propanol 98% heptane	NG (0); not retained
[d] 1978	μ-Bondapak C_{18} (Waters Assoc.) 30 cm × 3.9 mm)	UV (254 nm)	50% methanol 50% H_2O	EGDN (6), NG (~10) 2,4-DNT (~13) 2,6-DNT (~13)
[e] 1978	μ-Porasil (Waters Assoc.) 30 cm × 3.9 mm	UV (254 nm)	70% isooctane 30% methylene-chloride	2,6-DNT (~4) 2,4-DNT(~5) 2,4,6-TNT (~8)
[f] 1979	Lichrosorb (Merek) 25 cm × 6.3 mm	UV (254 nm)	n-hexane 1,2-dichloro-ethane 2-propanol (various proportions)	NG xylol 2,4-DNT and six others (various times)

[a] J. O. Doali, A. A. Juhasz: Application of High Speed Liquid Chromatography to the Qualitative Analysis of Compounds of Propellant and Explosives Interest. J. Chromatogr. Sci. *12*, 51 (1974)
[b] cf. Ref. [12]
[c] cf. Ref. [33]
[d] L. Svensson, Swedish National Laboratory of Forensic Science, unpublished results
[e] same as [d]
[f] cf. Ref [9]

Table B. 10. Mass spectrometry by direct inlet of explosives, 1968–1977

Ref. and year	Mode of operation (EI = electron impact, CI = chemical ionization) and experimental parameters	Explosives analyzed (no complete list) and evaporation temperature if published	The five most intensive peaks obtained (base peaks in italics, molecular peaks in bold face). Relative intensities given in brackets
[a] 1968	EI, 70eV	Nitroguanidin (80° C)	31 (77), 42 (86), *58*, 74 (43), **104** (90)
	ion source at 250° C	RDX (120° C)	*28*, 30 (77), 42 (62), 45 (62), **222** (0.04)
		HMX (160° C)	*28*, 30 (67), 42 (75), 46 (65), **296** (0.03)
		Tetryl (100° C)	30 (30), 181 (35), 224 (28), *241*, **287** (4)
		TNB (60° C)	30 (93), 74 (65), *75*, 120 (14), **213** (32)
		TNT (60° C)	76 (53), 89 (29), 134 (27), *210*, **227** (3)
[b] 1968	EI, 70eV	DEGDN	29 (11), 30 (13), 45 (11), *46*, 73 (38), no mol. peak
		NG	28 (6), 29 (15), 30 (24), *46*, 76 (9), no mol. peak
[c] 1972	EI, 2 eV negative ion MS	TNT	*46*, 120 (40), 151 (80), 181 (40), **227** (~2)
[d] 1974	CI: H_2, 500 eV	RDX (120° C)	*75*, 105 (8), 149 (30), **223** (MH^+, 4), 252 (2)
		HMX (230° C)	*75*, 105 (70), 149 (62), **297** (MH^+, 10), 326 (15)
		PETN (100° C)	*46*, 48 (49), 272 (10), **317** (MH^+, 75), 362 (10)
[e] 1974	CI: H_2O	RDX (170° C)	93 (52), 149 (16), **223** (MH^+, 14), *241*, 445 (19)
		HMX (265° C)	*75*, 93 (94), 223 (14), 244 (18), **297** (MH^+, 7)
		PETN (160° C)	46 (52), 64 (43), 268 (17), **317** (MH^+, 70), *362*
	CI: CH_4	RDX (170° C)	75 (62), 103 (60), 105 (41), *149*, **223** (MH^+, 16)
		HMX (265° C)	*75*, 102 (22), 104 (26), 149 (22), 223 (2), no mol. peak
		PETN (160° C)	30 (9), *46*, 254 (1), **317** (MH^+, <1)
	CI: isobutane	RDX (170° C)	30 (38), 104 (25), *131*, 176 (84), no mol. peak, 252
		HMX (265° C)	30 (26), 86 (24), 104 (36), *131*, 176 (27), no mol. peak
[f] 1975	CI: isobutane 300 eV	2,4-DNT (250° C)	153 (10), **183** (MH^+, base peak)
		4-Nitrotoluene (250° C)	**138** (MH^+, base peak) –
		Tetryl (250 °C)	242 (56), *243*, 283 (18), **288** (MH^+, 29)
		NG (room temp.)	73 (27), 88 (19), *117*, no mol. peak
		RDX (250° C)	149 (85), 117 (23), *149*, 178 (30), **223** (MH^+, 56)
		HMX (250° C)	149 (58), 117 (17), *149*, 189 (16), 223 (39), no mol. peak
		PETN (250° C)	76 (41), **317** (MH^+, 24), 346 (15), 359 (37), *362*
[g] 1976	CI: CH_4 70 or 300 eV	TNB (150–200° C)	184 (14), 185 (5), 200 (29), **213** (5), *214*
		TNT (150–200° C)	198 (2), 210 (24), 211 (9), 214 (4), **228** (MH^+, base peak)
		PA (150–200° C)	199 (8), 200 (18), 213 (12), **229** (19), *230*
		Tetryl (150–200° C)	224 (26), 241 (48), 242 (52), *243*, **288** (MH^+, 1)

Table B. 10 (continued)

Ref. and year	Mode of operation (EI = electron impact, CI = chemical ionization) and experimental parameters	Explosives analyzed (no complete list) and evaporation temperature if published	The five most intensive peaks obtained (base peaks in italics, molecular peaks in bold face). Relative intensities given in brackets
	CI: isobutane 70 or 300 eV	TNB (150–200° C)	184 (28), 185 (11), 200 (17), **213** (2), *214*
		TNT (150–200° C)	210 (22), 211 (7), 214 (6), **227** (4), *228*
		PA (150–200° C)	200 (75), 201 (6), 216 (7), **229** (1), *230*
		Tetryl (150–200° C)	242 (30), *243*, 272 (5), **288** (MH^+, 5)
[h] 1977	CI: NH_3	RDX (100° C)	176 (57), **240** ($M+NH_4^+$, base peak), 398 (37), 427 (22), 445 (9)
		HMX (100° C)	104 (10), 130 (12), 176 (44), **314** ($M+NH_4^+$, base peak), 315 (10)
		TNT (100° C)	134 (20), 210 (63), 211 (22), **228** ($M+NH_4^+$, base peak) 229 (13)
		NG (room temp.)	*76*, 118 (25), 149 (7), 165 (11), **228** ($M+NH_4^+$, 23)

[a] cf. Ref. [32]
[b] cf. Ref. [11]
[c] cf. Ref. [37]
[d] cf. Ref. [14]
[e] cf. Ref. [38]
[f] cf. Ref. [31]
[g] cf. Ref [39]
[h] cf. Ref. [33]

Table B. 11. Analytical results from a case of several burglaries by blasting

Exhibits[a]	Isolation method	Compounds or elements identified	Methods used	Probable type of dynamite (containing:)
1 and 2	Extraction with 1,2-dichloro-ethane-iso-octane	NG DEGDN Na, Ba, S	Quantitative HPLC Quantitative HPLC energy dispersive X-ray spectrometry in a scanning electron microscope	A (NG, DEGDN, $BaSO_4$, $NaNO_3$)
I–X	Collection by mechanical means			
XI	Extraction with 1,2-dichloro-ethane-iso-octane	NG DEGDN 2,4-DNT NH_4 NO_3	Quantitative HPLC Quantitative HPLC Quantitative HPLC wet chemistry wet chemistry	B (NG, DEGDN, NH_4NO_3, $NaNO_3$, 2,4-DNT)

Footnote see p. 82

Table B.11 (continued)

[a] The following exhibits were analyzed:

1:	red material from a scene of crime 1 (petrol station)
2:	material from a metal part belonging to a petrol pump (scene of crime 2)
I:	red material from the home of a suspect
II:	red material from the back seat of car no. 1 (rented by a suspect)
III–V:	material from different parts of car no. 2 (rented by another suspect)
VI and VII:	material from the clothes of a suspect
VIII:	material from the gloves of another suspect
IX and X:	material from the clothes of a suspect
XI:	material from under the driver's seat of car no. 1 (cf. above)

Table B.12. The nature of the eight exhibits and the results obtained by analyzing them (case c)

Desig-nation	Description of the material	Location of the material	Detection methods used[a]	Results obtained
A 1	Grey, sticky mass	At the blasted gasoline automat of a gas station (district no. 1)	1. Color 2. Consistency 3. PGC 4. SEM/X-Ray	Identical results of all 4 tests for A 1, B 4 and C 4
B 4	Grey, sticky mass	At the blasted gasoline automat of a gas station (district no. 2)		
C 3	Grey, sticky mass of the brand Bostik	At the home of the suspect		
A 2	Black, sticky material	At the blasted gasoline automat of gas station (district no. 2)	1. Color 2. Stickiness 3. PcGC	Identical results of the 3 tests for both materials
–	Black, sticky material	From the laboratory's reference collection of fuses		
B 1	Suspected residue of an explosive	At the blasted gasoline automat of a gas station (district no. 2)	1. Nitrate ion test 2. Ammonia test 3. HPLC	Identical results for the 3 tests for B 1, B 2, C 4 and the reference sample
B 2	Traces of suspected explosive on a detonating cap	On the ground close to a blasted gasoline automat (district no. 2)		
C 4	Three cartridges of a Swedish brand of dynamite (Dynamex B)	At the home of the suspect		
–	Dynamex B	From the laboratory's reference collection of explosives		

Table B. 12 (continued)

Designation	Description of the material	Location of the material	Detection methods used[a]	Results obtained
C 5	Pieces of compressed yellow sticks	At the home of the suspect	1. Melting point (81° C) 2. Yellow color 3. IR 4. HPLC	Identical results of all 4 tests for C 5 and the reference sample
	Trinitrotoluene (Trotyl)	From the laboratory's reference collection of explosives		

[a] The following abbreviations are used:
PGC: Pyrolysis gas chromatography
SEM/X-ray: Energy-dispersive X-ray analysis using a scanning electron microscope
HPLC: High-pressure liquid chromatography
IR: Infrared spectrophotometry

III. Polymers

1. Introduction

Polymers occur in nature and some of them have been used before the advent of the synthetic plastics industry. Among natural sources for plastics cellulose, rubber, casein, and various resins can be mentioned. The middle of the 19th century saw the introduction of synthetic polymers, but their industrial production was limited until the beginning of the present century. After the Second World War the development of new synthetic polymers accelerated and at present the world production of synthetic polymers has been estimated at around 50 million tons per years. This output places the plastics industry on the top of the synthetic chemical industry as far as bulk is concerned.

Table B.13 (p. 103) lists some of the major developments in the synthetic polymer field since 1839.

It is surprising that relatively few publications in the forensic science field are concerned with the investigation of plastic materials. The vast occurrence of these materials in industry, homes, automobiles, and applicances of countless nature should rightly be reflected in crime scene investigations and analysis in the forensic laboratories.

Due to the complexity of the field, only a short treatment of the analysis of synthetic polymers is possible in this book. Detailed treatises can be found in the ample literature on this subject, and the thorough analysis of particular compounds in important forensic cases usually requires the assistance of specialists in the plastics field.

We shall try to give a short survey of the classes of materials produced around 1980, and will discuss a few of the analytical methods available for the microanalysis of plastics. This discussion will also include some fiber polymers and thus touch on problems belonging to Chap. B. V. The synthetic methods used for the production of the polymers will not be covered in any detail but must be studied in other works.

2. Chemical Composition of the Polymers Proper

There exist numerous systems for classifying semisynthetic and synthetic polymers depending on the practical use of the system. We shall try to regard the various products from the point of view of the analytical chemist.

Polymers may be obtained by modification of natural products such as caoutchouc, cellulose, proteins or resins, or they may be synthesized from small molecules. Their physical properties are mainly determined by their size and spatial structure whereas their chemical properties depend largely on the active groups on the surface of the molecule. A rough group division of polymers is usually based on their mode of formation. In principle, they are formed by two processes, i.e., addition polymerization and condensation polymerization.

Addition polymerization is a chain reaction, initiated by activation of the reacting molecules by the formation of free radical or of an electric charge. This process is started by *initiators* (peroxides, azo compounds, photosensitizers and redox systems such as metal ions etc.). No byproducts are formed. The reaction is terminated by *terminators* or *inhibitors* (oxygen, iodine, quinones, polycyclic aromatic hydrocarbons etc.).

An example of a *free radical polymerization* is the formation of a vinyl polymer:

chain initiation:

$$ROOR \xrightarrow{100\,^{\circ}C} 2\,RO^{\cdot}$$

$$RO^{\cdot} + CH_2{=}CH{-}OCOCH_3 \longrightarrow RO{-}CH_2{-}\dot{C}H{-}OCOCH_3$$

vinyl acetate

chain propagation:

$$RO{+}CH_2{-}\underset{\textstyle OCOCH_3}{\underset{|}{CH}}{+}_n CH_2{-}\dot{C}H{-}OCOCH_3 + CH_2{=}CH{-}OCOCH_3$$

$$\longrightarrow RO{+}CH_2{-}\underset{\textstyle OCOCH_3}{\underset{|}{CH}}{+}_{n+1}{-}CH_2\dot{C}H{-}OCOCH_3$$

chain termination:

$$2P^{\cdot} \longrightarrow P{-}P$$

An example of *addition polymerization* through functional groups is the synthesis of polyurethanes: Dialcohols and diisocyanates react by additions and shifts of hydrogen atoms:

$$HO{-}R_1{-}OH + OCN{-}R_2{-}NCO + HO{-}R_1{-}OH + OCN{-}R_2{-}NCO{-}$$
$$\downarrow$$
$${-}O{-}R_1{-}O{-}CO{-}NH{-}R_2{-}NH{-}CO{-}O{-}R_1{-}O{-}CO{-}$$

The name is derived from the urethane moiety (–O–CO–NH–) formed in this process. Many polyurethanes contain high-molecular polyalcohols and only relatively few urethane groups – nevertheless these products are also designated as polyurethanes.

Condensation polymerization is a stepwise reaction where low-molecular compounds such as water, alcohols, hydrogen chloride, ammonia etc. are formed as byproducts.

An example of condensation polymerization is the formation of Dacron by condensation of ethylene glycol and the diethyl ester of terephthalic acid:

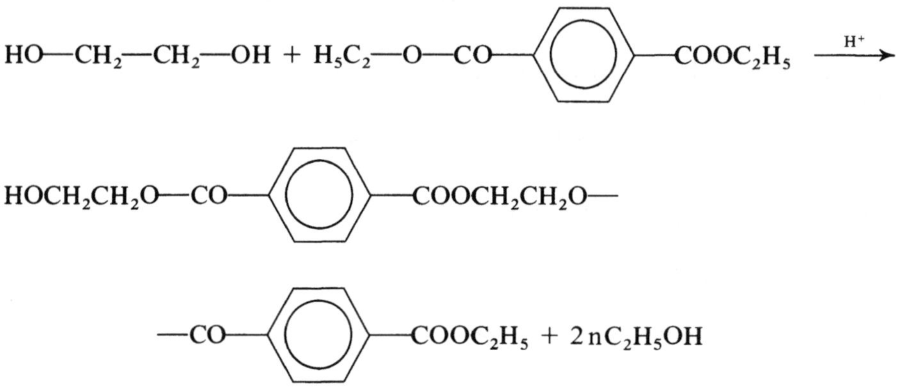

If the subunits of a polymer are all the same, one designates the product as a *homopolymer*. *Copolymers* consist of different monomers.

Linear polymers are usually soft, melt on heating and are soluble in organic solvents. *Cross-linked polymers* consist of monomers carrying more than two reactive groups. The cross-linkages lead to the formation of networks, and the resulting polymers are usually relatively hard, do not melt but decompose at high temperatures and are only slightly soluble in organic solvents.

The major groups of polymers can be classified as follows:

A. Polymerization Polymers

1. Natural rubber. Rubber consits essentially of a polymer of 1000–5000 subunits of cis-isoprene [$\leftarrow CH_2{-}\underset{\displaystyle CH_3}{\underset{|}{C}}{=}CH{-}CH_2 \rightarrow_n$.] It is usually treated with sulfur (vulcanized) in order to acquire the desirable elasticity. The sulfur content can vary between 2 and 35%.

2. Chloroprene rubber. This polymer, also known as Neoprene, is built up thus

$$n\,CH_2{=}CH{-}\underset{Cl}{C}{=}CH_2 \longrightarrow \leftarrow CH_2{-}CH{=}\underset{Cl}{C}{-}CH_2 \rightarrow_n$$

3. Butadiene rubber. This synthetic rubber, developed in Germany under the name of Buna, consists of polybutadiene:

$$n\,CH_2{=}CH{-}CH{=}CH_2 \longrightarrow {+}CH_2{-}CH{=}CH{-}CH_2{+}_n$$

4. Other diene rubbers. Various butadienes with methyl – or halogen – substituents have been synthesized but have not become products of major importance.

5. Butyl rubber. The starting material of this product is isobutene

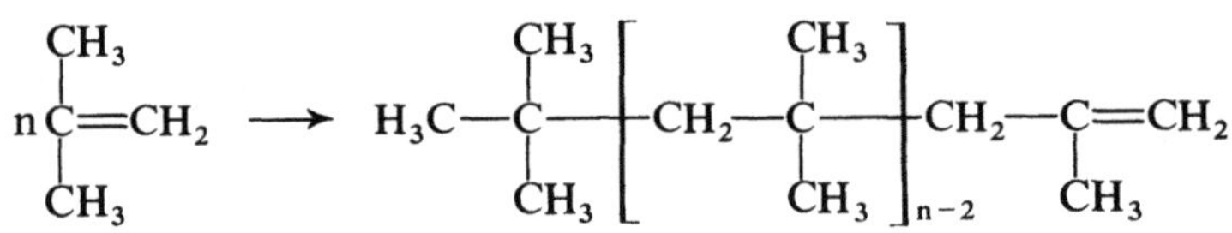

6. Polystyrene. The polymerization of styrene has been known since 1839 but production did not start until the 1930ies. Polystyrene is well suited for injection molding of various products. The polymer has the following structure

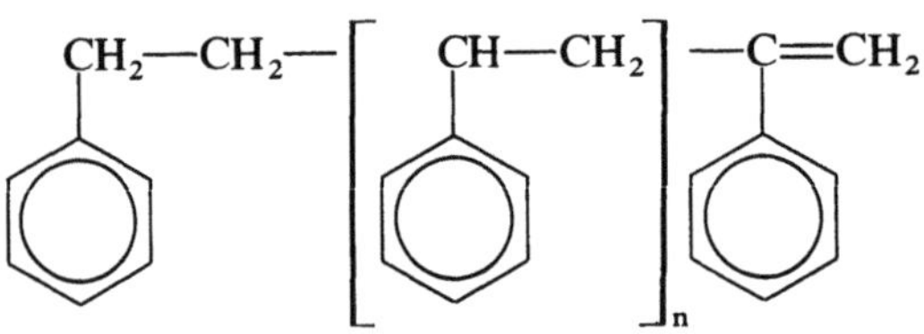

7. Poly (vinyl chloride) (PVC). A number of vinyl derivatives, such as vinyl chloride, vinyl bromide, or vinyl acetate are polymerized in the presence of plasticizers such as tricresyl phosphate or dioctyl phthalate. Vinyl chloride is the most common starting material.

$$n\ CH_2{=}CHCl \longrightarrow {-}CH_2{-}\underset{\displaystyle Cl}{CH}{-}CH_2{-}\underset{\displaystyle Cl}{CH}{-}$$

High proportions of plasticizer (up to 40%) give elastic products.

8. Vinylidene chloride polymers. Vinylidene chloride polymers have the following structure:

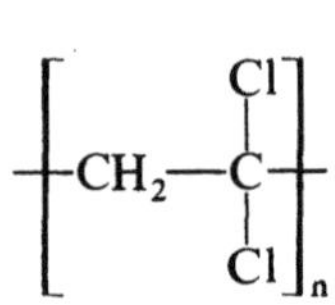

Vinylidene chloride and vinyl chloride give a copolymer used for fiber manufacture (Saran) and containing about 90% vinylidene chloride and about 10% vinyl chloride subunits.

9. Vinyl acetate polymers (PVAc). Vinyl acetate can be polymerized by itself as a homopolymer:

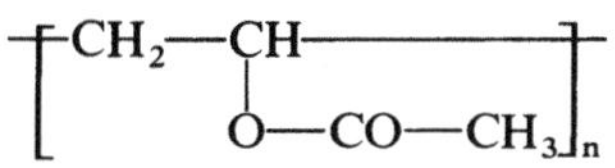

or it can be copolymerized with vinyl chloride to give a mixed product (Vinylite V).

10. Propylene polymers. The polymerization of propylene, CH_3–$CH=CH_2$ can proceed by different steric pathways, resulting in "isotactic" or "atactic" polymers. Completely atactic polypropylene is amorphous, soft and weak. With increasing content of isotactic configuration, strength and elasticity increase.

11. Acrylonitrile polymers. Acrylonitrile (vinylcyanide) polymerizes to give an excellent fiber material (Orlon)

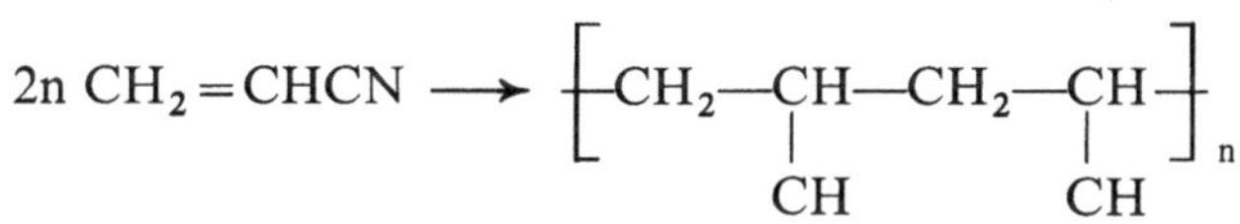

12. Acrylic resins. Acrylic acid, and especially β-substituted acrylic acid esters, can be polymerized to yield glass-like resins.

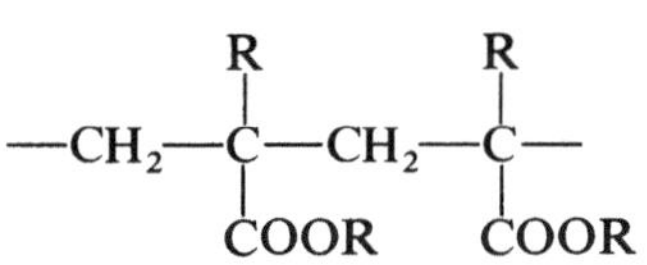

The methyl methacrylate polymers have trade names such as Plexiglas or Lucite.

13. Polyurethane polymers. The synthesis of polyurethane polymers from polyalcohols and isocyanates has been mentioned above. Dialcohols and low-molecular diisocyanates form linear polymers. Polyfunctional reaction partners yield branched and netted resins useful as elastomers. An excess of isocyanate moieties and water yields foams. Combinations of polyurethanes and polyesters give products with adhesive properties, lacquers, elastic fibres and textile coatings.

14. Ethylene polymers. Liquid ethylene at high temperatures and pressures is polymerized to polyethylene (polythene), $(CH_2CH_2)_n$.

15. ABS plastics. This group of polymers is built up of acrylonitrile, butadiene, and styrene (ABS) and these polymers are very tough and weather resistant. The properties of these products can be varied in many ways in order to meet specific purposes.

B. Polycondensation Polymers

1. Polyesters (alkyd resins). Among the many polyesters, the ethylene glycol-terephthalic acid polymer is of special importance since it is marketed as a successful fiber material (Dacron):

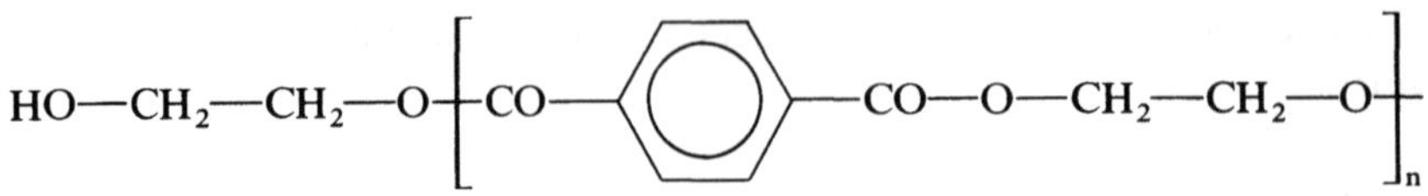

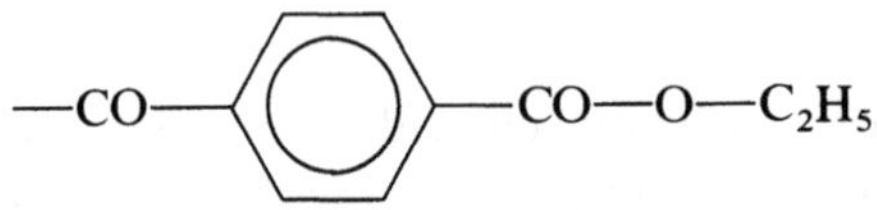

Another member of this group is the glycerolphthalic acid polymer (glyptal).

2. Polyamides. The best known example of a polyamide resin is again a synthetic fiber material, Nylon, introduced in 1935. Several chemical varieties of Nylon are produced. One of them is Nylon 6,6, i.e. the hexamethylene-diamine-adipic acid polymer:

$$H_2N{-}(CH_2)_6{-}NH{+}CO{-}(CH_2)_4{-}CONH{-}(CH_2)_6{-}NH{+}_n CO(CH_2)_4{-}COOH$$

3. Polyethers. Polyethers are usually prepared from cyclic ethers. The simplest example is the synthesis of polyethylene glycol from ethylene oxide in the presence of acids or bases:

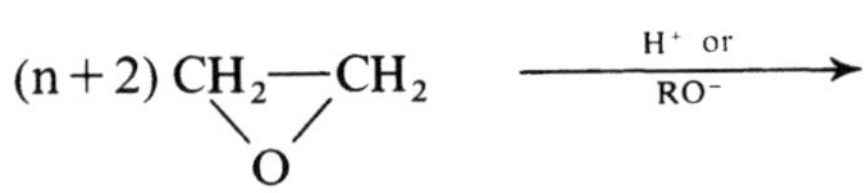

$$\longrightarrow \; ROCH_2CH_2{+}OCH_2{-}CH_2{+}_n O{-}CH_2{-}CH_2OH$$

4. Urea-formaldehyde resins. The reactants (urea or polyfunctional amides) and formaldehyde first condense to a liquid polymer which on heating (thermosetting) forms secondary cross linkages giving a hard final product

$$H_2N{-}\underset{\underset{O}{\|}}{C}{-}NH_2 + HCHO \longrightarrow HOCH_2NH\underset{\underset{O}{\|}}{C}NH(CH_2NH\underset{\underset{O}{\|}}{C}NH)_n{-}CH_2OH$$

$$\xrightarrow{\text{thermosetting}}$$

5. Melamine-formaldehyde resins. This type of polymer is related to the ureaformaldehyde group:

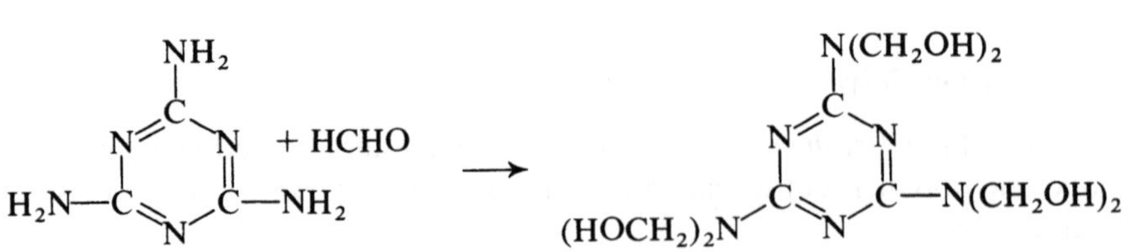

6. Phenol-formaldehyde resins. The simplest representative of this class of polymers is Bakelite, formed from phenol itself and formaldehyde by acid or alkaline catalysis:

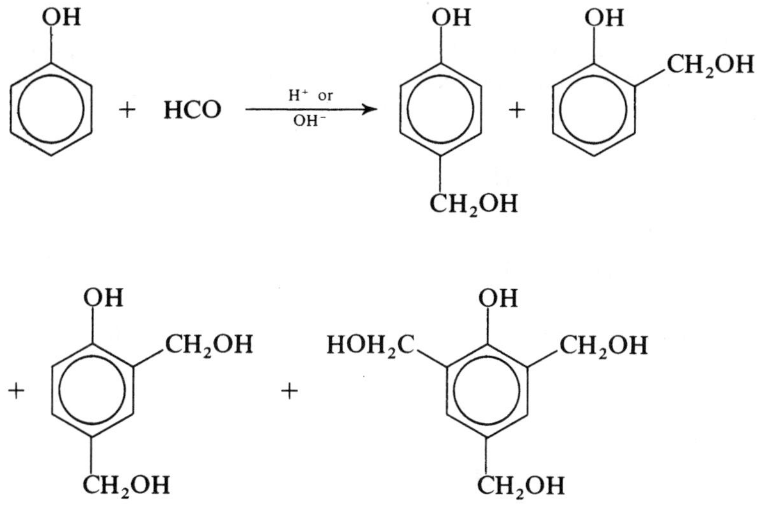

These intermediates condense further, giving products of the following type:

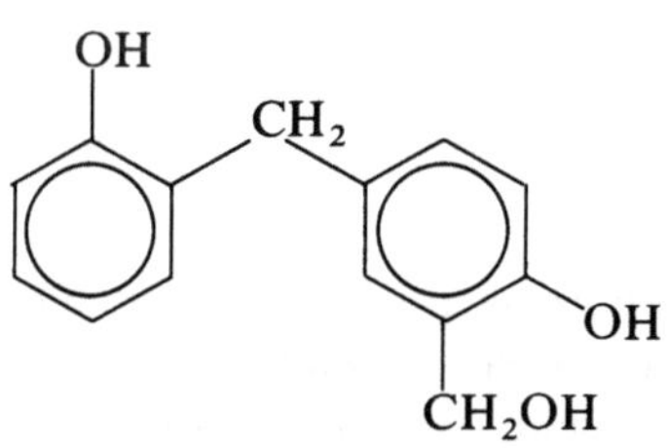

(liquid)

Thermosetting produces the solid final product by the formation of cross-linkages with loss of water.

7. Polyphenylene oxide (PPO). This polymer is formed by oxidative coupling between 2,6-dimethylphenol monomers:

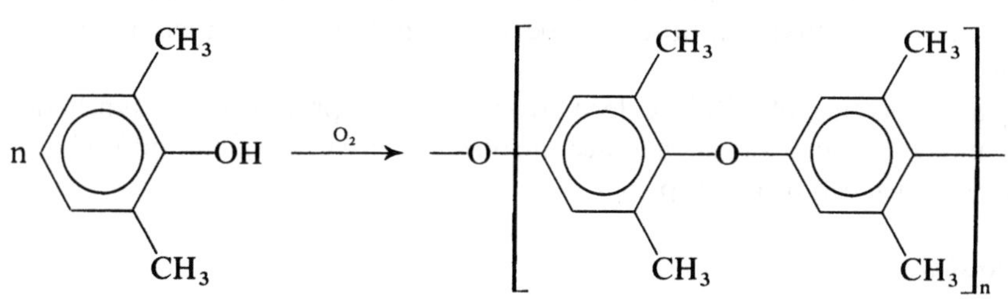

The resin has a high resistance to cold and heat and is self-extinguishing.

8. Polysulfone. This group of polymers is heat-resistant and self-extinguishing.

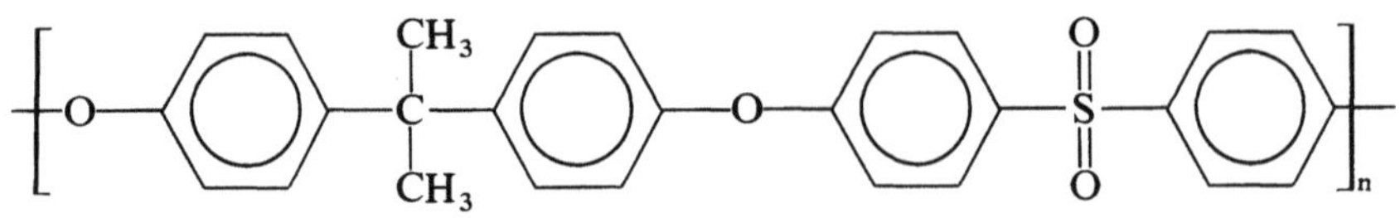

9. Polyacetals. It is well – known that formaldehyde can polymerize. A special process is required, however, to make this reaction irreversible and to form a stable resin, polyoxymethylene or polyacetal:

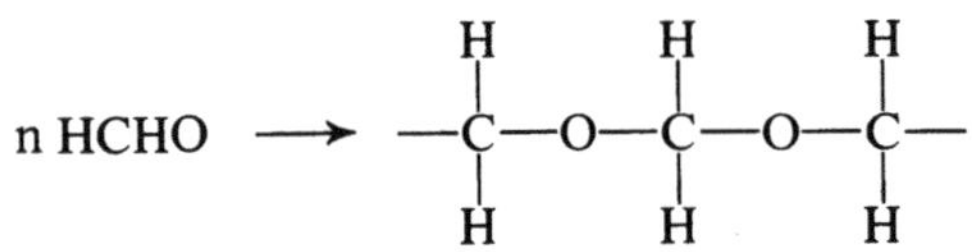

10. Polycarbonate polymers. Polycarbonates are made by reacting phenoxyalkanes with phosgene, e. g.:

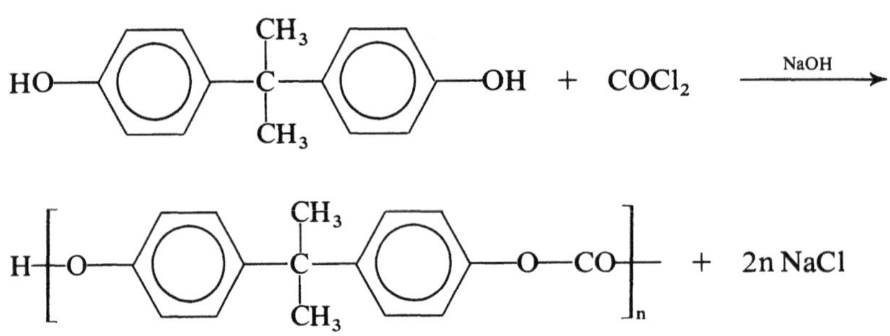

11. Cellulose derivatives. Cellulose, a naturally occurring polymer, forms the basis for a number of plastics. Polymerization has thus already been provided by nature, and the chemical processes for the production of cellulose plastics consist of introducing reactive proups into the matrix.

In 1869, the Hyatt brothers in the United States patented a resin prepared by blending nitrated cellulose with camphor. The resin, under the name of Celluloid, was the first plastic of commercial importance.

In 1905, cellulose acetate became commercially available. A mixed ester, cellulose acetate butyrate, was introduced in 1935. Cellulose propionate came into use in 1945. Finally, ethyl cellulose can be mentioned as a product of minor importance.

The main importance of cellulose derivatives lies in the production of synthetic (or rather semi-synthetic) fibers such as acetate fibers or rayon (artificial silk). These products will be discussed in Chap. B.V.

C. Epoxy Resins

Epoxy Resins are formed by condensation of polyvalent phenols or alcohols with compounds incorporating an ethylene oxide moiety with a vicinal reactive group capable of forming a second ethylene oxide configuration.

A typical reaction sequence starting from epichlorohydrin and p-xylol is the following:

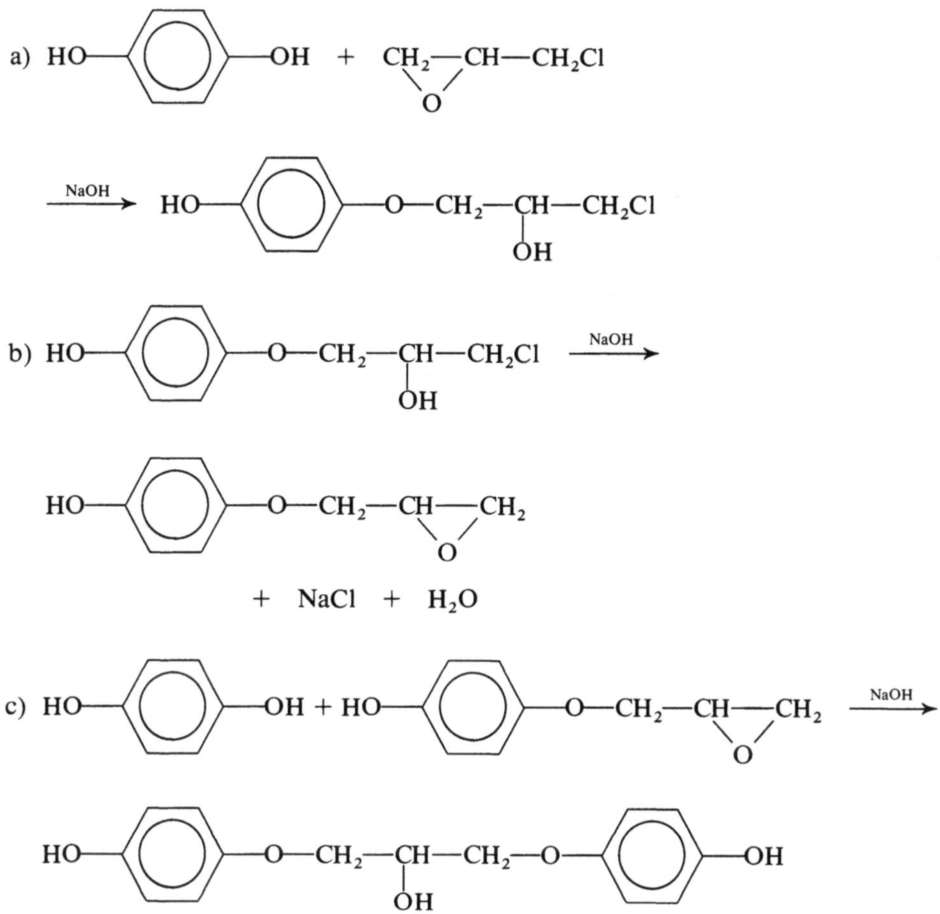

and polymerization to

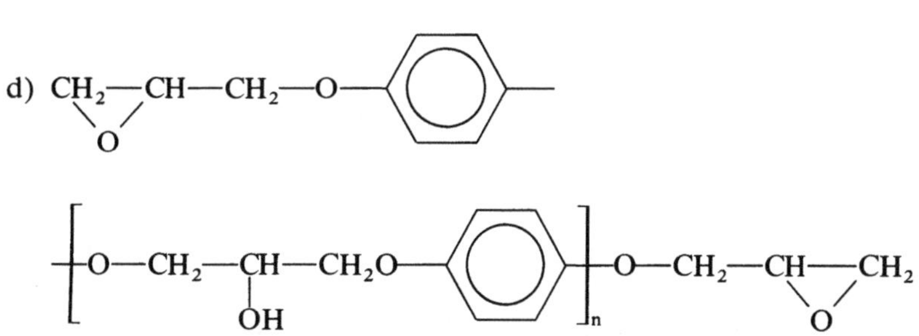

The polymerization is brought about by mixing the reaction products with hardeners. The most common hardeners are polyamines such as diethylenetriamine. One phenolic hydrogen atom reacts with one epoxy group and forms a new reactive hydroxy group, thus promoting a chain reaction, even at room temperature. Other hardeners are polyamine amides, ketimides, boron trifluorid derivatives as well as anhydrides of phthalic acid and similar poly(carboxylic acid)s.

D. Silicon Polymers

Silica, when heated with methyl chloride, forms a mixture of methylchlorosilanes from which dimethyldichlorosilane is isolated. Treatment with water yields hydroxy-dimethylsilane. This can be polymerized to silicon oils of the general structure

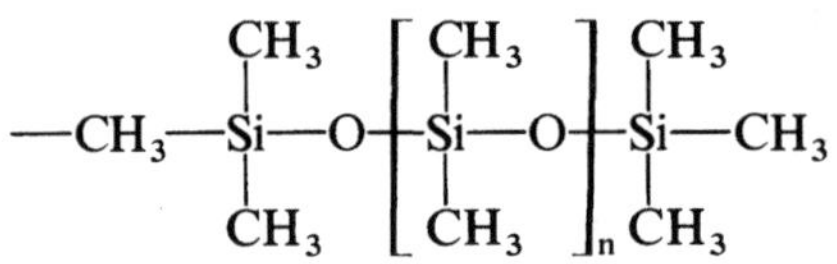

(The reaction is terminated by trimethylchlorosilane)

E. Fluorocarbon Plastics

1. Polytetrafluoroethylene (PTFE). Pyrolysis of chlorodifluoromethane followed by emulsion polymerization yields polytetrafluoroethylene (Teflon):

$$2\,CHClF_2 \longrightarrow F_2C{=}CF_2 + 2\,HCl$$

$$n\,F_2C{=}CF_2 \longrightarrow -(CF_2CF_2)_n- (n \sim 1000)$$

PTFE is a polymer with very high resistance to corrosive chemicals and to heat.

2. Polychlorotrifluoroethylene (CTFE). This is a more expensive fluoroplastic for special purposes:

$$-(CF_2CFCl)_n-$$

3. Other fluoroplastics. Less frequently used are fluorinated ethylene – propylene copolymers, as well as poly(vinyl fluoride) and poly(vinylidene fluoride) polymers.

3. Additives in Commercial Plastics

In the manufacture of plastic products, the polymers are modified by a number of additives. The most important classes of additives are *plasticizers*, *antioxidants* and *hardeners* (especially for epoxy resins). Also, *fillers* (adding bulk to the product), dyes and pigments are added to give the product the desired properties. *Lubricants* are used for facilitating the removal of polymer casts from their molds.

Plasticizers act like high-boiling solvents, providing greater freedom of chain movement within the polymer. They should be reasonably cheap but should *not*: migrate during use, increase the flammability of the product, be toxic, result in color bleeding, cause blocking of coating fabrics, reduce chemical resistance nor increase water resorption. Hundreds of plasticizers are now available most of which are used as additives in vinyl polymers. Examples are listed in Table B.14 (p. 103). The more frequently used *antioxidants* and *hardeners* are listed in Tables B.15 (p. 104) and 16 (p. 104).

An enormous variety of *fillers* are employed in the plastics industry. These cannot be listed in this book, but it may be useful to know a few of the components of such fillers. They are cellulose, synthetic polymers (other than the main product), carbon, powdered metals (e. g. Al, Cu, Fe, Pb) and a large variety of minerals containing e. g. Al, B, Ba, Mb and Si.

It would be useless to list pigments and dyes used in the manufacture of plastics since their variety is obviously enormous. A list of plastic products is collected in Table B.17 (p. 104).

4. Analytical Methods

a. Introduction

Procedures for the analysis of polymers are needed in industry for controlling the purity and composition of the starting materials as well as the physical and chemical properties of the industrial products. Typical features of these analytical procedures are the availability of large samples so that faulty or inconclusive analyses can be repeated, and a pretty good advanced knowledge of the main composition of the substances to be analyzed.

In forensic chemistry the situation is usually quite different. Large samples are the exception rather than the rule, and the composition or even the type of polymer is rarely known in advance. In some cases, inquiries at the manufacturer – when known – may help. In addition, the forensic chemist is often concerned with comparative analyses where a detailed knowledge of the composition of a product is not always required. These comparative analyses may go beyond the quantitative composition and may include queries concerning batch variations of commercial products. Finally, it is rarely possible to repeat a doubtful analysis due to the paucity of sample available. For these reasons, many methods used in product control laboratories cannot be applied to forensic analysis. Nevertheless, there are a number of techniques available which are well suited for forensic use. These include:

Ultraviolet spectrophotometry (UV)
Infrared spectrophotometry (IR)
- conventional IR
- IR of pyrolysis products

Differential thermal analysis (DTA)
Mass spectrometry (MS)
- conventional MS (direct probe)
- pyrolysis MS
- pyrolysis GC/MS

Solubility tests (ST)
Thin-layer chromatography (TLC)
Gas chromatography (GC)
- conventional GC
- capillary GC
- pyrolysis GC (Curie point, laser, stepwise heating
- photolysis GC

Gel permetation chromatography (GPC)
Luminescence photometry (LP)

Numeros publications on most of these techniques have appeared, and review articles are cited in the bibliography [AA-AP].

b. The Analysis

In the forensic science literature, relatively few publications on the analysis of polymers have appeared to-date. In view of the rapidly increasing use of plastics, this situation is bound to change.

Thin-layer chromatography has not been discussed by criminalists but has probably been used in some cases. Two comprehensive TLC review papers (AB and AI) are good guidelines for the forensic chemist. In the case of vinyl plastics, Janca et al. [15] have described a TLC system.

Pyrolysis gas chromatography (PGC) is now probably the most frequently used technique in forensic laboratories. It has the advantages of small sample requirements and of the production of "patterns", generally well suited for comparative investigations and frequently also useful for identification purposes within certain limits.

The major problem of the use of PGC in polymer analysis is the difficulty of getting reproducible results within a laboratory and exchangeable comparative data between several laboratories. Critical parameters of the pyrolysis steps include:

pyrolysis temperature,
pyrolysis duration,
sample size,
sample geometry,
rate of heat transfer to the sample,
trapping of solvents or pyrolysis products in the sample matrix.

Strassburger et al. [25] were among the first to use PGC for polymer analysis but only in 1965 Kirk et al. [14] introduced the PGC technique into forensic science. In recent years, the Curie-point method has won widespread acceptance. Its inherent temperature reproducibility is of great help in comparative analytical work. If the experimental conditions of the gas chromatography step also become standardized in several laboratories, common reference libraries of pyrograms can be used. This has been done within the network of the English Home Office regional forensic science laboratories. Curie-point PGC has been described by Wheals and Noble [27] and by Noble et al. [22].

The PGC chromatograms obtained can be compared by pattern recognition techniques as shown by Küllik et al. [20] and – for forensic work – by de Forest [9].

Pyrolysis gas chromatography of polymers by laser was introduced by Folmer et al. [7, 8] in 1971 but seems not to have been applied to forensic studies. The main drawback of this method is the need to mix transparent samples with carbon in order to achieve the hight light absorption necessary for succesful pyrolysis. Degradation of polymers by photolysis followed by gas chromatography was carried out by Juvet et al. [16] who irradiated samples with a Hg-lamp. Their technique does not seem to be very suitable for forensic purposes. An interesting approach to PGC of polymers was worked out by Kaljurand and Küllik [17] who used cross-correlation chromatography (CCC). The principle of the method is a continuous heating of the sample while analyzing the pyrolysis products at regular intervals (e. g. every 4 seconds) by GC. A microcomputer is needed for plotting the sets of chromatograms produced. This method has not been used for forensic purposes.

Pyrolysis Gas Chromatography Used in the Authors Laboratory

Pyrolysis gas chromatography is carried out using two different systems. In System A, a Pye-Unicam 104 gas chromatograph with a Curie point pyrolyser is employed. A wire with a Curie point of 610° C and a heating period of 10 second was chosen. A glass column (1,7 m × 2 mm i. d.) is filled with Porapak Q (80–100 mesh). The injection port temp. is 200° C, the detector temp. 290° C (FID). The oven temp. is programmed from 100° C (initial) to 200° C (final) at 7° C min $^{-1}$.

In System B, the gas chromatograph is a Σ 2 by Perkin-Elmer equipped with an exit splitterleading to an FID and NPD. The column is made from 1/8 in. copper tubing (inner diameter 0.8 mm) of 3 m length and filled with Porapak Q, 80–100 mesh. The carrier gas is N_2 at about 30 ml min^{-1}.

Pyrolysis is carried out in a CDS-pyroprobe 120, where the sample is placed into a silica tube, heated by a platina coil (rate of heating ~75° C ms^{-1}).

Luminescence and phosphorescence spectroscopy of plastics have been described by Allen et al. [1, 2] who also cite the earlier literature (94 ref.). For a monograph of this technique cf. [AK]. It seems that the method has not been applied to forensic analyses.

Infrared spectrophotometry is extensively used for product control in the plastics industry. Haslam and Willis [D] published a collection of 188 IR-spectra of polymers and 44 IR-spectra of plasticizers. Monographs were written by Krimm [AJ] and by Tosi and Ciampelli [AM]. A prerequisite for the use of IR spectrophotometry is obviously the absence of opaque or colored additives in the polymer. This drawback as well as the requirement for relatively large amounts of sample restricts the use of this technique in forensic work.

However, in some cases this technique can be used for comparing polymer specimens in order to find their possible common origin. Cleverley [4, 5] has thus used IR spectrophotometry for investigating the common origin of low density polyethylene bags which are widely used for packaging and distributing drugs of addiction. In some cases, they found even batch-to-batch differences in bags from the same manufacturer using IR analysis of weak bands in the fingerprint region.

The use of a diamond cell for the IR-analysis of small samples of plastics for forensic purposes was described by Tweed et al. [26].

In 1952, Kruse and Wallace [19] reported the identification of polymers by IR-analysis of their pyrolysis products. In the following year, Harms [11] published a detailed paper on this method where he reproduced 41 IR-spectra of pyrolyzed plastics. In 1969, Smalldon [23] applied this technique to forensic investigations. A systematic approach to PIR shows promising results [10].

Finally, pyrolysis in combination with mass spectrometry (with or without gas chromatography as an intermediate step) has become an important technique for the forensic analysis of polymers. Kistemaker et al. [18] as well as Maynard et al. [21] and others have used PGC/MS. In the forensic field, Hughes et al. [13] and later Hickman and Jane [12] have reported the use of PMS. The latter authors have also compared the reproducibility of three different pyrolysis systems in forensic analysis of polymers.

Proton-Nuclear Magnetic Resonance Spectrometry was applied to polymer analysis by Evans et al. [6]. It has, to our knowledge, not been used in forensic science.

Ambler [3] described the application of gel permeation chromatography to polymer analysis.

Finally, Smalldon et al. [24] published the use of thermogravimetry combined with mass spectrometry for the forensic investigation of polymers.

5. Practical Examples

a. Identification of Kitchen Ventilator

In the investigation of a fire the police suspected that the fire was caused by the malfunction of a kitchen ventilator. Debris from the scene of the fire were sent to the laboratory with the aim of identifying pieces of plastic found there.

According to the manufacturer of the ventilator in question, the filter housing was made of polycarbonate and the fan of polyamide.

Analysis of a molten particle in the fire debris by pyrolysis gas chromatography infrared spectrophotometry gave diagrams corresponding to polycarbonate (cf. Fig. B. 20).

Another particle was analyzed in the same manner and shown to consist of a mixture of polycarbonate and polyamide.

The conclusion reached was that these two types of polymers could stem from the kitchen ventilator.

b. Pollution from Burning Discarded Electric Cables

The local health authorities accused a person of repeated violation of the anti-pollution laws by burning used isolated cables. The police sent pieces from four different types of cables to the laboratory with the request to determine the kinds of plastic used in the isolation, and to report which gases or fumes these form on burning.

The outermost layer from each of the four cable pieces, as well as a few randomly chosen samples from the deeper layers of the insulation, were analyzed by pyrolysis gas chromatography and pyrolysis infrared spectrophotometry. The results were the same within the limits of experimental error. Figures B. 21 and 22 show the diagrams obtained from one of the insulation samples and from a poly (vinyl chloride) reference sample.

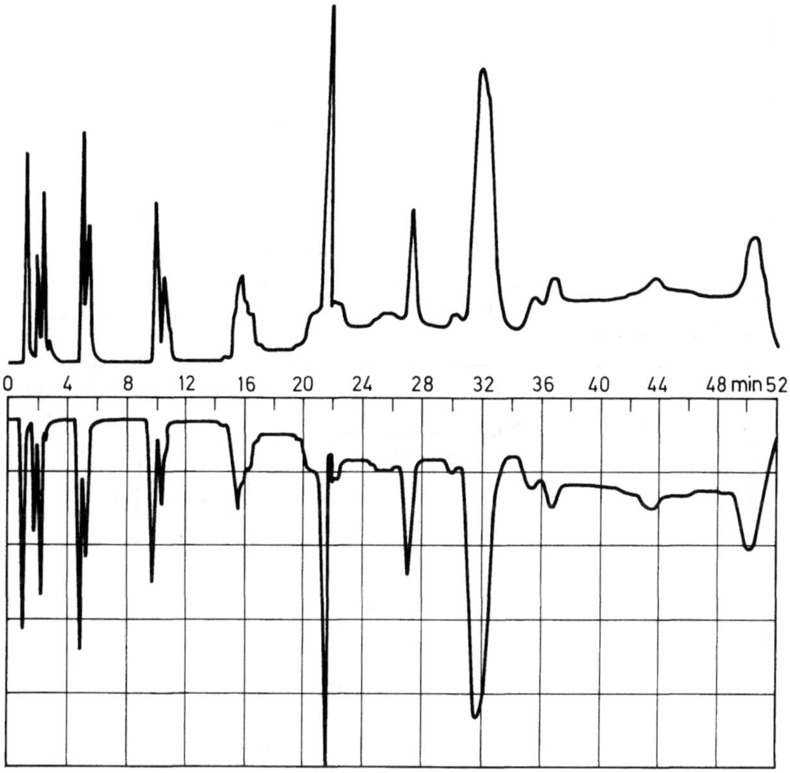

Fig. B. 20. The pyrograms for a polycarbonate reference sample (upper trace) and for a molten particle (probably from the filter housing of a kitchen fan) mentioned in case a. The experimental conditions are those described in the text for method B. Due to the absence of nitrogen, no NPD signal was recorded. The pyrograms were not recorded on the same day and the retention times are therefore not exactly identical, probably due to slightly different N_2 – flow rates

The conclusion read as follows:

1. All analyzed samples of insulation from the cables sent in were composed of poly (vinyl chloride) (PVC).
2. On burning, PVC yields *inter alia* fumes of hydrochloric acid which have a corrosive effect. For details of the dangers involved, other government agencies will have to be contacted.

The "cable burner" was sentenced to a stiff fine.

6. General References (In Chronological Order)

A) Cantow et al. (eds.): Advances in Polymer Science, Berlin – Heidelberg – New York: Springer 1958–1978

B) F. W. Billmeyer: Textbook of Polymer Science, New York – London: Interscience 1962

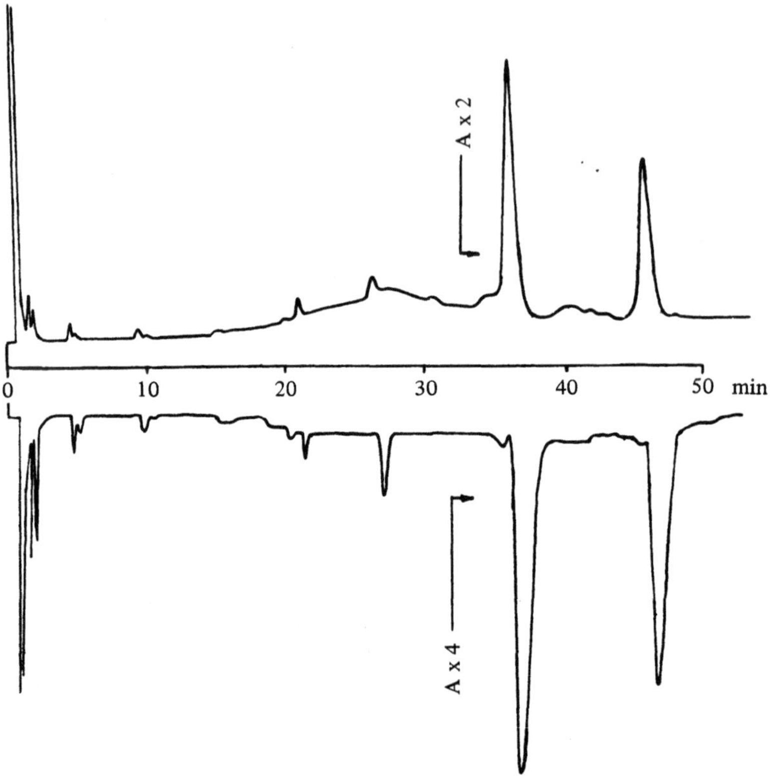

Fig. B. 21. The pyrograms for a polyvinyl choride (PVC)-reference sample (upper trace), and for the insulation of one of the cables in case b (lower trace, mirror image). The experimental conditions are those described in the text for method B. However, the NPD trace showed no response due to the absence of nitrogen containing groups

C) J. Becke: Leichtverständliche Einführung in die Kunstoff-Chemie, München: Hanser 1964

D) J. Haslam, H. A. Willis: Identification and Analysis of Plastics. London: Iliffe Books Ltd. and Princeton: D. van Nostrand Co. Inc. 1965

E) D. C. Miles, J. H. Bristow: Polymer Technology. London: Temple Press 1965

F) K. J. Saunders: The Identification of Plastics and Rubbers. London: Chapman & Hall 1966

G) Hj. Saechtling, A. Krause: Kunstoff-Bestimmungstafel, 5th edit. München: Hanser 1966

H) R. W. Lenz: Organic Chemistry of Synthetic Polymers. New York – London – Sydney: Interscience 1967

I) K. Stoeckhert: Kunststoff-Lexikon, 4th edit. München: Hanser 1967

J) D. O. Hummel, F. Scholl: Atlas der Kunststoff-Analyse, Vol. 1. München: Hanser 1968

K) E. R. Yescombe: Sources of Information on the Rubber, Plastics and Allied Industries. London: Pergamon 1968

L) R. V. Milby: Plastics Technology. New York and 15 other cities: Mc Graw-Hill 1973

M) B. Vollmert: Polymer Chemistry' Berlin – Heidelberg – New York: Springer 1973

N) J. Brandrup, E. H. Immergut: Polymer Handbook, 2nd edid. New York: Wiley-Interscience 1975

O) J. Urbański et al.: Handbook of Analysis of Synthetic Polymers and Plastics, Engl. edit. New York – London – Sydney – Toronto: John Wiley 1977

P) Hj. Saechtling: Kunststoff-Taschenbuch, 21st edit. München – Wien: Hanser 1979

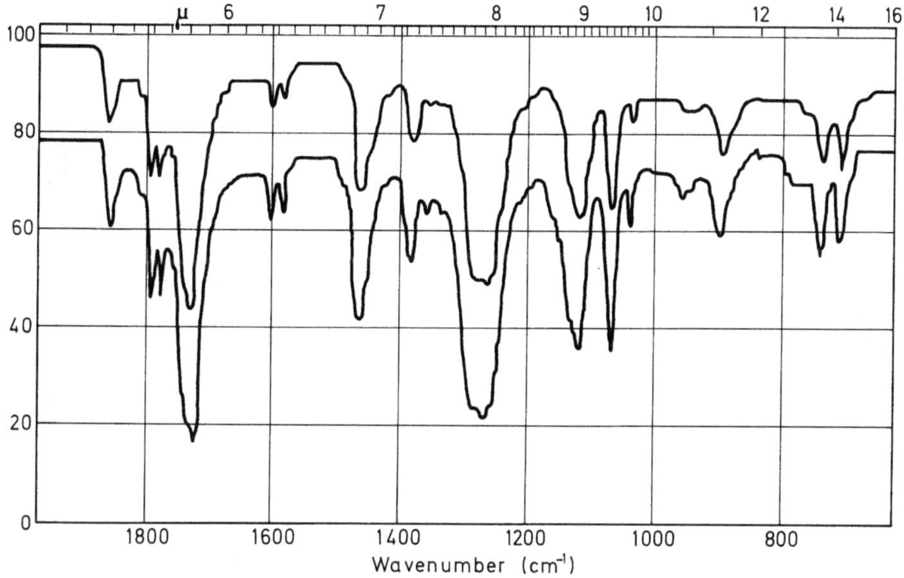

Fig. B. 22. The pyrolysis-infrared spectra of a PVC reference sample (upper trace), and of the insulation from one of the cables in case b (lower trace, displaced by 20% T). The experimental conditions are described in the text

7. References on Reviews on Analytical Methods

AA) G. Agrey: The Use of Isotopes in Polymer Analysis. Adv. Polym. Sci. *6*, 128 (1969)

AB) B. G. Belenkii, E. S. Gankina: Thin-Layer Chromatography of Polymers. J. Chromatogr. *141*, 13 (1977) with 123 ref.

AC) F. A. Bovey, G. V. D. Tiers: The High Resolution Nuclear Magnetic Resonance Spectroscopy of Polymers. Adv. Polym. Sci. *3*, 139 (1961/64)

AD) J.-M. Braun, J. E. Guillet: Study of Polymers by Inverse Gas Chromatography. Adv. Polym. Sci. *21*, 107 (1976)

AE) D. T. Clark: ESCA Applied to Polymers. Adv. Polym. Sci. *24*, 125 (1977)

AF) J. G. Cobler, C. D. Chow: Analysis of High Polymers. Anal. Chem. *51*, 287R (1979)

AG) R. B. Fox: Photodegradation of High Polymers. Progr. Polym. Sci. *1*, 45 (1967)

AH) P. J. Hendra: Laser-Raman Spectra of Polymers. Adv. Polym. Sci. *6*, 151 (1969)

AI) H. Inagaki: Polymer Separation and Characterization by Thin-Layer Chromatography: Adv. Polym. Sci. *24*, 189 (1977) with 106 ref.

AJ) S. Krimm: Infrared Spectra of High Polymers. Adv. Polym. Sci. *2*, 51 (1960/61)

AK) G. Oster, Y. Nishijima: Fluorescence Methods in Polymer Science. Adv. Polym. Sci. *3*, 313 (1961/64)

AL) W. P. Slichter: The Study of High Polymers by Nuclear Magnetic Resonance. Adv. Polym. Sci. *1*, 35 (1958/60)

AM) C. Tosi, F. Ciampelli: Applications of Infrared Spectroscopy to Ethylene-Propylene Copolymers. Adv. Polym. Sci. *12*, 87 (1973)

AN) W. Wrasidlo: Thermal Analysis of Polymers. Adv. Polym. Sci. *13*, 1 (1974)

AO) H. Yoshida, K. Hayashi: Initiation Process of Radiation-induced Ionic Polymerization as Studied by Electron Spin Resonance. Adv. Polym. Sci. *6*, 401 (1969)

AP) H. G. Zachmann: Das Kristallisations- und Schmerzverhalten hochpolymerer Stoffe: Adv. Polym. Sci. *3*, 581 (1961/64)

8. Specific References

1) N. S. Allen, J. Homer, J. F. Mc Kellar: The Use of Luminescence Spectroscopy in Aiding the Identification of Commercial Polymers. Analyst. *101*, 260 (1976)
2) N. Allen, J. F. Mc Kellar: Luminescence Applications in Commercial. Chem. Ind. *23*, 907 (1978)
3) M. R. Ambler: Recent Advances in Polymer Characterization by GPC, Intern. Laboratory May/June 35–48 (1979)
4) B. Cleverley: Comparison of Plastic Materials and Paint Films Using Infra Red Spectroscopy. Med. Sci. Law 7, 148 (1967)
5) B. Cleverley: The Comparison and Matching of Low Density Polyethylene Plastic Bags by Infrared Spectroscopy. J. Forens. Sci. *24*, 339 (1979)
6) D. L. Evans et al.: Compositional Determination of Styrene-Methacrylate Copolymers by Pyrolysis Gas Chromatography, Proton-Nuclear Magnetic Resonance Spectrometry, and Carbon Analysis. Anal. Chem. *50*, 857 (1978)
7) O. F. Folmer Jr., L. V. Azarraga: A Laser Pyrolysis Apparatus for Gas Chromatography. J. Chromatogr. Sci. 7, 665 (1969)
8) O. F. Folmer Jr.: Laser Pyrolysis Gas Chromatography – Application to Polymers. Anal. Chem. *43*, 1057 (1971)
9) P. R. de Forest: The Potential of Pyrolysis – Gas Chromatography for the Pattern Individualization of Macromolecular Materials. J. Forens. Sci. *19*, 113 (1974)
10) L. Haeger, A. C. Maehly: Unpublished experiments
11) D. L. Harms: Identification of Complex Organic Materials by Infrared Spectra of Their Pyrolysis Products. Anal. Chem. *25*, 1140 (1953)
12) D. A. Hickman, I. Jane: Reproducibility of Pyrolysis – Mass Spectrometry Using Three Different Pyrolysis Systems. Analyst *104*, 334 (1979)
13) J. C. Hughes, B. B. Wheals, M. J. Whitehouse: Pyrolysis Mass Spectrometry. A Technique of Forensic Potential? Forens. Sci. *10*, 217 (1977)
14) N. C. Jain, C. R. Fontan, P. L. Kirk: Identification of Paints by Pyrolysis – Gas Chromatography. J. Forens. Sci. Soc. *5*, 102 (1965)
15) J. Janča et al.: Separation of Vinyl Chloride-Vinyl Acetate Copolymers According to Their Chemical Composition by Thin-Layer Chromatography. J. Chromatogr. *171*, 81 (1979)
16) R. S. Juvet, Jr., J. L. S. Smith, K-P. Li: Polymer Identification and Quantitative Determination of Additives by Photolysis-Gas Chromatography: Anal. Chem *44*, 49 (1972)
17) M. Kaljurand, E. Küllik: Continuous Thermal Volatilization Analysis of Polymers by Gas Chromatography with Pseudo-Random Injection of Samples. J. Chromatogr. *171*, 243 (1979)
18) P. Kistemaker, A. J. H. Boerboom, H. L. C. Meuzelaar: Laser Pyrolysis Mass Spectrometry: Some Aspects and Applications to Technical Polymers. Dyn. Mass. Spectrom. *4*, 139 (1975)
19) P. F. Kruse, Jr., W. B. Wallace: Identification of Polymers by IR Analysis of Their Pyrolysis Products: Anal. Chem. *24*, 2015 (1952)
20) E. Küllik, M. Kaljurand, M. Koel: Analysis of Pyrolysis Gas Chromatograms Using Pattern Recognition Techniques: J. Chromatogr. *126*, 249 (1976)
21) J. B. Maynard, J. E. Twichell, J. Q. Walker: Determination of Carboxyl-Terminated-Butadiene-Nitrile Polymer in Commercial Cresol-Novalac Epoxy Resin Systems by Pyrolysis Capillary Gas Chromatography and Mass Spectrometry. J. Chromatogr. Sci. *17*, 82 (1979)
22) W. Noble, B. B. Wheals, M. J. Whitehouse: The Characterisation of Adhesives by Pyrolysis Gas Chromatography and Infrared Spectroscopy: Forens. Sci. *3*, 163 (1974)
23) K. W. Smalldon: The Identification of Paint Resins and Other Polymeric Materials from the Infra Red Spectra of Their Pyrolysis Products. J. Forens. Sci. Soc. *9*, 135 (1969)
24) K. W. Smalldon, R. E. Ardrey, L. R. Hullings: The Characterization of Closely Related Polymeric Materials by Thermogravimetry – Mass Spectrometry: Anal. Chim. Acta *107*, 327 (1979)
25) J. Strassburger et al.: Analysis of Methyl Methacrylate Copolymers by Gas Chromatography. Anal. Chem. *32*, 454 (1960)
26) F. T. Tweed et al.: The Forensic Microanalysis of Paints, Plastics and Other Materials by an Infrared Diamond Cell Technique. Forens. Sci. *4*, 211 (1974)
27) B. B. Wheals, W. Noble: Forensic Applications of Pyrolysis Gas Chromatography. Chromatographia *5*, 553 (1972)

9. Tables B. 13–17

Table B. 13. Some historical dates on the development of synthetic polymers

Year	Event
1839	Observation of styrene polymerization
1843	Vulcanization of caoutchouc
1844	Production of linoleum
1846	Nitration of cellulose
1869	Acetylation of cellulose. Celluloid from nitrocellulose und camphor
1872	Poly(acrylic acid)
1873	Poly(isobutylene)
1877	Poly(methacrylic acid)
1878	Poly(vinyl chloride)
1892	Cellulose xanthogenate
1897	Casein-formaldehyde polymer
1905	Cellulose acetate
1909	Bakelite (phenol-formaldehyde polymer)
1911	Production of cellophane
1915	Production of dimethylbutadiene rubber
1920	Urea-formaldehyde polymers
1924	Poly(vinyl alcohol)
1927	Acryloid
1928	Production of poly(methyl methacrylate) poly(vinyl acetate)s
1930	Production of poly(vinyl acetate). Epoxy polymers from phenol-polyalcohols
1931	Production of neoprene (polychloroprene), poly(vinyl acetal) and polyisobutylene
1933	Teflon (polytetrafluoroethylene)
1935	Nylon (hexamethylenediamine and adipic acid)
1938	Perlon fibers. Polyamides. Epoxy resins
1939	Polyurethane fibers
1946	Poly(vinyl chloride) fibers
1942	Saran fibers (poly(vinylidene chloride))
1950	Orlon (poly(acrylonitrile))
1960	Polycarbonates
1960	Polyacetal
1965	Poly(phenylene oxide)
1965	Polysulfone

Table B. 14. Some groups of plasticizers frequently encountered

A. Esters
  1. Dialkyl phthalates
  2. Dialkyl sebacates
  3. Dialkyl adipates
  4. Higher glycol esters of oleic, stearic, adipic, and benzoic acid
  5. Ethyl palmitate
  6. Butyl lactate

B. Alkyl phosphates
  1. Tritolyl phosphate
  2. Trixylenyl phosphate
  3. Trioctyl phosphate

C. Toluenesulfonamides

D. Dialkyl tin compounds

E. Castor oil

F. Camphor

Table B. 15. Some frequently employed antioxidants

1. Bis(2-hydroxy-3-methylcyclohexyl-5-methylphenyl)methane
2. Butylated hydroxyanisole
3. Dicresylolpropane
4. 2,6-Di-tert-butyl-4-methylphenol
5. N,N'-Di-β-naphthyl-p-phenylenediamine
6. Phenyl-α-naphthylamine
7. Phenyl-β-naphthylamine

Table B. 16. Some more frequently ecountered hardeners[a]

A. *Aliphatic primary amines*
   Diethylenetriamine (DETA)
   Triethylenetetramine (TETA)
   Triethylamine
B. *Aromatic amines*
   m-Phenylenediamine (MPDA)
   Diaminodiphenyl sulfone (DADS)
   Piperidine
   Tridimethylaminoethylphenol (DMP-30)
   Dimethylaminophenol (DMP-10)
C. *Acid anhydrides*
   Phthalic anhydride (PA)
   Pyromellitic dianhydride (PMPA)
   Methyl nadic anhydride (MNA)
   Hexahydrophthalic anhydride (HHPA)

[a] Essentially from (L)

Table B. 17. List of some products made of plastics with indication of the type of polymer usually employed[a]

Adhesives, general	EP
Adhesives (for wood)	EP, MF, UF
Advertising signs	Acr
Air-space industry products	FP, Si, Su
Ashtrays	PF
Automobile parts	ABS, Acr, Ce, P, PU, PVC
Automobile body parts	ABS, PE
Balloons	PE
Ball point pens	Ce
Bearings, oilless	PF
Bicycle parts	PVC
Billiard balls	Ce
Binders (for molding and laminating)	Ce, MF, UF
Boat hulls	ABS
Body implants (incl. foams)	Si
Book covers	Ce
Boots	Si
Bottles	E
Bushings	PF
Camera parts	PF
Carpeting (incl. outdoors)	P
Cement (organic)	EP

Table B. 17 (continued)

Chemical equipment	DP
Coatings, general	Am, Ce, EP, MF, PE, Si
Coatings of fabrics	PVC
Coatings of glass	EP
Condensers	PF
Containers (also baskets)	Ce, E, PVC
Cups	St
Cushions (foam)	PVC
Dice	Ce
Dinnerware	MF (CF)
Drink vending machine parts	Ca
Electric shaver housings	Ce, MF (CF)
Electronic circuits	FP
Enamel (for appliances)	Su, UF
Eyeglass frames	Ce
Face shields	Ce
Fibers (textile)	Am, Ce, PE, V
Film (non-inflammable)	Ce
Films and sheets, general	E, PE, PVC
Flash tubes	St
Floats (foam)	PVC, St
Food containers	St
Frying pan coatings	FP
Furniture	PU
Gaskets	E
Gears	PF
Glass coatings	EP
Glue	EP
Gunstocks	PF
Heat insulating foam	PU, St
Insulators (electrical)	Ca, Ce, E, FP, P, PF, Su, UF
Ironing iron handles	PF
Knobs	PF, St
Laboratory equipment	E, PF, PVC, PMP
Lacquers	Ce
Lenses and watch-glasses	Ce
Light guides	St
Lucite	Acr
Luggage	ABS
Lunch boxes	St
Matrasses and cushions (foam)	PU
Medical equipment	Acr, PPO
Microphone housings	Ce
Models	Ce
Molds	EP
Oilless bearings	PF
O-rings (inert)	FP
Packaging material	PVC, St
Panel laminate	DP, MF
Paper laminate	DP, MF
Patching compounds	PVAc
Photo darkroom equipment	PF
Ping pong balls	Ce
Pipe linings	FP, PU

Table B. 17 (continued)

Pipes and pipe fittings	Ce, E, PVC
Plexiglas	Acr
Plastic models (machined)	Acr
Playing cards	Ce
Plumbing hardware	ABS
Pot handles	PF
Primers	PE
Profiles	E
Pump parts	E, FP, PF, Su
Radio tube bases	PF
Radomes	PE
Recording tape	Ce
Refrigerator parts	ABS, Ca, FP
Relays	PF
Rubber substitutes	D, Si
Sheets and films	E, PE, PVC
Skylights	Acr, PVC
Spacers	PF
Sport applicances	ABS, PVC
Surface coatings	Am, Ce, EP, PE, Si
Switches	MF, (GF), PF
Syringes	E
Tank linings	FP, PU
Textile fibers	Am, Ce, PE, V
Tool handles	Ce, MF, (CF), PF
Tooth brushes	Ce
Tow ropes	P
Toys	Ce, E, PVC, St
Trailer components	ABS, DP
Truck cabs	DP
Tubing	E, FP, Si
Typewriter keys	Ce
Umbrella handles	Ce
Vacuum cleaner parts	Ce, P
Valves	E, FP, Su
Wall tiles	St
Washing machine parts	P, PF
X-ray equipment parts	PF

[a] Abbreviations used in Table B. 17

ABS	acrylonitrile-butadiene-styrene copolymers (A. 14)
Acr	acrylic polymers (A. 11)
Am	polyamide resins (B. 2)
Ca	polycarbonate resins (B. 10)
Ce	cellulose derivatives (B. 11)
CF	cellulose-filled products
D	diene polymers (A. 2–3)
DP	diallylphthalate polymers
E	polyethylene resins (A. 13)
EP	epoxy polymers (C)
FP	fluorocarbon plastics (E)
GF	glass-filled products
KF	casein-formaldehyde polymers
MF	melamine-formaldehyde polymers (B.5)
P	polypropylene resins (A. 9)
PE	polyesters (B. 1)
PF	phenol-formaldehyde polymers (B. 6)
PMP	polymethylpentenes
PPO	poly(phenylene oxide)resins (B. 7)
PU	polyurethane resins (A. 12)
PVAc	poly(vinyl acetate) (A. 8)
PVC	poly(vinyl chloride) (A. 6)
Si	silicones (D)
Su	polysulfone resins (B. 8)
St	styrene polymers (A. 5)
UF	urea-formaldehyde polymers (B. 4)
V	vinyl polymers (A. 6–8)

IV. Fibers

1. Introduction

The forensic investigation of fibers has traditionally been carried out by biologists since the classical fiber materials all had biological origins. The methods employed were accordingly morphological in character. The physical appearance of the material was noted and compared with standards whenever possible, both by macroscopic and microscopic means.

The appearance of synthetic fiber material did not, to begin with, change this mode of investigation. However, the introduction of chemical and physical micromethods in the middle of this century soon had a pronounced influence also on the examination of fibers in the forensic laboratory. At present, a large number of morphological, physical, and especially chemical methods are available – and are used – for fiber identification and comparison.

Historically, flax fibers were probably the first to be used by man. Both in ancient China and Egypt flax-woven clothing was common. Cotton came to Europe from Egypt in the 6th century A. C. Jute had been used in India for centuries but reached Europe only in the 1860's. Silk is said to have been introduced as fiber material around 2650 B. C. in China.

The first half-synthetic fiber material, nitrocellulose, was produced in Switzerland by Audemar in 1885, but a safe cellulose fiber appeared only in the 1890's. The first fully synthetic fiber, nylon, was made by Carothers in 1927. Since then the production of new types of synthetic fibers has been accelerating at at tremendous rate.

Table B. 18 (p. 123 and 124) gives an overview of the most important types of fibers used today. The list is already outdated since new types of fibers are constantly synthesized and marketed. Also, it is important to realize that many fabrics are made of fiber blends, often consisting of both natural and synthetic fibers.

From the chemical point of view it is important to realize that the original fiber is subjected to a number of treatments so that the final product can contain up to 50% of non-fiber materials such as fillers, stabilizers, hardeners, antistatics, bleaching agents, waterproofing agents, tanning agents, and dyeing chemicals.

Dyeing is a complicated procedure since the various natural and synthetic fibers have widely different affinities for the different classes of dyes available. Numerous methods for pretreatment are therefore necessary to ensure the proper uptake of dyes. For the forensic chemist the reverse operation, i. e. the removal of the dyes for chemical analyses is of special importance. A short discussion of the analysis of textile colors is therefore included in this chapter since it forms an integral part of textile fiber investigations.

A large number of textbooks and monographs on the analysis of textile fibers are available. Many of the methods described in these texts are not suitable for the forensic scientist or are superseded by more modern microtechniques. It is nonetheless advisable to read pertinent chapters of these texts. The general references are listed at the end of this chapter (p. 121) as [A-R]. References [F] and [I] are the only review articles on forensic fiber analysis known to the authors at this time. One of the very useful publications in this field is the treatise "Identification of Textile Materials"[Q] which is revised and enlarged every few years.

The remainder of this chapter will be concerned with the analytical methods used at present for the forensic examination of fibers. We have not tried to separate the chemical methods from morphological and physical techniques since a thorough fiber examination makes use of a variety of approaches.

Instead, the methods are roughly grouped into non-destructive and destructive procedures. This enables the forensic scientist to choose a correct sequence when progressing from one method to another. No general guidelines can be given about the number of methods required for a comparative investigation since this obviously depends on the nature of the case, the availability of other exhibits with high evidenciary value, the level of ambition, the resources of the laboratory etc.

The non-chemical methods mentioned in Sects. B. IV. 2 and 3 will not be treated in great detail. These details will be found in the general forensic literature (cf. C. IV) and in the general references listed at the end of this chapter [A-R].

In a recently published proficiency testing program for American forensic science laboratories [19] a number of tests were carried out for fiber examinations. The frequency of the methods used for this purpose is listed in Table B. 19 (p. 124).

The data of Table B. 19 show how insignificant the impact of modern instrumental methods is at present, at least in the majority of American laboratories. This does not mean that the classical techniques should be discarded but that they should be supplemented by chemical and physical micromethods available today and in the immediate future.

2. Non-Destructive Micromethods of Analysis

a. Introduction

The most frequently used non-destructive fiber analytical methods are listed in Table B. 20 (p. 125). They are, somewhat arbitrarily, grouped into visual and instrumental methods. We have regarded the magnifying glass and the conventional light microscopes as an extension of the human eye but have considered the scanning electron microscope as an "instrument".

In the great majority of cases the forensic scientist need not *identify* a fiber in detail but is mainly concerned with a *comparative* investigation. The forensic importance, as well as the correct collection of fiber evidence will be described in Chap. C. I. Great care must be taken in keeping the samples collected at the scene of the crime, or from a victim, apart from the reference samples obtained from the suspect(s) or his (their) belongings. The slightest uncertainty pertaining to the identification or origin renders the whole effort useless. Of course this is always true of forensic comparison work, but especially important in fiber, hair, paint chip, and glass examinations.

b. Visual Comparisons

An examination of the fiber material with respect to amount, general appearance, type, and color is first done by the naked eye and under a binocular microscope at low magnification and in white light. At this point labeled containers (covered Petri dishes or the like) are kept in readiness for unequivocal sorting and identification of the material. The material is then transferred to a high quality microscope or a comparison microscope for transmitted light.

c. Microscopic Examination

A selection of literature references on microscopic methods is found in Table B. 21 (p. 125). Fibers collected on an adhesive tape are observed in the dry state, but in most other cases it is preferable to immerse them in a liquid with a similar refractive index as that of the fiber itself. Table B. 22 b (p. 126) lists the refractive indices of a number of fibers and Table B.22 a (p. 125) of a number of mounting media. The ratio of the refractive indices should not exceed 1.06. It is often more practical, however, to use water as a mounting medium.

The most useful and also the easiest way for observing and comparing fibers is viewing them longitudinally. The observation of fiber cross sections is a good complement for comparing fibers, but cross sectioning requires some experience in microtome techniques. It also leads to the destruction of part of the material. Microscopic fiber identification is greatly helped by using collections of micro photographs such as those reproduced in Ref. [H, Q, R] and others. However, the direct comparison of exhibit and reference fibers can best be accomplished if the laboratory has access to a good and updated reference collection of microscopic slides.

The study of the morphology of colored fibers is not always easy but can be considerably improved by casting the fibers in an appropriate medium. For this purpose the use of a 20% solution of poly (vinyl acetate) in benzene has been suggested [R].

The microscopic examination is not limited to a comparison of the morphological features, including the measurement of fiber diameters and a subjective color comparison. A number of special optical techniques are available for further comparative investigation. Thus, the refractive index and – if present – the birefringence of the fibers can be measured using polarized light and interference microscopy. In ultraviolet light, the presence and the nature of fluorescence can be observed. The comparison of colored fibers may be carried out subjectively in white light but is helped by observation in filtered or monochromatic light.

The double refraction or birefringence is the numerical difference between the index of refraction along the longitudinal axis of a fiber, $n\gamma$, and the index of refraction perpendicular to this axis, $n\alpha$. The birefringence is called positive for $n\gamma > n\alpha$, and negative for $n\gamma < n\alpha$. Some birefringence values are listed in Table B. 22 b (p. 126).

d. Density Determinations

Density measurements are an interesting complement to other physical examinations. They can either be carried out by placing fibers in mixed solvents of known densities or in specially prepared density gradient tubes. It is important to ascertain the complete absence of minute air bubbles in the material which would lead to erroneous results. More often than not, however, the small amount of available material makes this technique difficult to apply to forensic work. Good descriptions of density determination techniques can e. g. be found in Ref. [Q].

e. Microspectrophotometry

In recent years a number of microspectrophotometers have become commercially available, and this allows for objective measurements of absorption spectra from the near ultraviolet to the near infrared. The advantages and pitfalls of this technique have been discussed in the literature (cf. Table, B. 23, p. 127).

Whenever color comparisons are made one has to keep in mind that differences in hue and/or intensity between two fiber colors may be due to different histories of the samples in question, such as exposure to light, water, heat, washing, bleaching etc.

f. X-Ray Diffraction and ESCA

Blends of cotton and polyester have been examined by X-ray diffraction [18] and by electron spectroscopy chemical analysis but these methods require at present rather large samples exceeding those usually met in forensic practice.

g. Infrared Spectrophotometry

The use of infrared spectrophotometry (IR) is usually also destructive (cf. below). If enough material is available, recovery is possible by using a special technique described by Carlsson et al. [3]. In other cases the fibers must be cut into small pieces and their retrieval will be virtually impossible.

3. Destructive Micromethods of Analysis

a. Introduction

Judging from Table B.19 (p. 124) the most widely used chemical methods for fiber examinations are, at present, solubility tests and the flame test. Other chemical methods are only used sporadically in American laboratories. This situation will rapidly change in the near future. Modern instrumental techniques are being constantly introduced and will increase in importance. The most important destructive methods are listed in Table B. 24 (p. 127).

b. The Flame Test or Burning Test

This classical test consists of four groups of observations:

(i) the odor of the smoke,
(ii) the rate and character of melting, shrinking, curling and burning,
(iii) the action after removing the flame, and
(iv) the properties of the ash.

Obviously, this test can only be used if the amount of material suffices after all other experiments have been carried out.

The burning characteristics of a large numer of fibers are listed in Ref. [Q].

The drawbacks of this technique are the relatively large amounts of fiber material required, the subjective character of the observations, and the limited value of the test in examining fiber blends which are rather common in modern fabrics.

c. Solubility Tests

Solubility tests can be useful if

(i) enough material is available
(ii) there is a blend of different fibers where the components show differences in solubility
(iii) and the examiner has acquired a good deal of practical experience prior to using the method in an actual case.

There are extensive tables of solubility data for textile fibers available in the literature. Some of these are in the form of stepwise differentiation procedures [G, Q, 2].

Suchenwirth and Brück [25] have pointed out some pitfalls for the use of solubility tests, especially when these are to be used in comparison work. The clothing of the criminal can have been subjected to changes brought about by wear (dirt, oil etc.) or – on the contrary – by thorough washing or drycleaning.

d. Hot-Stage Microscopy

Controlled heating of fibers on the hot stage of a microscope can give some information on the nature and composition of the fiber material. One observes the contraction, softening and melting regions (a melting *point* is rare) as well as other phenomena such as gas bubble formation, decolorization, sublimation, formation of droplets from water or other compounds etc. as a function of temperature. An "exhibit fiber" and a reference fiber may be placed next to each other to make a direct comparison, but in this case the material may not be used for further tests because of contamination risks. Some melting points of man-made fibers are listed in Ref. [Q].

e. Staining Techniques

Staining techniques cannot be recommended for forensic purposes. They are primarily designed for testing untreated fibers and give unreliable results on treated and/or dyed fiber material since such material must first be stripped of any chemicals added.

f. Microscopy of Cross Sections

The preparation of cross sections of fibers always leads to small losses of material. The sectioning can be done directly on the fibers or after imbedding them in a suitable medium. Again, Ref. [Q] gives very useful technical details for this purpose. For identification of fibers, several collections of photomicrographs are available, e. g. in [H, Q], and others. However, in comparison work a good reference collection is always preferable. Exhibit fibers and comparison fibers should be cut in the same fashion.

g. Scanning Electron Microscopy

The rapid spread of scanning electron microscopy (SEM) to forensic science laboratories has made this technique available to an ever-growing community of criminalists. In fiber investigation, this instrument can give detailed pictures of the surface characteristics and morphology of single synthetic fibers [16]. In addition, the elemental composition of the outer layers of fibers can be measured and compared. The presence of a wide variety of elements in the surface layers is mainly due to various chemical treatments of the textile material such as stabilization, antistatic treatments, water-proofing, sizing, finishing by a variety of processes, and the treatments involved in dyeing (application of dye carriers, pigment dispersants and the dyes themselves). The examination of fibers in the SEM is theoretically a non-destructive process, but it will be difficult in most cases to retrieve a fiber from its SEM-mount without damaging it.

h. Extraction of Dyes from Textile Fibers

Not all fiber materials are suitable for extraxtion nor can all types of dyes be extracted into solvents. In addition, some solvents destroy or change the dye so that further analysis cannot be carried out.

Madan and Khan [15] have described extraction procedures for dispersed dyes on poly(ethylene terephthalate) using chlorobenzene. Macrae and Smalldon [14] have studied the extraction of dyestuffs from wool fibers for forensic purposes. They recommend a 3-step procedure:

(i) extraction of the dye with pyridine-water (4+3 v/v), first at room temperature (30 min) then at 90° C (15 min),
(ii) pretreatment with 2% aqueous oxalic acid for 20 min at room temperature followed by extraction as above, and
(iii) extraction with dimethyl formamide – 4% aqueous ammonia – 4% aqueous 1,2-diaminoethane.

Appleyard [R] has collected the literature on dye extraction from fibers prior to 1969. Cook and Paterson [4] have recently described the extraction of dyes for thin-layer chromatography. They also recommend pyridine – water (4+3 v/v) for wool fibers.

The extracted dyes can be analyzed efficiently by thin-layer chromatography which has adequate sensitivity for this purpose (10–100 ng), especially when microplates are employed. Macrae et al. [14] recommend the following conditions: "The dye solution was applied to the origin of the plate (Merck DC-Alufolien Kieselgel 60 F_{254}) to give a spot about 2 mm in diameter. The plate was warmed at 50° C on a hot plate in a stream of air during spotting and for about 10 min afterwards. When a suitable number of samples had been applied to the origin the plate was developed for 2 or 3 mm in methanol and then dried once more. The plate was suspended in a cylindrical tank for 5 min to allow equilibration with the solvent vapor. Development was then carried out for about 5 cm with the solvent system pyridine/amyl alcohol/10% aqueous ammonia (4+3+3 v/v). The developed plate was finally dried for 15 min as previously described."

Alternative solvents recommended by these authors are:

(2) Chloroform-methanol-0.88 d ammonia-water = 11+7+1+1, and

(3) Sec. butanol-acetone-0.88 d ammonia-water = 5+5+2+1

i. Analysis of Extracted Dyes by Instrumental Methods

Dyes extracted from fibers by one or another of the solvents mentioned above can also be analyzed by spectrophotometry in the visible region. The amount of dye available in forensic cases is, however, usually a limiting factor. Macrae et al. [14] have used a specially constructed microcell in connection with a beam condenser to overcome this difficulty.

Even more difficult, due to small sample sizes, is the use of X-ray fluorescence. Macrae and Smalldon [13] have applied this technique to show the presence of Cr in a chrome dye. The situation is improved when scanning electron microscopy (SEM) is employed instead of X-ray analysis [16].

j. Pyrolysis Gas Chromatography

The application of pyrolysis gas chromatography (PGC) to the analysis of polymers has been discussed in some detail in Chap. B. III of this book. PGC is a very useful and highly discriminating method for man-made fibers. It is, however, of limited use for the analysis of natural fibers which often give uncharacteristic and – to boot – varying pyrograms from sample to sample. It is advisable to separate fiber blends (if this is possible) by mechanical means before analyzing them by PGC.

It seems that PGC was introduced into fiber analysis in 1964 by Groten [8] who applied this technique to over 150 polymer materials, among them some synthetic and natural fibers. In 1968, Janiak and Demerau [12] used the method for forensic analyses, and further forensic applications can be found in the literature collected in Table B. 25 (p. 127).

The Home Office Forensic Science Service in England has assembled a very useful reference collection of pyrograms based on the use of a commercial Curie-point pyrolyzer, and standardized conditions for pyrolysis and chromatography. The minimum amount of fiber necessary for PGC has been given as 40 μg by Bortniak et al. [1] and corresponds to a few millimeters of a single fiber.

k. Thermal Instrumental Methods

A survey of the forensic literature has not shown any publication on the use of differential thermal analysis or thermogravimetric analysis for comparative fiber examinations even is the technique has been used in forensic laboratories. Possibly, the limited amounts of fiber material have been considered a problem. However, modern instruments can handle samples of less than 1 mg.

On the other hand, there are two publications on the use of thermomechanical analysis in forensic fiber examination [17, 24]. The latter authors used this method in conjunction with a mass spectrometer and their paper will be discussed below.

l. Infrared Spectrophotometry

Infrared (IR) spectra give detailed information on the structure of organic molecules and should at first sight appear to be an ideal criterion for fiber analyses. There are, however, several difficulties which limit the value of IR-spectra for this purpose. These are

(i) the limited amounts of material usually available to the criminalist,
(ii) the wide-spread occurrence of fiber blends, giving complex IR-spectra
(iii) the similar IR-spectra given by natural fibers,
(iv) the loss of material, common to all destructive methods, and
(v) the interference of dyes usually present on the fibers.

In spite of these difficulties IR-spectrophotometry has been widely used for forensic fiber examination. In particular, the interference from dyes seems to be minimal [5]. Selected publications on IR methods are listed in Table B. 26 (p. 128).

Sample preparation is a crucial step in IR-procedures. If enough material is available the fibers can be mounted in special holders and can be recovered (cf. B. IV. 2 g). In most cases other techniques must be used. One of these could be the "film-method", where a solution of the fiber is evaporated on a KBr-disk [6]. Better suited for work with very small amounts of material is the use of a KBr micro-pellet, with or without intermediate grinding with a suitable solvent. Fox and Schuetzman [5] published such a procedure and illustrated it with 12 IR-spectra of fibers. Rouen [21] gave another version of this technique and reproduced 22 IR-spectra. Finally, Grieve et al. [6, 7] described a third way of working and published 16 IR-spectra. These authors recommend the following procedure.

The fiber is placed in a well of a spot plate and ground with a drop of solution of 10 g of phenol in 100 ml of cloroform, followed by a small amount (enough for a micropellet) of spectroscopic grade dry KBr powder. If needed, undissolved fibers are cut with a scalpel. The mixture is dried for 1 hour at 100° C and immediately transferred to the press for a micropellet (they used a Perkin-Elmer Ultra-Micro Die 186-0007 producing a KB-pellet of 1.5 diameter). A 1 by 4 refracting beam condenser was employed for recording the spectra.

Finally, Cook and Paterson [4] described in detail two microtechniques for obtaining IR-spectra of fibers, i. e. a variation of the solvent cast technique, and a new method, the lead foil technique.

m. Mass Spectrometry

Mass spectrometry (MS) has only found limited use in fiber analysis. This is not surprising since the molecular weights of the polymers are very large, and since the fragmentation patterns cannot be expected to be very discriminating. The combinations pyrolysis-MS or gas chromatography-MS may be more promising since the pyrolysis products are better defined entities.

Hughes et al. [9–11] have published three papers on pyrolysis gas chromatography.

Smalldon et al. [24] coupled a mass spectrometer to a thermobalance. They named this method TG-MS (thermogravity – mass spectrometry). The sophisticated instrumentation which must be studied in the original paper, yields among other data two kinds of diagrams: mass spectra, and mass thermograms. The latter are plots of total ion current as a function of temperature.

4. Practical Examples

a. A Case of Rape

A practical application of some of the methods described above was published by us in 1979 [16].

A 19-year old girl was attacked by two men and was raped on her own bed by one of these. The police arrested a 20-year old suspect on the following day. The medical examiner found a small tuft of yellow-brown fibers (A) under the prepuce of the suspect's penis.

On the scene of the crime the police found similar fibers on three fabrics, i. e. (B) the panties of the victim, (C) the ankle socks of the victim, and (D) the bedspread.

All four textile samples were sent to the laboratory for examination. Microscopy and solubility tests showed that (B) consisted of nylon, (C) of cotton, whereas (A) and (D) were based on a non-identified synthetic fiber material. Both (A) and (D) had a similar profile, hardness, and very similar color (strongly yellow-brown) and had a similar structure as seen under a comparison microscope in transmitted white light. After imbedding in epoxy resin, the cross sections were compared and found to be similar and of unusual shape.

Small samples of (A) and (D) were now examined in the scanning electron microscope (SEM). The profiles of the fibers seen in the SEM are shown in Fig. B. 23.

The specimens were then subjected to a burning test (giving similar results) and the ashes subjected to energy-dispersive X-ray analysis and yielded practically identical spectra (cf. Fig. B. 24). The non-identified peaks to the right stem from Cu and are mostly due to the material from the sample holder. The difference in Na-concentration stems probably from slight differences in the ashing procedure.

Pieces of single fiber (about 5 mm long) were subjected to pyrolysis gas chromatography (Curie-point apparatus), yielding very similar pyrograms. These are shown in Fig. B. 25 in a schematic way since the original pyrograms differed considerably in intensity (some peaks had to be attenuated).

The dye was eluted from small pieces of fibers (A-D) with pyridine-water $=4+3$ and subjected to thin-layer chromatography (Fig. B. 26). The chromatograms for A and D are in good mutual agreement.

The results of the various analyses described above led to the conclusion that fibers A and D could not be discriminated between.

b. Robbery of a Money Transport

A messenger carrying a large amount of money from an automobile company to a bank was driving across a bridge in Stockholm when a large Mercedes edged him off

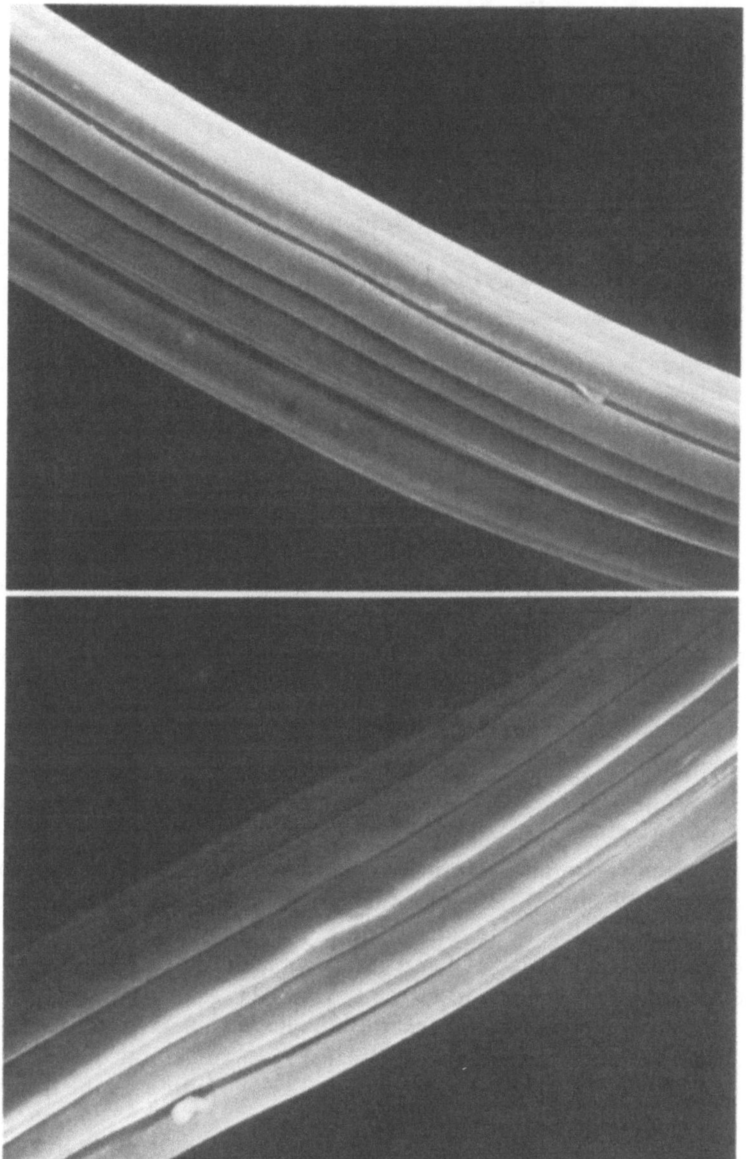

Fig. B.23. The structure of the fibers from the penis sample, above, and the reference sample (bed overcast), below as seen in the scanning electron microscope

to the sidewalk. Three people rushed him, took his case with the money and sped away. The robbers drove to a restaurant, locked the car and celebrated their crime. They left no fingerprints on the car nor on the dinnerware which was recovered.

The Mercedes had been stolen shortly before the crime. It was towed away to the technical squad of the Stockholm police and opened. Since no usuable fingerprints

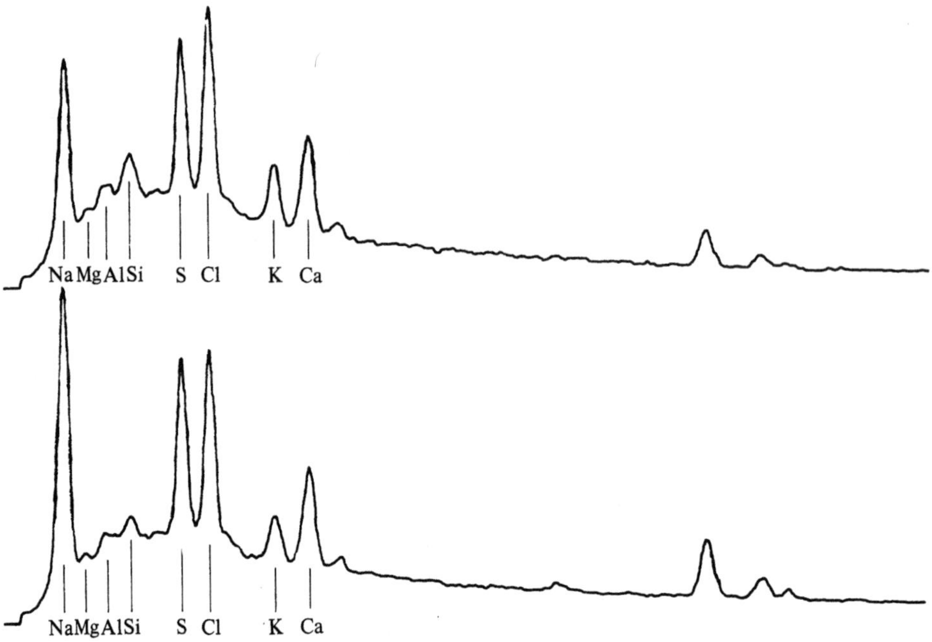

Fig. B. 24. The energy – dispersive X-ray analyses of the incriminated fibers A (above) and the reference fibers D (below). The same elements occur at virtually identical relative concentrations

were found, the seats were taped with adhesive tape. The 12 tapes used for taping the sitting surfaces and backs of each car seat separately were sent to the laboratory.

Three days after the crime, a suspect was arrested in Northern Sweden and his clothes sent in. On three tapes from the back seats of the car very unusual blue leatherlike fibers were found (cf. Fig. B. 27). Among the suspect's clothes a blue-dyed leather jacket was located. Fibers from this jacket had a similar size, shape, and color as the blue fibers from the tapes.

Due to the rare occurrence of blue leather jackets of this particular type, the evidential value of the fiber comparison was high, and the court decided that the suspect could be kept in custody. He later admitted to the crime.

c. Murder or Suicide?

A person was found dead, strangled with a scarf tied in three loops around the neck. At autopsy, death by strangulation was established. The circumstances at the scene of the death pointed to suicide.

In order to exclude murder, both hands of the deceased were taped off and the tapes sent to the laboratory together with the scarf.

The scarf was made of green and orange synthetic fibers. The fibers of both colors had two different physical shapes: one type had a smooth surface and a bean-shaped cross section, the other a pitted surface and a round cross section.

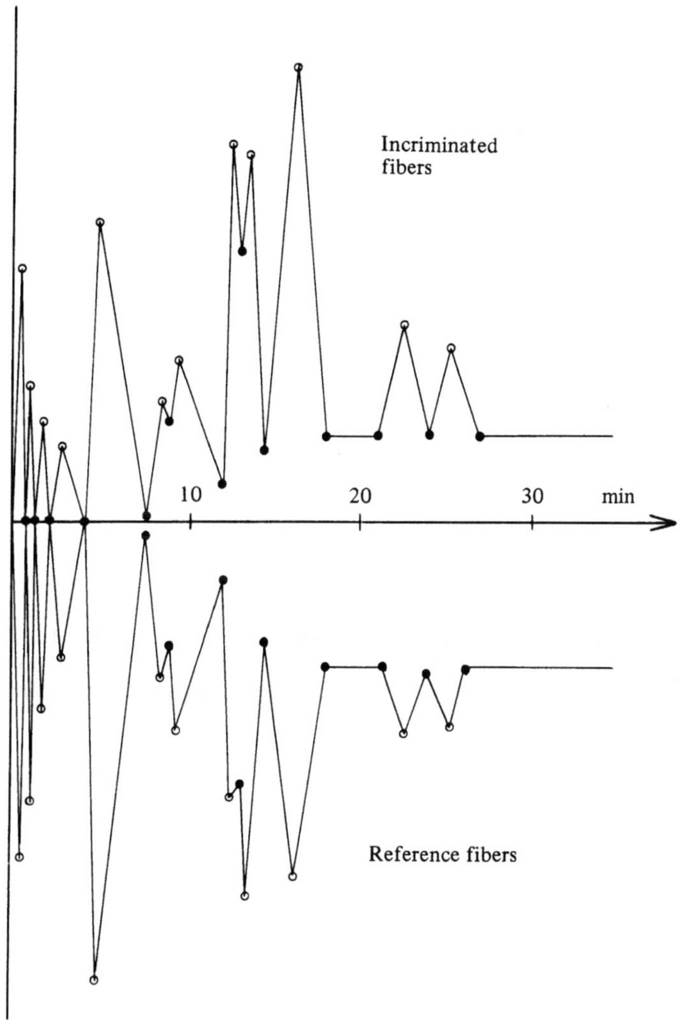

Fig. B. 25. Pyrolysis-gaschromatograms of the incriminated fibers A (above) and the reference fibers D (below). The original pyrograms had very different intensities and in addition, several attenuations were used. The pyrograms maxima and minima were therefore measured and their scales adjusted. A mirror image of the reference pyrogram was then drawn. Experimental conditions: Unicam 104 gaschromatograph, equipped with a Curie-point pyrolysing unit. Column: 1,5 m glass, packed with Parapak Q (80–100 mesh). Injector and detector temperature: 220°. Oven temperature: 100–200° (8° min^{-1}), then 200° (27,5 min). Curie-wire: 610° (10 seconds)

Four green fibers could be isolated from the tape marked "right hand". Two of these were smooth with a bean-shaped cross section and the other two pitted with a round cross section.

One fiber, 25 mm long, could be picked from the tape marked "left hand". It was smooth and had a bean-shaped cross section. The colors from this fiber and from a similar reference fiber were extracted with pyridine-water-acetic acid. The extracts were thin-layer chromatographed and gave indistinguishable color patterns.

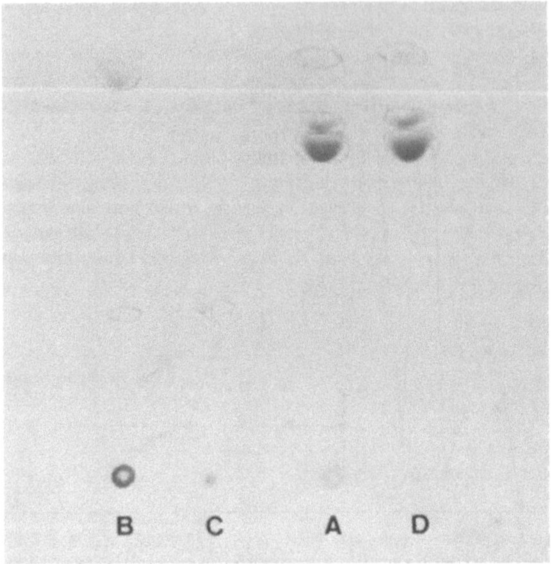

Fig. B. 26. Thin layer chromatogram of extracts from fibers B, C, A and D. Plates: DC-Alufolien Kieselgel 60, ohne Fluoreszenzindikator (Merck). Developing solvent: Chloroform + methanol + 0,88 d ammonia + water = 55 + 35 + 5 + 5. On the original plate, also the colors of the spots for A and D were the same

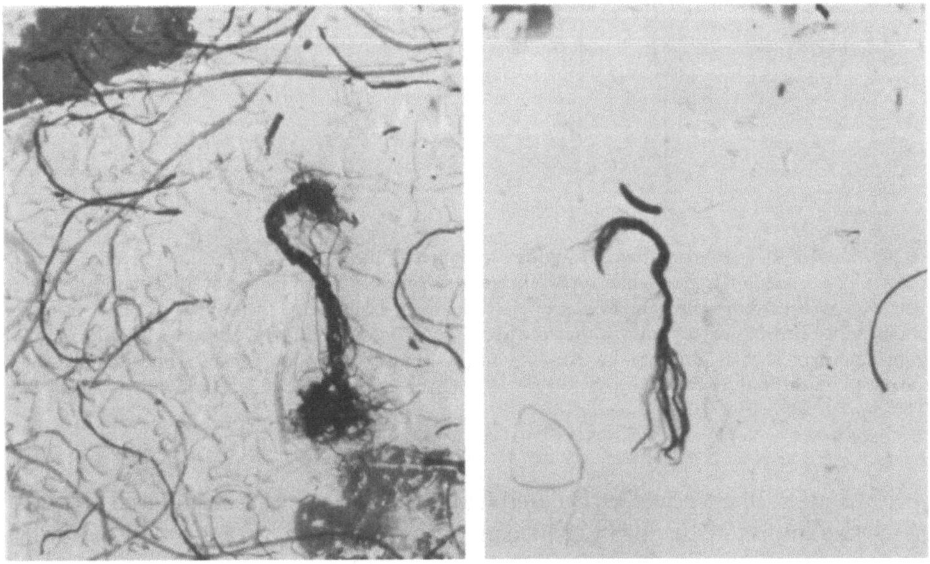

a b

Fig. B. 27. a A microphotograph showing part of the adhesive tape from the back seat of the getaway car. The unusual bunch of fibers was blue and about 3 mm long, **b** A microphotograph of a fiber bunch (about 3 mm long) taken from the blue-colored suède jacket of the suspect

The conclusion was that:

Fibers were found on the tapes which corresponded well with the reference fibers from the scarf in respect to their morphology and color. The fibers could therefore have had their origin from the scarf sent in.

These findings supported the suicide theory.

5. General References (Arranged in Chronological Order of Publication)

A) A. Longhetti, G. W. Roche: Microscopic Identification of Man-Made Fibers from the Criminalistics Point of View. J. Forens. Sci. *3*, 303 (1958)

B) J. J. Press (ed.): Man-Made Textile Encyclopedia. New York: Interscience 1959

C) H. M. Appleyard: Guide to the Identification of Animal Fibres. Leeds: Wool Industries Research Association 1960

D) E. Clayton: The Identification of Dyes on Textile Fibres. Bradford, England: The Society of Dyers and Colourists 1963

E) G. H. Rettie, C. G. Haynes: Thin-Layer Chromatography and its Application to Dyes. J. Soc. of Dyers and Colorists *80*, 629 (1964)

F) M. Frei-Sulzer: Coloured Fibres in Criminal Investigation with Special Reference to Natural Fibres. In: Methods of Forensic Science, Vol. 4. Curry, A. S. (ed.). London: Interscience 1965

G) C. E. M. Hall, M. S. T. Davis: Identification of Textile Materials, 5th ed. Liverpool-London-Prescot: The Textile Institute 1965

H) W. C. Mc Crone, R. G. Draftz, J. G. Delly: The Particle Atlas. Ann Arbor: Ann Arbor Science Publishers 1967

I) M. G. Grieve: Forensic Examination of Textile Fibres. Police Res. Bull. *2*, 26 (1967)

J) W. Garner: Textile Laboratory Manual, 3rd edit., 6 Vols. New York: American Elsevier 1967

K) H. Schweppe: Synthetic Colouring Materials. In: Thin-Layer Chromatography. Stahl, E. (ed). New York: Springer 1967, pp. 612–630

L) J. E. Booth: Principles of Textile Testing. London: Heywood Books 1968

M) J. Gordon-Cook: Handbook of Textile Fibers. Vol. 2: Man-Made Fibers. Watford, England: Merrow 1968

N) J. W. Weaver: Analytical Methods for a Textile Laboratory, 2nd edit., AATCC Monograph No. 3. Research Triangle Park N. C.: American Association of Textile Chemists and Colorists 1968

O) J. F. Feeman: An Introduction to Modern Methods of Dye Identification – Chromatography and Spectrophotometry. Cand. Text. J. *83*, 1970

P) Anonymous: Glossary of Generic Names for Man-Made Fibres BS 4815. London: British Standards Institution 1972

Q) Anonymous: Identification of Textile Materials, 7th edit. Manchester: The Textile Institute 1975

R) H. M. Appleyard: Fibre Identification and Analysis of Fibre Blends. A Review. Analyst *101*, 593 (1976)

6. Specific References

1) J. P. Bortniak, S. E. Brown, E. H. Sild: Differentiation of Microgram Quantities of Acrylic and Modacrylic Fibers Using Pyrolysis Gas-Liquid Chromatography. J. Forens. Sci. *16*, 380 (1971)

2) M.-L. Bumiller: Identifizierung synthetischer Fasern gemäß ihrer Löslichkeit: Chemiefasern 7, 488 (1963)

3) D. J. Carlsson, T. Suprunchuk, D. M. Wiles: Fiber Identification at the Microgram Level by Infrared Spectroscopy. Text. Res. J. *47*, 456 (1977)

4) R. Cook, M. D. Paterson: New Techniques for the Identification of Microscopic Samples of Textile Fibres by Infrared Spectroscopy. Forens. Sci. Intern. *12*, 237 (1978)

5) R. H. Fox, H. I. Schuetzman: The Infrared Identification of Microscopic Samples of Man-Made Fiber. J. Forens. Sci. *13*, 397 (1968)
6) M. C. Grieve, J. A. Kearns: Preparing Samples for the Recording of Infrared Spectra from Synthetic Fibers. J. Forens. Sci. *21*, 307 (1976)
7) M. C. Grieve, T. M. Kotowski: The Identification of Polyester Fibers in Forensic Science. J. Forens. Sci. *22*, 390 (1977)
8) B. Groten: Application of Pyrolysis-Gas Chromatography to Polymer Characterization. Anal. Chem. *36*, 1206 (1964)
9) J. C. Hughes, B. B. Wheals, M. J. Whitehouse: Simple Technique for the Pyrolysis – Mass Spectrometry of Polymeric Materials. Analyst *102*, 143 (1977)
10) J. C. Hughes, B. B. Wheals, M. J. Whitehouse: Pyrolysis Mass Spectrometry. A Technique of Forensic Potential? Forens. Sci. *10*, 217 (1977)
11) J. C. Hughes, B. B. Wheals, M. J. Whitehouse: Pyrolysis – Mass Spectrometry of Textile Fibres. Analyst *103*, 482 (1978)
12) R. A. Janiak, K. A. Damerau: The Application of Pyrolysis and Programmed Temperature Gas Chromatography to the Identification of Textile Fibers. J. Crim. Law Criminol. Police Sci. *59*, 434 (1968)
13) R. Macrae, K. W. Smalldon: The Extraction of Dyestuffs from Single Wool Fibers. J. Forens. Sci. *24*, 109 (1979)
14) R. Macrae, R. J. Dudley, K. W. Smalldon: The Characterization of Dyestuffs on Wool Fibers with Special Reference to Microspectrophotometry. J. Forens. Sci. *24*, 117 (1979)
15) G. L. Madan, A. H. Khan: Determination of Dye on Textile Fibers. Part I. Disperse Dyes on Polyethylene Terephthalate. Text. Res. J. *48*, 481 (1978)
16) A. C. Maehly, M.-A. v. Wachenfelt: Textilfasern unter der Vorhaut – kriminaltechnischer Beweis bei einem Sexualverbrechen. Z. Rechtsmed. *82*, 237 (1979)
17) M. R. Martinelli, S. W. Mayer, P. F. Jones: Thermomechanical Examination of Fabric Composed of Synthetic Polymers. J. Forens. Sci. *24*, 130 (1979)
18) D. Mitcham, R. J. Berni, V. W. Tripp: An X-Ray Diffraction Method for the Analysis of Cotton/Polyester Blends. Text. Res. J. *47*, 386 (1977)
19) J. L. Peterson et al.: Crime Laboratory Proficiency Testing Research Program. Washington: National Institute of Law Enforcement and Criminal Justice 1978
20) H. Rein et al.: Fasern und Borsten. In: Ullmanns Encyklopädie der technischen Chemie, vol. 7, Foerst, W. (ed.). München-Berlin: Urban & Schwarzenberg 1956, pp. 234–374
21) R. A. Rouen, V. C. Reeve: A Comparison and Evaluation of Techniques for Identification of Synthetic Fibers. J. Forens. Sci. *15*, 410 (1970)
22) E. Schwartz: Fasern. In: Ullmanns Encyklopädie der technischen Chemie, Ergänzungsband, Foerst, W. (ed.). München – Berlin – Wien: Urban & Schwarzenberg 1970, pp. 333–357
23) K. W. Smalldon: The Identification of Acrylic Fibers by Polymer Composition as Determined by Infrared Spectroscopy and Physical Characteristics. J. Forens. Sci. *18*, 69 (1973)
24) K. W. Smalldon, R. E. Ardrey, L. R. Mullings: The Characterization of Closely Related Polymeric Materials by Thermogravimetry-Mass Spectrometry. Anal. Chim. Acta *107*, 327 (1979)
25) H. Suchenwirth, H.-J. Brück: Über den Aussagewert von mikrospektralphotometrischen Messungen an Textilfaserspuren. Arch. Kriminol. *142*, 16–25, 111 (1968)

7. Tables B. 18–26

Table B. 18. Classification of textile fibers (according to Rein et al. [20] and Schwartz, E. [22])

A. *Natural fibers*
- I. Fibers of botanical origin (cellulose)
  - 1. *Seed hair fibers*
    - a. Cotton (from Gossypium)
    - b. Kapok (e. g. from Ceiba)
  - 2. *Bast fibers*
    - a. Flax (from Linum)
    - b. Hemp (from Cannabis)
    - c. Sun hemp (from Urticacea)
    - d. Sisal hemp (from Agave)
    - e. Manila hemp (from Musa)
    - f. Gambo hemp (from Hibiscus)
    - g. New Zealand hemp (from Phormium tenax)
    - h. Mauritius hemp (from Fourcroya)
    - i. Jute (from Corchorus)
    - j. Ramie (from Urticacea)
    - k. Burning bush fiber (Dictamus)
    - l. Yucca fiber
    - m. Kendyr (from Apocynum)
    - n. Cocos fiber
    - o. Broom fiber (from Genista)
    - p. Typha fiber (from Reeds)
- II. Fibers of zoological origin
  - 1. *Wool and hair from* (keratin)
    - a. Sheep
    - b. Goat
    - c. Mohair (from Angora goat)
    - d. Kashmir (from Kashmir goat)
    - e. Cattle
    - f. Camel
    - g. Camel goat (Llama, Alpaca, Hua naco)
    - h. Horse
    - i. Rabbit
    - j. Human
  - 2. *Silks* (fibroin)
    - a. Silk (from Bombyx mori)
    - b. Wild silk (from certain butterflies)
    - c. Spider silk (from Nephila)
- III. Mineral fibers
  Asbest (hydroxylated CaMg silicates)
- IV. Man-made fibers
  - 1. *From biological sources*
    - a. Regenerated cellulose fibers (viscose f., copper f., nitrate f., saponified f.)
    - b. Cellulose acetate fibers
    - c. Regenerated protein fibers (from casein, corn, peanuts, soja beans, cotton seeds, hair and wool)
    - d. Alginate fibers (from cell walls of algae)
  - 2. *From mineral raw materials*
    - a. Graphite
    - b. Boron
    - c. Boron nitride
    - d. Silicium carbide
    - e. Silicium oxide (silica)
    - f. Glass
- V. Synthetic fibers
  (Names of commercial brands are given in parenthesis)
    - a. Polyethylene fibers (Courlene, Dawbarn DLP, Drylene 3, Fortiflex, Hostalen G, Levilen, Nishikalon, Norfil, Northylen, Nymplex, Pex, Platilon, Tanikalon, Tory-Pylen E, Trofil, Tuff-Lite-L, Vectra, Vestolen A, Vislene)
    - b. Polypropylene fibers (Avisum, Beamette, Cetryl, Courlene Py, Chisso Poripuro, Daiwabo, DLP, Drylene 6, Fil Pak, Heraclon, Herkulon, Hostalen HD, Hostalen PP, Lustre Lite, Lus Trus, Marwess, Meraklon, Nyfil, Olane, Polycrest, Polypro, Polystress, Prolene, Proplon, Pylen, Ribofil, Trofil P, Vectra, Velon, Vestolen P, Voplex)
    - c. Poly(vinyl chloride) fibers (Clevyl, Crinovyl, Envilon, Fibroyl, Isovyl, Khlorin, Krehalon, Leavin, Movil, PeCe, Piavicid, Rovyl, Texvil, Thermovyl, Valtren, Viclon)
    - d. Poly(vinylidene chloride) fibers (Bolta-Saran, Clorène, Dawborn, Draka-Saran, Rovana, Saran, Sowiden, Velon, Vectra E, Vectra K)
    - e. Vinyl chloride-vinyl acetate copolymers (Avisco-Vinyon, Bistrand, MP-Faser, Nipla, Niplon, Teviron, Voplex)
    - f. Vinylidene cyanide-vinyl acetate copolymers (Darvan)
    - g. Vinyl chloride-vinylidene chloride-acrylonitrile copolymers (Diorid, PeCe-120)
    - h. Poly(vinyl alcohols) (Alprona, Biolan, Cremona, Fosfol, Kuralon, Letilan, Manryo, Mewlon, Titand, Vi-lon, Vinilon, Vinylan, Vinylon, Wodon)
    - i. Polystyrene fibers (only for insulating materials: Emprene, Polyfiber, Styroflex)

Table B. 18 (continued)

j. Polyacrylonitrile (Acribel, Acrilan, Beslon, Cashmilon, Celacryl, Courtelle, Crylor, Dolan, Dralon, Euroacril, Exlan, Greslan, Krilion, Leacril, Nitron, Nymcrylon, Orlon, PAN, Redon, Sniacrile, Tacryle, Toraylon, Velicren, Vonnel, Wolpryla, X 51, Zefkrome, Zefran)
k. Acrylonitrile-vinyl chloride copolymers (Dynel, Vinyon N)
l. Polyethylene-terephthalate polymers (Dacron, Diolen, Elana, Encron, Fiber 200, Fortrel, Grisuten, Lawsan, Tergal, Terital, Terlenka, Terylene, Tetoron, Trevira, Wistel)
m. Copolyesters (Fiber K, Grilen, Kodel, Kuraray-polyester, NRC-polyester, Toyobo-polyester, Vestan, Vitel, Vycron)
n. Hexamethylenediamine-adipic acid polymers (Amyd, Anid, Antron, Blue-C-Nylon, Bri-Nylon, Cantrece, Nailon, Nygon, Nylfrance, Nylon 66, Nylor, Nylsuisse, Nylon-Fabelta, Nylon Type 8, Permalon, Promilan, Velon.
o. Polycaprolactam fibers (Akulon, Amilan, Bodanyl, Caprolan, Celon, Dederon, Enka-Nylon, Grilon, Kapron, Lilion, Mirlon, Nivion, Nylenka, Ortalion, Perlon, Polana, Silon, Steelon, Velion, Velon-NF)
p. Other polyamide fibers (Qiana, Rilsan)
q. Polyurethane (only for fibers with high elasticity)
r. Polyurea (Urilon)
s. Polytetrafluoroethylene fibers (Algoflon, Klingerflon, Polifen, Polyflon, Teflon, Toyoflon, Vulflon)

Table B. 19. Fiber examination tests carried out by 117 American forensic science laboratories. The methods are listed in the order of frequency in percent of the number of the laboratories taking part in the test (ref [19])

x	Microscopy	100%
x	Macroscopy	72%
	Solubility test	48%
x	Birefringence	40%
(x)	Melting point	17%
x	Refractive index	17%
x	Fluorescence	11%
(x)	Infrared spectrophotometry	9%
	Flame test	2%
x	Density	1%
	Thin-layer chromatography	1%
	Staining	1%
x	Thermal depolarization	1%
x	Color test	1%
(x)	Ultraviolet spectrophotometry	1%
x	Diameter measurement	1%

x Physical methods (x) Physico-chemical methods

Table B. 20. Non-destructive micromethods for fiber examinations

I. Visual observations
   A. Measurement of thickness and yarn structure
   B. Subjective color comparison
   C. Microscopic examination
      1. In air
      2. In an immersion liquid
         a. In white light
         b. In monochromatic light
         c. In polarized light (birefringence)
         d. In ultraviolet light (fluorescence)
      3. Refractive index measurement
   D. Density determination

II. Instrumental techniques
   A. Microspectrophotometry
   B. Scanning electron microscopy
         a. Detailed morphology
         b. Energy-dispersive X-ray analysis
   C. X-ray diffraction
   D. Infrared spectrophotometry by non-destructive techniques

Table B. 21. Literature on microscopic examination of textile fibers

I. General references
Cf. [A, C, G, H, Q]

II. Selected references

1952 A. N. J. Heyn: Observations of the Birefringence and Refractive Index of Synthetic Fibers with Special Reference to Their Identification. Text. Res. J. *22*, 513

1953 A. N. J. Heyn: The Identification of Synthetic Fibers by Their Refractive Indices and Birefringence. Text. Res. J. *23*, 246

1963 B. J. Culliford: The Multiple Entry Card Index for the Identification of Synthetic Fibres. J. Forens. Sci. Soc. *4*, 91

1970 R. A. Rouen, V. C. Reeve: A Comparison and Evaluation of Techniques for Identification of Synthetic Fibers. J. Forens. Sci. *15*, 410

1979 M. C. Johari, D. P. Jatar: Identification of Some Synthetic Fibers by Their Birefringence. J. Forens. Sci. *24*, 692

Table B. 22 a. Refractive indices of mountants for microscopy (acc. to Ref. [R])

Mountant	Refractive index (n_{20}^{D})	Mountant	Refractive index (n_{20}^{D})
Water	1.33	Ethyl salicylate	1.525
n-Heptane	1.39	Canada balsam	1.53
n-Decane	1.41	2-Phenylethyl alcohol	1.533
Butyl stearate	1.445	Methyl salicylate	1.537
Glycerine jelly	1.45	o-Dichlorobenzene	1.549
Liquid paraffin	1.47	Tricresyl phosphate	1.556
Olive oil	1.48	Bromobenzene	1.560
Cedarwood oil	1.513–1.519	l-Bromonaphthalene	1.658
Anisole	1.515	Methylene iodide	1.74
ALP I immersion oil	1.524		

Table B. 22 b. Refractive indices of fibers (acc. to Ref. [R])

Fiber	$n_{\parallel}$	$n_{\perp}$	Birefringence Δn
Acetate			
Diacetate (Dicel)	1.476	1.473	0.003
Triacetate (Tricel)	1.469	1.469	0
Acrylic			
Acrilan 36	1.511	1.514	−0.003
Courtelle	1.511	1.514	−0.003
Orlon 42	1.511	1.515	−0.004
Aramid			
Kevlar	> 2.000		
Asbestos			
Chrysotile	1.50–1.56	–	varies
Amosite	1.64–1.69	–	varies
Crocidolite	1.68–1.71	–	varies
Chlorofibre			
Fibravyl	1.541	1.536	0.005
Cupro			
Cuprammonium rayon (Bemberg)	1.553	1.519	0.034
Glass			
A-glass	1.542	–	–
E-glass	1.550	–	–
S-glass	1.523	–	–
C-glass	1.541	–	–
Modacrylic			
Dynel (unrelaxed)	1.535	1.533	0.002
Teklan	1.520	1.516	0.004
Nylon			
Nylon 11 (Rilsan)	1.553	1.507	0.046
Nylon 6 (Enkalon)	1.575	1.526	0.049
Nylon 6.6 (ICI nylon)	1.578	1.522	0.056
Polyester			
Terylene	1.706	1.546	0.160
Polyolefin			
Polypropylene (Ulstron)	1.530	1.496	0.034
Polyethylene (Courlene X3)	1.574	1.522	0.052
Viscose			
Normal-tenacity viscose	1.542	1.520	0.022
High-tenacity viscose	1.544	1.505	0.039
High-wet-modulus viscose	1.551	1.513	0.038
Wool	1.557	1.547	0.010
Cotton			
Tanguis	1.577	1.529	0.048
Silk			
Degummed	1.591	1.538	0.053
Flax	1.58–1.60	1.52–1.53	0.06

Table B. 23. Literature on microspectrophotometry of fibers

1959	H. Amsler: Die Mikro-Spektralphotometrie, ein wichtiges Hilfsmittel für den Farbvergleich kleinster corpora delicti. Arch. Kriminol. *124*, 85
1968	H. Suchenwirth, H.-J. Brück: Über den Aussagewert von mikrospektralphotometrischen Messungen an Textilfaserspuren. Arch. Kriminol. *142*, 16, 111
1973	R. Halonbrenner, J. Meier: Mikrospektralphotometrische Untersuchungen an Textilfasern. Kriminalistik *27*, 344
1978	R. Halonbrenner: Microspectrophotometric analysis of textile fibres. Intern. Crim. Police Rev. *33*, 7
1979	R. Macrae, R. J. Dudley, K. W. Smalldon: The Characterization of Dyestuffs on Wool Fibers with Special Reference to Microspectrophotometry. J. Forens. Sci. *24*, 117

Table B. 24. Destructive micromethods for fiber examinations

I. Visual observations
   A. Burning tests (flame test)
   B. Solubility tests (microscope)
   C. Hot-stage microscopy, observing: contraction, softening, melting
   D. Staining techniques
   E. Microscopy of cross sections
   F. Extraction of dyes followed by thin-layer chromatography

II. Instrumental techniques
   A. Extraction of dyes followed by visible spectrophotometry and/or X-ray fluorescence
   B. Pyrolysis-gas chromatography
   C. Differential thermal analysis
   D. Thermogravimetry
   E. Thermomechanical analysis
   F. Infrared spectrophotometry (KBr-disks, mulls, solventcast films or diamond cells)
   G. Mass spectrometry
      1. After pyrolysis
      2. After pyrolysis – gas chromatography
      3. After thermogravimetry

Table B. 25. Literature on pyrolysis gas chromatography of fibers

Year	Authors
1960	J. Strassburger, G. M. Brauer, A. F. Forziati: Analysis of Methacrylate Copolymers by Gas Chromatography. Anal. Chem. *32*, 454. 7 pyrograms
1964	B. Groten: Application of Pyrolysis-Gas Chromatography to Polymer Characterization. Anal. Chem. *36*, 1206. 26 pyrograms
1966	U. Gokcen, D. M. Cates: Thermal Analysis of Fibers and Fiber Blends. Part II. Pyrolysis-Gas Chromatography. Applied Polymer Symposia nr. 2, 15–24. 13 pyrograms
1968	R. A. Janiak, K. A. Damerau: The Application of Pyrolysis and Programmed Temperature Gas Chromatography to the Identification of Textile Fibers. J. Crim. Law. Criminol. Police Sci. *59*, 434. 13 pyrograms
1970	R. A. Rouen, V. C. Reeve: A Comparison and Evaluation of Techniques for Identification of Synthetic Fibers. J. Forens. Sci. *15*, 410. No pyrograms
1971	J. P. Bortniak, S. E. Brown, E. H. Sild: Differentiation of Microgram Quantities of Acrylic and Modacrylic Fibers Using Pyrolysis Gas-Liquid Chromatography. J. Forens. Sci. *16*, 380. 6 pyrograms
1972	B. B. Wheals, W. Noble: Forensic Applications of Pyrolysis Gas Chromatography. Chromatographia *5*, 553.

Table B. 26. Literature on infrared spectrophotometry of fibers

Year	Authors
1968	R. H. Fox, H. I. Schuetzman: The Infrared Identification of Microscopic Samples of Man-Made Fiber. J. Forens. Sci. *13*, 397. 13 IR-spectra
1970	D. J. Carlsson, T. Suprunchuk, D. M. Wilses: The Uses of ATR Infrared Spectroscopy in Fibre Science. Canad. Text. J. *87*, 73, 78, 84. 6 IR-spectra
1970	R. A. Rouen, V. C. Reeve: A Comparison and Evaluation of Techniques for Identification of Synthetic Fibers. J. Forens. Sci. *15*, 410. 24 IR-spectra
1973	K. W. Smalldon: The Identification of Acrylic Fibers by Polymer Composition as Determined by Infrared Spectroscopy and Physical Characteristics. J. Forens. Sci. *18*, 69. 7 IR-spectra
1976	D. J. Carlsson, F. R. S. Clark, D. M. Wiles: Infrared Spectra of Monofilaments. Textile Res. J. *46*, 318. 3 IR-spectra
1976	M. C. Grieve, A. J. Kearns: Sample Preparation for the Recording of Infrared Spectra from Synthetic Fibers. J. Forens. Sci. *21*, 307. 10 IR-spectra
1976	P. B. Iyer, K. R. K. Iyer, N. B. Patil: An Infrared Technique for the Quick Analysis of Cotton-Polyester Blends. J. Appl. Polym. Sci. *20*, 591. No spectra
1977	D. J. Carlsson, T. Suprunchuk, D. M. Wiles: Fiber Identification at the Microgram Level by Infrared Spectroscopy. Textile Res. J. *47*, 456, 5 IR-spectra
1977	M. C. Grieve, T. M. Kotowski: The Identification of Polyester Fibers in Forensic Science. J. Forens. Sci. *22*, 390. 8 IR-spectra
1978	R. Cook, M. D. Paterson: New Technique for the Identification of Microscopic Samples of Textile Fibres by Infrared Spectroscopy. Forens. Sci. Intern. *12*, 237. 6 IR-spectra
1978	L. K. Read, R. J. Kopec: Analysis of Synthetic Fibres by Diamond Cell and Sapphire Cell Infrared Spectrophotometry. J. A. O. A. C. *61*, 526. 9 IR-spectra

V. Paints, Varnishes and Lacquers

1. Introduction

Modern paints and lacquers are very complex systems, incorporating an enormous variety of ingredients. It is impossible to give even a simple survey of the many types of paints and their composition. The general bibliography at the end of this chapter can help the interested reader to penetrate into this field where practical experience and know-how often still prevail over theoretical and systematic knowledge.

In principle, paints and varnishes, contain 3 classes of ingredients, i.e.

(i) the carriers or binders,
(ii) the pigments, dyes and additives, and
(iii) solvents that allow the application of the first two groups of substances to surfaces to be covered.

The combination of binder (carrier) and solvent is often called the vehicle or medium. Lacquer is a term usually restricted to clear solutions of cellulose derivatives and a few other polymers. The forensic analyst is normally faced with a comparative investigation of finished products, and is therefore rarely called upon to analyze for solvents. Consequently, the latter will not be further discussed. Carriers, binders or film formers are macromolecular substances in which the additives are dissolved or suspended. They must possess the property of hardening after application of the paint. This hardening can be a *chemical* or a *physical* process. The major *physical* processes are:

1) evaporation of the solvent, leaving a hard and dry residue,
2) evaporation of the water from a suspension of a carrier,
3) cooling of molten carriers.

The major classes of physically hardening vehicles are:

1) natural resins,
2) synthetic resins,
3) cellulose derivatives (nitrocellulose, cellulose esters and ethers),
4) vinyl polymers,
5) chloroprene polymers,
6) polysaccharides (gums),
7) proteins (casein, glues),
8) asphalt- and tar-based vehicles,
9) water-soluble binders.

The hardening of these vehicles is mainly the result of the evaporation of the solvent, but in some cases, such as with the water-soluble binders, coagulation occurs simultaneously.

Water-soluble paints have become widespread and are increasingly used also for exterior painting. The older type of these paints were based on casein but modern paints use latex as the base.

Chemical hardening proceeds mainly by oxidation or polymerization. Binders can be classified into the following groups:

1) oxidizable binders containing unsaturated oils and resins,

2) condensation polymers,
   a) phenolic-aldehyde resins,
   b) urea alkyds,
   c) melamine alkyds,
   d) epoxy resins,
   e) polyurethane resins,
   f) silicone alkyds,
   g) aldehyde resins,
   h) polyester alkyds (alkyds, maleic resins).

3) addition polymers,
   a) polyethylene resins,
   b) polyvinyl resins,
   c) polyacrylates,
   d) styrene polymers,
   e) copolymers of various kinds.

Table B. 27 (p. 151) gives a short overview of the historical development of paint vehicles.

The pigments, dyes and additives give paints and lacquers the desired properties for their practical application. The more important types of additives are:

1) plasticizers,
2) extenders or fillers,
3) driers,
4) skinformation preventors,
5) viscosity-control substances,
6) pigment dispersants and dispersing resins.

Plasticization is needed to give the finished film the desired flexibility and toughness. This effect may be achieved either by *internal* plasticization by chemical modifications of the monomers before polymerization, or by *external* plasticization, i. e. the addition of plasticizers. Typical plasticizers include phthalates (dibutyl, dioctyl, butyl benzyl), phosphates (tricresyl, trioctyl), and polyesters. *Extenders* are used to "dilute" the expensive dyes and pigments without interfering with the coloring properties. *Driers* are needed to speed up the drying of unsaturated oils. The most important driers are the naphthanates, tallates and octoates of cobalt, lead and manganese. *Skin-forming preventors* are e. g. butyraldoxime or methyl ethyl ketoxime. These compounds are volatile and do therefore not inhibit the drying process. Viscosity is controlled by the addition of *thinners* such as butyl glycol, cyclohexanone or other high boiling solvents.

The settling of pigments and other heavy components can be counteracted by the addition of finely powdered silica, aluminium salts of colophonium or long-chain fatty acids. Lately, charged detergents have been increasingly used. They act by inducing ionic charges, by hydrogen bonding, by dipole-dipole interaction, and as protective colloids.

Silicone oils are added to prevent the separation of pigments into layers which would give the applied paint a streaky appearance. If glossy paints are to be dimmed, an increased amount of settling preventors can be added. Also waxes, aluminium colophonate or talcum can be used for this purpose.

Crown [A] lists a large number of pigments and dyes used in paints and lacquers which are schematically grouped in Table B. 28 (p. 152). Thus over 400 types of coloring matter are available to the paint manufacturers. Crown [A] also gives extensive lists of vehicles and of various types of additives.

The complex nature of paint and varnish composition needs not frighten the forensic analyst too much. In actual practice, the comparative investigation of paint samples reduces to three relatively basic operations, i. e.

(i) the visual (or photographic) recording of the color and the sequence of the paint layers in the samples,
(ii) the comparative analysis of the vehicles, and
(iii) the corresponding (semi) quantitative analysis of the various elements (other than C, H, O and N) forming parts of the pigments, fillers, additives etc. in each layer, or several layers at once.

2. Decoration and Protection Paints of Buildings

The type of paint used in buildings depends on the surface (wood, metal, plastic etc.) and on the nature of the surroundings, especially interior and exterior use.

Wood is traditionally subjected to three operations, using primers, undercoats and finishes. The primer is adapted to seal the pores of the wood, the undercoat brings the surface to an even color and provides a suitable ground for the finish or top coat, and the latter provides the desired color and glossiness.

Metals, especially iron, are first protected against corrosion. Redlead primers are being gradually replaced by primers containing lead or zinc chromates.

3. Modern Automobile Surface Coatings

by Inge Horkeby, AB Volvo, Göteborg, Sweden

a. The Phosphating Process

The painting of a car starts really at the steel mill where the sheet metal is made. Precoated steel is widely used today on the interior parts of the car which are heavily exposed to corrosion. This is, of course, of minor interest in the forensic analysis of car-body-paint where the exterior paint is of particular interest.

The bottom layer in a car paint is a coat of phosphate. The best pretreatment of sheet metal used today is a phosphating process which, when properly done, gives not only a clean surface but also a good adhesion for the following paint coats.

The phosphating process is a multistep procedure including alkaline degreasing, rinsing, phosphatizing, rinsing, passivating (chromic acid rinse) and finally rinsing with deionized water. Both iron and zinc phosphatizing is used but the latter dominates. In the zinc phosphate process two types of crystals are formed: hopeite and phosphophyllite. The thickness of the coat is about 1–3 μm (1–3 g/m^2). If a chromic acid rinse is used (Cr^{III} and Cr^{VI}) it forms an iron chromate which will fill the holes between the zinc phosphate crystals.

b. Primers

The first paint coat is today almost always an electrocoat (EC) primer. In the 1970's the EC-primers became very popular. Before the EC a diprimer in an organic solvent was used, which had many disadvantages for the working environment. In order to reduce the use of solvents, a water-borne EC-primer was introduced as an inevitable replacement. Electrocoating had the additional advantages of being fully automized and of giving a uniform coat thickness on different parts of the car (inside, outside, on the edges etc.).

Electrocoating is used by almost all car producers in Europe today. Only FIAT uses in one work a powder paint as a primer on the outside of the body instead of EC. In electrocoating the body can be the anode or the cathode in an electrochemical cell and is immersed in a tank, with an EC-primer containing about 10–15% of solids in water. The paint particles, which are electrically charged will be transported to the body of the car and are precipitated on the metal.

Anodic electrocoating, where the paint is precipitated on the anode, is the oldest and the most common method in Europe. The use of cathodic electrocoating is growing very fast and is the most widely employed method in the USA since about 1978.

c. Electrocoating Plant

Prior to dipping, auxiliary electrodes are attached to the inside of the body. The dipping time varies but is usually 2–3 minutes. The voltage is in the range of 100–400 V. On the outside of the body, a dried coat of 15–35 µm will be precipitated. Cathodic EC-paint gives a lower coat thickness than anodic EC. The paint coat on interior parts is much thinner than on the outside (5–15 µm). The content of solids of the EC-paint tank is low, often 10–15% w/w in water.

When the body is immersed into the tank, and the current is switched on, precipitation will start on all exterior parts. As the coat is growing in thickness precipitation will decrease as the resistance of the formed paint coat increases until a limiting coat thickness is reached. The precipitation on the interior parts then increases. This process is enhanced by auxiliary electrodes.

d. Anodic Electrocoating

In anodic electrocoating the body is attached as the anode. The binder contains carboxy groups and has been neutralized with a suitable base, usually diethylethanolamine or potassium hydroxide. On the body (anode) oxygen is formed and the hydrogen activity will increase. At sufficiently low pH-values the binder will start precipitating on the anode. On the cathode (e. g. the tank walls or separate cathode boxes) hydroxide ions will be formed. The binders (resins) used in anodic electrocoating paints are most often polybutadiene or epoxy esters while the alkyds are no longer used.

e. Cathodic Electrocoating

In cathodic electrocoating the body is used as the cathode. In this case the binder is alkaline and is made hydrophilic with e. g. formic, acetic, citric, phosphoric or hydrochloric acid. The paint particles are transported to the body surface and precipitate at sufficiently high pH-values.

The paint materials for cathodic electrocoating of car bodies are based upon an epoxy resin, e. g. an epoxy amine adduct dissolved in a suitable acid. The use of cathodic electropaint increases every year and is already dominating in the car industry of the USA. All new tanks in Europe will also be filled with cathodic paint. The advantages of cathodic electrocoating are e. g. good corrosion protection, good resistance to mechanical damage and good alkaline resistance.

f. Surfacer – 3 or 4 Coats

The differences between the methods used by the various producers of automobiles are greater in the following steps of the painting process than in the choice of phosphating and EC-priming. The painting processes used are often described as 3- or 4-coat painting depending on the number of separately cured paint coats (incl. the EC-primer). On the parts of the body prone to stone-chipping or severe abrasion, the number of paint coats is increased by many producers.

A few car models from Japan and Italy have a smaller number of paint coats. On the phosphated body a powder coat is applied in a dip- or spray-process followed by electrocoating of the inside. A top coat is then sprayed on the powder coat.

The choice of the second paint coat, the surfacer, depends very much on the process used. If sanding is employed to give a good finish the surfacer must have good sanding properties. The cross-linking of the binder is of a higher magnitude where sanding is used. The pigmentation of the surfacer will also influence these properties. The resin in a surfacer is often an alkyd. To improve corrosion protection the alkyd is often modified with an epoxy resin. To cross-link the alkyd a heat reactive amino resin, e. g. melamine- or benzoguanoamino-resin, is added. Depending on the amino resin the reaction between the alkyd and amino resin will take place at 110–170° C during 10–20 minutes. The surfacer coat is about 30–50 µm thick. While the surfacer is sprayed those parts of the body which are exposed to stone chipping are protected. On the sides of the body a stone chip resistant paint is applied in a thin coat (15–25 µm), by three European car producers, Mercedes, Saab and Volvo. This paint has a poor cohesion which will prevent chipping to the bare metal. On sills, bottom plate and wheel houses most car manufacturers spray a thick coat (300–800 µm) of an abrasive resistant paint. Elastic polyurethanes or PVC can be used.

g. Sealer in 4-Coat

In a four-coat process a sealer is sprayed on the surfacer before the top coat is applied. The sealer is usually sprayed in a thin coat (15–25 µm) and is cured at the same temperature range as the top coats (110–150° C for 15–20 minutes). The sealer is usually of the same quality as the top coat used but is differently pigmented. Some manufacturers spray almost the same color of sealer as the top coat in order to get a better hiding. Others use a uniform color for the sealer. A red-greyish color will provide the best color to be hidden by different top coat colors. The color deviation between surfacer and sealer in such a case can be difficult to examine in a dried paint coat. On a surfacer one almost always uses a light grey color, which ensures that a good coat is applied (when manually sprayed). If black color is used there is a tendency to spray a thinner coat.

h. Top Coats

Top coats in car painting can be divided into two major groups, solid colors and polychromatic colors (metallics). The top coat is sprayed to a thickness of 35–50 µm.

Totally, the body will have a coating of about 100–120 µm, but large deviations occur depending on the location of the surfaces and the number of repainting operations.

The choice of top coat quality is delicate since it is very important for the durability of the whole surface coating. The specifications for outdoor durability differ among the car producers and hence also the top coat qualities. The biggest suppliers of top coat car paints in Europe are e. g. BASF Lackchemie (Glasurit) and Herberts GmbH in the BRD, Corona and Valentine in France, Levis in Belgium, Sikkens in Holland, ICI and International in Great Britain, AB Wilhelm Becker in Sweden, and IVI in Italy.

i. Solid Top Coats

Solid top coats have been dominating the car painting up to the 1970's. The last decade has shown a tremendous increase in the interest in metallics. Nitrocellulose was the most common resin in Europe for top coats in the 1950's but is no longer of any importance. Alkyd-melamine resin top coats took the market in Europe and are used by Ford in the USA, but GM, USA, has started using thermoplastic acrylates. Ford, USA, uses today thermosetting acrylate-melamine resin top coats. The acrylate resin consists of polyacrylates, polymethacrylates, copolymers of acrylates, methacrylates and other monomers e. g. acrylonitrile. Thermosetting acrylic resins for car painting are flexible and have excellent weather durability, but the chemical resistance is not as good as alkyd-melamine resin enamels. In car painting, thermosetting alipathic acrylates, dispersed in an alipathic organic solvent, NAD-paints (Non-Aqueous-Dispersion) are often used. These NADs have a high solids content and less agressive solvents. Alkyds for car paints are polyesters of dicarboxylic acids, e. g. phthalic acid, polyalcohols, e. g. trimethylolpropane, and monocarboxylic acids, e. g. coconut oil fatty acid or pelargonic acid (synthetic). For car top coats less than 40% of fatty acid in the alkyd is used. If no fatty acid is used the binder is called a polyester.

j. Polychromatic Top Coats (Metallics)

Metallics are becoming more and more popular in Europe. With many car producers the metallics represent more than 50% of the production.

Two types of metallics are used, 1-coat and 2-coat paints. In the USA the 1-coat metallic is most common. This paint has low hiding power and must be sprayed in several coats.

Two-coat metallics, first used by Volvo and Mercedes, are more and more used mostly due to superior gloss and outdoor durability. In a two-coat metallic a pigmented base metallic coat is sprayed first (10–20 µm). This coat dries quickly (physically in 1–3 minutes) and a clear coat is then sprayed on top. Both 1-coat and 2-coat metallics are based on acrylate resins. Some base coats are made of polyester resins. The clear coat is sometimes a mixture of an alkyd and an acrylic resin. In the clear coat UV-additives and quenchers are used to give better weather durability.

k. Pigments

Not only the resin is important for giving good durability of the top coat. The choice of pigments is extremely important. Earlier, mostly inorganic pigments were used. The choice of inorganic pigments is now limited due to environmental regulations. Chromate, lead and cadium pigments are limited or forbidden to use. Organic pigments are therefore becoming more important for car top coats. The preparation of these pigments often implies the use of extensive organic syntheses to give large complicated molecules.

l. Future Car Paints

Our environment is constantly deteriorating and hence higher demands are put on the car paint to be resistant to weather and chemicals.

One of the most important goals of car painting today is to reduce the emission of organic solvents from the paint shops at the plant. The content of solids of the paints has to be increased. The future will probably bring about a wide use of:

1. Paints with less organic solvents,
2. Powder coatings,
3. Water-borne paints.

4. Visual Comparison of Paints, Varnishes and Lacquers

The first step in comparing paint flakes is the determination of the number, sequence and color of the various layers. The subjective perception of color is here a big problem. About 10% of all men are more or less severely color-blind, but even people with "normal" color perception have difficulties in accurately describing the hue, intensity and tone of a color.

Hue is the attribute of a color according to common speech and is equivalent to wavelength. *Intensity* is the "strength" of a hue; the denser the color, the more distinct the hue. *Tone* is related to the grey scale which overlays color of a given hue and intensity.

In order to overcome the subjectivity in forensic color comparisons, it is advisable to use standard charts. Fouweather et al. [13] have recommended using the "Methuen Handbook of Colour" [D] for this purpose. This book contains 30 numbered color plates, each representing one distinct hue. Each plate is composed of 48 samples, arranged in 8 horizontal rows of decreasing intensity [1–8]. Within each row, the tone increases in 6 steps [A–F]. For example, a medium intense blue of intermediate tone could have the designation 21 D 4.

Fouweather et al. [13] conducted some very interesting experiments for testing the errors made by a number of investigators using the book for classification of automobile paints. They also surveyed a 287 × 573 m area of a London satellite town with respect to door and window paint colors.

The Swedish Standard Commission has published a color atlas [K] based on a somewhat different principle.

It can be hoped that the frequency distribution of paint colors can be used in the evaluation of evidence with the help of color standards of the type mentioned above.

Tippett [48] has made a thorough effort to study the statistical distribution of the colors of 2000 paint samples from buildings in Wales (UK). His investigation started with noting the number, sequence and colors (8 only) of the paint layers of each sample. In this work, paint samples that could not be discriminated were also subjected to chemical tests (solubility in solvents, pyrolysis gas chromatography).

There have been numerous efforts to establish reference collections (physical, written or computer-based) of the paints and lacquers of automobiles. The task of collecting even approximately complete information on this subject is obviously tremendous. One of the first efforts in this direction was made by Tippett in 1964 [48], Later, Gothard [15] made a similar survey in Australia. Many forensic science laboratories have their own collection of car paint data. Recently, a Royal Canadian Mounted Police laboratory has started a very extensive collection and data project [2] on car paints. The forensic analysis of body primers was discussed by Deacon [11].

Various techniques are available for observing and measuring the appearance of paint flakes. Some investigators advocate imbedding the chips vertically in a suitable medium; after shaving or cutting by microtome and, in some cases, polishing, the block can be easily handled, and the chips investigated. Others prefer to observe the samples as such. This can be done by placing them on a microscope slide between cover glasses or other slides. If essentially optical methods only are used, the former technique has some advantages. However, if the optical examination is followed by instrumental analysis it seems better to skip the embedding step.

The visual observations start with recording the number, sequence, thickness, and color of the paint layers under low magnification. Obviously, some layers may be missing in one of the samples (usually the questioned one). In such cases, observation under higher magnification or in a scanning electron microscope may reveal traces of a subsequent layer on the outside of the "sandwich".

Valuable information can also be obtained from the topography of the various paint layers. Mc Queen et al. [29] have investigated the appearance of such layers by using a scanning electron microscope. Of particular importance is the viewing of the top layer which can carry typical particles from the environment or from surface treatments such as waxes and polishes.

Before proceeding to other techniques it is useful to take color pictures of the paint chips, either by macro- or by microphotography. The chips can then observed in UV-light (fluorescence), in IR-light (luminescence) or through suitable filters. If clear results are obtained, a photographic record can be made. In the absence of photographs, sketches of the chips should be made for future reference in conjunction with scanning electron microscopy. In principle, the recording of the reflectance spectrum of a paint sample would be desirable, since unbiased, objective color data could be obtained in this way. It must be emphasized, however, that the human eye is still more sensitive to small differences in hue, tone and intensity between two very similar colors than present-day spectrophotometers. Paul et al. [32] have used an integrating sphere for enhancing the reflectance spectra of small paint samples. The results were not encouraging.

Recently, microspectrophotometers with adequate performance for the measurement of small surfaces in reflected light have become available. Such instruments have previously been used in the forensic investigation of colored textile fibers, but they operated only in transmitted light. With the advent of reflected light microspectrophotometers unbiased, objective color data can be obtained within the limits indicated above.

Irrespective of the technique for color measurement employed, the forensic chemist must be aware of the inherent pitfalls in the examination of small and very small objects. The smaller the sample, the greater the danger of contamination. The signal-to-noise ratio decreases, and artifacts can falsify the analytical result. These difficulties start already in the collection of the evidence at the scene of the crime or accident. Great care must be taken during packing and transportation of the samples, and in the laboratory contamination must be kept at a minimum. In spite of all precautions, the possibility must be kept in mind that questioned sample(s) and the reference sample(s) could have acquired different contaminations.

5. Inorganic Paint Analysis

a. Introduction

In a recently published Crime Laboratory Proficiency Testing Research Program [33], 110 American laboratories carried out comparative analyses on paint samples from automobiles and from buildings. The methods used by these laboratories are listed in Table B. 29 (p. 152). The table makes also reference to some techniques which were not used by the laboratories mentioned above but are of general importance. Some of the inorganic methods from Table B. 29 will now be discussed.

b. Emission spectroscopy

Emission spectroscopy is the classical method for qualitative and semiquantitative inorganic analysis of small specimens. After the Second World War, several improvements of this technique were introduced. Diffraction gratings of good quality and reasonable cost became available. High purity graphite electrodes of sophisticated shapes were introduced. The arc was allowed to burn in a nitrogen-free atmosphere to prevent the formation of molecular bands from cyanide and related compounds. The photographic plate was, in some instruments, replaced by electron multiplier tubes placed at the location of selected emission lines. Ellen [12] described an apparatus for producing compressed graphite pellets suitable for examining paint flakes weighing 50 µg by emission spectroscopy. A similar device was used by the Basle criminalist Martin in 1963 according to Ellen. Commercial graphite electrodes are now available which can accommodate similar sample sizes.

And finally, the laser-excited emission spectroscopy was found to be well suited for forensic analysis of paint. Its principle is the following: a high voltage is applied between two graphite electrodes which are brought close to the sample. A laser beam is directed to a small area of the object with the help of a microscope. The beam volatilizes some material and the plasma cloud formed ignites a spark between the graphite electrodes. Many forensic science laboratories are now using this technique.

The first forensic application was apparently reported by Witte [57]. Neuninger [31], Butterworth [8] and Sullivan et al. [46] published more extensive reports on the forensic analysis of small samples by laser emission spectroscopy. Manura and Saferstein [26] tested the method on 112 paint chips from wrecked late model automobiles.

The main disadvantages of emission spectroscopy are the destruction of a fair amount of sample (50 μg are about the lower limit for a successful analysis), the tricky preparation of the sample, the time-consuming evaluation of the results via a photographic plate or film, and the great difficulty of analyzing single layers of paint samples. If two chips are to be compared, they should each contain the same amount of layers. The main advantages of emission spectroscopy are its very high sensitivity and the fact that all elements are represented on one photographic plate.

c. Electron – induced X-ray spectroscopy

The advent of the electron microprobe has made it possible to examine paint chips layer by layer and to obtain semiquantitative data of the elements in each one. The first forensic applications were published by Schöntag in 1966 [37] and by Butler in 1967 [7] but this method is rather time-consuming, two to three days. Later, Smale [40] as well as Labana and Wheeler [22] described similar procedures. While the microprobe is very sensitive and can detect elements in the ppm range (roughly the same as emission spectroscopy), its complicated sample preparation makes it awkward for routine use. Also, the need for embedding the samples precludes a subsequent analysis for organic paint constituents. Finally, the instrument is quite expensive.

The real breakthrough for a rapid, reliable, reasonably sensitive and semi-quantitative procedure for the inorganic microanalysis of multilayered paint flakes came with the introduction of scanning electron microscopy (SEM) into forensic science. Initially, the SEM was only used as a high-power microscope with a remarkable depth of focus [4, 16]. Then, attachments for the wavelength – or energy-dispersive analysis of electron-induced X-rays became available.

Pioneers in this field were the English forensic scientists, especially the Metropolitan Forensic Science Laboratory in London. Williams [54] could already in 1971 report the first results obtained with a laboratory owned SEM equipped with an X-ray analyzer. In the following years several other authors [6, 34, 39, 51, 55] published SEM/X-ray data. From 1974 there was an almost explosive increase in the use of the new forensic technique, and it would be ungainly to cite all of the literature that appeared on the subject. Special symposia on the forensic applications of SEM-methods were arranged within the framework of the IIT Research Institute Scanning Electron Microscopy Meetings of the years 1971, 1975, and 1979. In the proceedings from these meetings useful articles can be found.

The analysis of automobile paint top coats by SEM/X-ray was described by Reeve and Keener [34], and that of household paints by Howden et al. [18]. Mogami [30] described the analysis of paint chips by SEM/X-ray carried out at the National Research Institute of Police Science in Tokyo which also had acquired a SEM at the beginning of the 1970's.

d. *Other methods*

Neutron activation analysis (NAA) was first described as a means for discriminating between different samples of black paint [42]. Later, colored paints [23, 43] and white paints [21] were investigated by NAA.

Black and white paints have often quite similar compositions among themselves but can show differences with respect to trace elements. Among other methods for comparative inorganic analysis spark source mass spectrometry [9 and 5] and the Weisz ring oven technique [25] may be mentioned.

6. Organic Paint Analysis

a. *Introduction*

Out of the methods for comparative forensic analysis of the organic components of paints and varnishes listed in Table B. 29 (p. 152), only pyrolysis gas chromatography, infrared spectrophotometry, pyrolysis mass spectrometry and spectrophotometry will be discussed. Other methods are either not specific enough or have not been properly tested.

Since the binders used in the paint industry usually are natural or synthetic polymers, it is necessary to pyrolyze the samples prior to analysis by gas chromatography, infrared spectrophotometry or mass spectrometry. In other words, the instrumental methods used for the analysis of polymers (cf. Chap. B. III.) are also applicable to paints and varnishes.

b. *Pyrolysis Gas Chromatography*

Already the grand old man of American forensic science, Paul L. Kirk, had investigated the identification of paints by pyrolysis gas chromatography (PGC) [20]. An important contribution to this technique appeared in 1973 from the Home Office Central Research Establishment (HOCRE) in England [27]. The importance of this work was discussed in Chap. B. III.

Stewart [44] analyzed a number of top coats from 1973 American automobiles by PGC and showed that the pyrograms obtained could be grouped according to the four main car companies in the United States. Independently, Wheals and Noble [53] carried out similar work on a variety of British vehicles but they pyrolyzed the whole flake. They used somewhat different experimental conditions than those of their British colleagues at Aldermaston. These two different conditions are compared in Table B. 30 (p. 153).

A collaborative study of automobile paint analysis by PGC was carried out by 8 American forensic science laboratories in 1974 [45]. Table B. 31 (p. 153) shows some of the experimental conditions used by these laboratories. This study showed that the same paint give vastly different pyrograms, depending on the conditions used by the 8 laboratories. It confirms the value of standardized conditions such as those suggested by May et al. [27]. On the other hand, it is of course feasible to use a given system within a specific laboratory if conditions are "frozen" for a reference collection of pyrograms.

Schultz and Perros [38] investigated the possibility of discriminating between similar types of paint with respect to the manufacturer or even batch-to-batch variations. Using PGC they analyzed 79 black paints from 27 manufacturers in the United States. They found relatively small differences between the resulting pyrograms.

c. *Infrared Spectrophotometry*

The other principal method for the analysis of the organic substituents of paints is infrared spectrophotometry (IR). The amount of sample required is usually larger in IR than in PGC but since it is always desirable to apply several different techniques to the same sample, IR analysis in an important tool for the forensic chemist also in this field.

The first application of IR to paint investigation is probably due to Heilman [17]. Later, Cleverley [10], Smalldon [41], May and Porter [28] as well as Gothard [15] reported the use of this method in analytical paint comparisons.

The KBr pellet technique, even when micropellets and beam condensers are used, requires sometimes up to 50 µg of sample for giving good results. In fiber examinations, sample sizes as low as 3 µg have been reported [14], but satisfactory analysis of such small amounts is not always possible for paint samples. A big step toward smaller sample weight requirements was the application of the diamond cell IR technique. This technique was described by Lippincott et al. in 1961 [24]. The heart of the procedure is a (commercially available) high-pressure device: Two diamonds, set in holders and having flat surfaces can be pressed together with a force of up to 50 kilobars: The sample is placed directly on one of the diamond surfaces which are about 0.25–1.50 mm^2 in area. After applying slight pressure, the device is placed in an IR spectrophotometer equipped with a beam condenser. The diamonds of type II have two wide absorption bands between about 2700–1750 cm^{-1}, leaving the important areas above and below essentially transparent. However, the diamonds transmit only a few percent of the total energy due to unspecific absorption. Details of the procedure for obtaining useful IR spectra can be found, e. g. in the paper by Tweed et al. [50] who introduced the method into forensic science. The same group of workers from the Royal Canadian Mounted Police laboratory in Ottawa published two articles on the classification of automotive paints by diamond cell IR spectrophotometry [35]. A group from another RCMP laboratory presented an interesting procedure combining PGC and IR spectrophotometry [1]. They first prepared 0.5 mm diameter KBr micropellets from 3–5 µg of paint (some authors have not been able to use so little, cf. above). After recording the IR spectrum using a beam condenser, the whole micropellet was inserted in the quartz tube of a pyrolyzer attached to gas chromatograph. Pyrolysis was carried out at 1000° C for 20 seconds, giving good pyrograms. In this way, one and the same sample can be investigated by both methods.

d. *Pyrolysis Mass Spectrometry*

A third instrumental technique introduced into forensic science is pyrolysis mass spectrometry (PMS). Two different approaches to this methodology have been published in the same year. Hughes et al. [19] used a Curie-point pyrolyzer attached to an empty glass column at 200° C which, in turn, was connected to a magnetic-sector mass spectrometer, using electron impact. They point out that a quadropole instrument with its faster scanning speed would have been preferable.

Saferstein and Manura [36] on the other hand used a quarz-tube pyrolyzer, taking 0.5 mg of sample and attaching it to a mass spectrometer in the chemical ionization mode. They used isobutane as the reactive gas and published a list of the mass peaks obtained for 30 different automobile paints.

7. Discussion of Methodology

The various methods for forensic paint analysis have been repeatedly compared and evaluated in the literature. In recent years such publications have e. g. been compiled by Ceccaldi and coworkers in Paris [3], by May and Porter at the Home Office Central Research Establishement in England [28], by Gothard [15], and by West [52] in Australia and finally by Tauchel and Helbig in East-Berlin [47].

Aufroix et al. [3] discuss the following investigative methods:

photography (black and white, color),
sectioning of paint flakes,
reflection spectrophotometry,
X-ray spectrometry,
X-ray diffraction,
gamma spectrometry,
IR spectrophotometry.

May and Porter [28] examine the following techniques:

arc emission spectroscopy,
laser-induced spark emission spectroscopy,
solvent tests,
IR spectrophotometry,
X-ray diffraction,
pyrolysis gas chromatography,
pyrolysis IR spectrophotometry.

Gothard [15] takes up the following four procedures:

microscopic examination,
IR spectrophotometry,
emission spectroscopy,
pyrolysis gas chromatography.

West [52] discusses all sorts of materials containing polymers such as paint, textile fibers, plastics, insulating material, adhesives, buttons etc. Among the methods for analyzing the inorganic constituents he mentions spectrography, X-ray fluorescence spectrometry, infrared spectrophotometry and SEM-energy dispersive X-ray analysis. Organic analytical methods discussed include IR, PIR, PGC, and GC of derivatized products.

Finally, Tauchel and Helbig [47] first give a general introduction into paint technology and the forensic importance of paint examinations (14 pp.). They proceed to discuss the following analytical methods:

microscopy,
emission spectroscopy,
absorption spectrometry,
gas chromatography,
neutron activation analysis.

Their bibliography (35 refs.) contains many references to East-European publications which are little known in the West.

In the authors' laboratory, paint chips are investigated mainly by comparative microphotography in color, followed by energy-dispersive X-ray analysis in a SEM and pyrolysis gas chromatography. In certain cases, are emission spectroscopy in oxygen followed by semiquantitative measurement of the emission lines by a microprocessor is used. Pyrolysis-infrared spectroscopy is now a routine method for polymers and can be recommended as a stand-by procedure.

In general, the difficulties of comparative paint examinations are not due so much to experimental parameters as to the correct interpretation of the experimental results.

8. Practical Examples

a. The Case of the Tanker that Ran Aground

On December 20, 1973, a tanker was reported leaking crude oil in Öresund, the passage between Denmark and Sweden. According to witnesses, the tanker had run aground in this passage, but backed out and continued her trip. Considerable damage by the oil leakage was caused to the beaches.

Divers investigated the suspected region for the ground contact. They brought up oil samples but also stones which seemed to have been in contact with the vessel. Since the master of the ship denied any ground contact it was necessary to obtain forensic evidence.

The oil samples and reference samples from the vessel tanks were analyzed by the Swedish Institute for Water and Air Pollution Research. Infrared spectrophotometry and gas chromatography gave virtually the same results for all samples. However, the ratio of the vanadium and nickel concentrations gave values of $V/Ni = 9.5$ and 10 respectively for two of the incriminated samples but $V/Ni = 16.9$ for the oil from the tanker. Since this deviation was uncomfortably large, a comparative analysis of the paint was deemed necessary.

At our laboratory, one paint chip was found on one of the stones brought up by the divers. Eight comparison paint samples from the vessel were obtained from the wharf where she was being repaired. One of the latter having 7 layers is pictured in Fig. B. 28.

The paint chip from the stone had only 4 layers: dark grey, brown, dark grey again, and finally reddish brown.

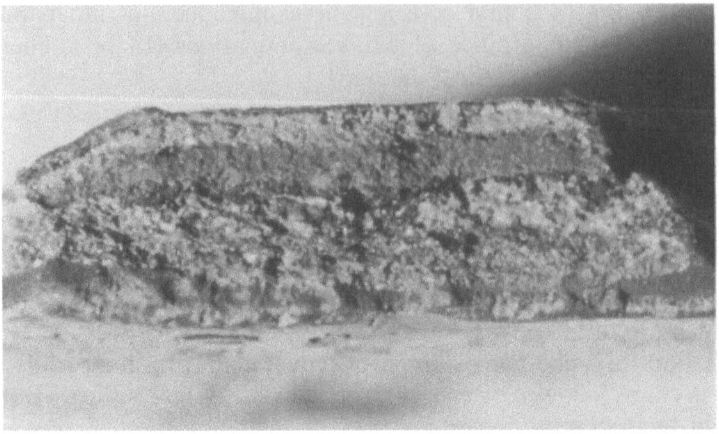

Fig. B. 28. A paint chip taken from the tanker by the police when the vessel was lying in dock for repairs

Visual comparison between these layers and corresponding layers in the reference chips showed no observable difference, neither in daylight nor in ultraviolet light. Analysis by energy-dispersive X-ray in the scanning electron microscope gave layer by layer the same results. A piece of a reference chip as it appeared in the SEM is illustrated in Fig. B. 29 and one pair of analyses is shown in Fig. B. 30.

Two-layer fragments (reddish brown and dark grey) from the incriminated chip and a reference chip were analyzed by pyrolysis gas chromatography using a wire with a Curie point at 610° C. Very similar pyrograms were obtained. The result of these investigations resulted in the following statement:

On the stone no. 3, a paint fragment with 4 layers was found. The layers were: dark grey, brown, dark grey, and reddish brown. These layers were very similar to 4 corresponding layers on one of the reference paint samples.

b. A Case of an Automobile Collision

Whereas the case of the tanker was rather unique, there are many instances of collision between automobiles where one of the drivers is away from the scene of the accident. One of these cases, a rather typical one, will be related here.

A member of the CID in a Swedish city found his parked car, a yellow Saab 96 standing at a large angle half-way on the sidewalk. The car's left side was bashed in. A smear of brown paint was found in this region and pieces from an automobile headlight were located on the street. A witness had seen another car with two people crashing into the parked automobile. She described this car as an Opel station-wagon of yellow-brown color having the registration number 358, 538 or 835 (the remainder is made up of three letters on Swedish registration plates).

With the help of the computerized number register this car was found by the police (number XXX 458)*) in spite of the small error of the witness. The owner was

*) The letters have been changed

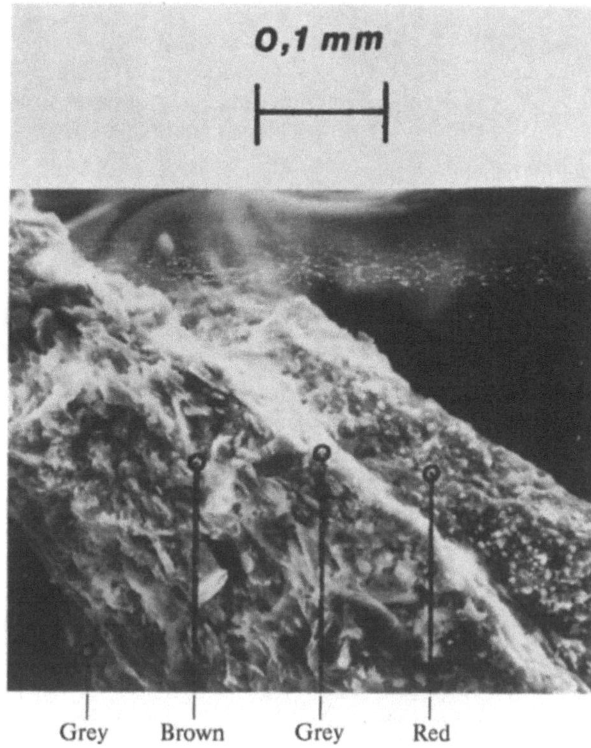

Fig.B. 29. Part of the paint chip pictured in the scanning electron microscope. The colors indicated are based on a sketch of the chip made under the light microscope

heard but remembered only having left the key of the car to an acquaintance after a 3-day party. It was now necessary to prove that XXX 458 was responsible.

The laboratory received the following material:

A. Brown paint chip from the yellow car.
B. Parts of an automobile headlight from the scene of the accident.
1. Paint sample from the car XXX 458.
2. Parts of broken right-handside headlight from XXX 458.

The questions posed were

a. Is A identical with 1?
b. Can parts of B and 2 belong together?

The investigation showed that A as well as 1 had 5 layers of paint: brown – light grey (thin) – yellow – light grey – black (cf. Fig. B. 31). Energy-dispersive X-ray analysis in a SEM gave layer for layer identical results within the limits of error. Pieces of the two flakes were then investigated by pyrolysis – gas chromatography using exit-splitting and detection by flame ionization (FID) and nitrogen-phosphorous detection (NPD), a new technique developed in our laboratory [25 a]. The pyrograms were identical within the limits of error. One of these is pictured in Fig. B. 32.

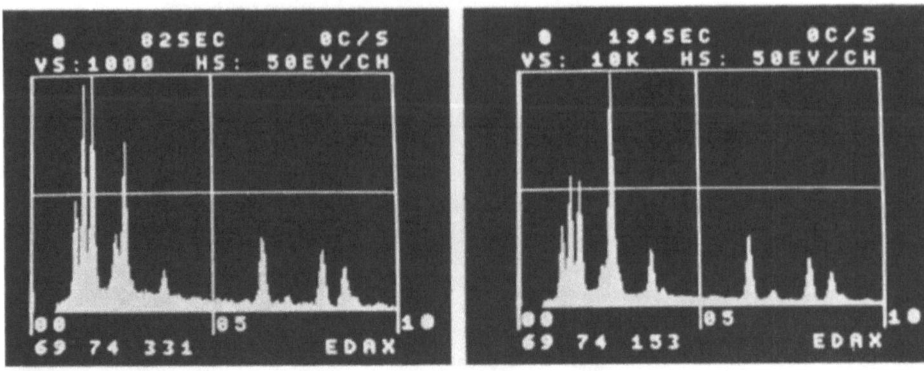

Fig. B. 30. Energy dispersive X-ray analyses of one of the corresponding layers of paint (brown) from the crime sample (left) and reference sample (right). The relative intensities of the signals are closely similar but not identical. This is probably due to a certain extent of cross-contamination between the paint layers. Close similarities between several other paint layers confirmed the theory of a common origin of the paint chips

Among the parts from headlights B and 2 respectively were pieces of some plastic material. One piece from B and one piece from 2 could be fitted together exactly.

Based on these investigations it was stated that:

(i) The five-layer paint samples A and 1 could not be distinguished from each other by the methods used. This strongly supports the assumption that they have the same origin.

(ii) At least one of the pieces of plastic B must, with virtual certainty, have belonged to one of the pieces of plastic 2.

Fig. B. 31. Paint chips from the parked car (crime sample, left) and the hit-and-run car (reference sample, right). The chips are shown glued side by side on to a cylindrical carbon rod for SEM-investigation

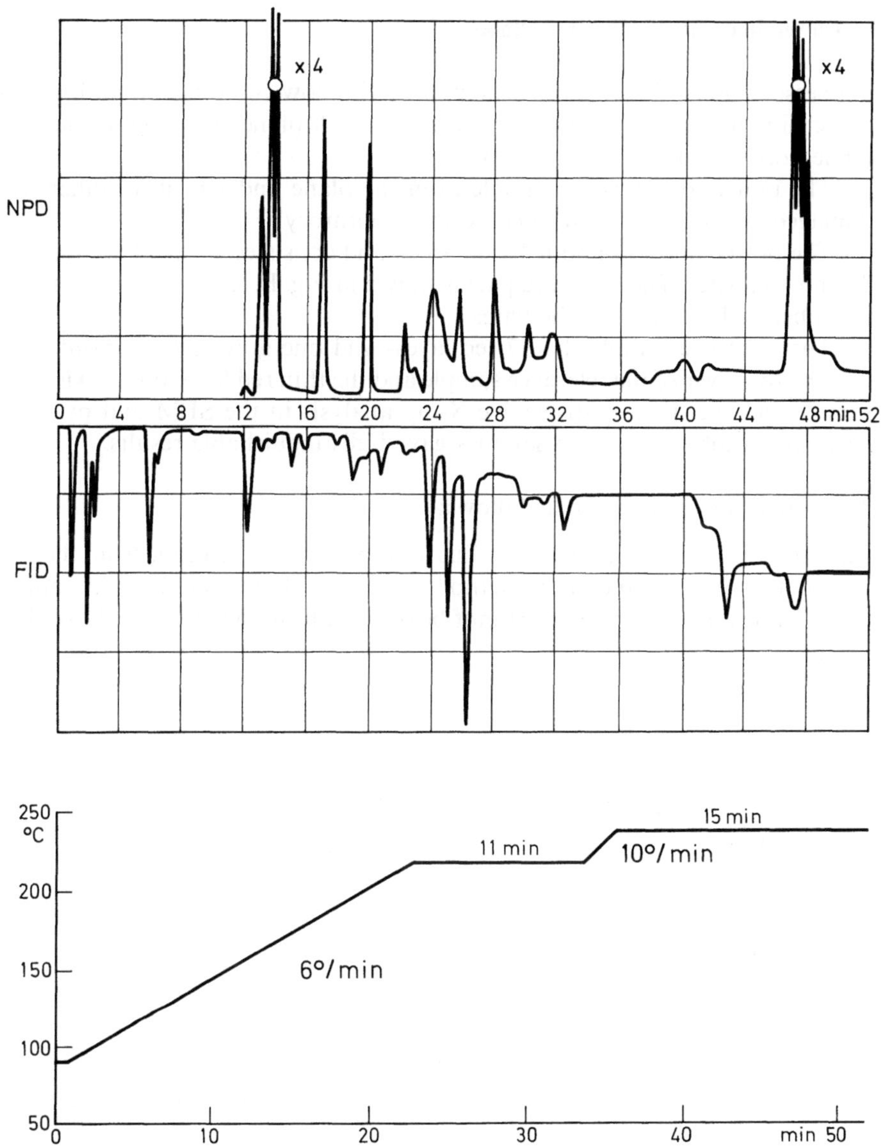

Fig. B.32. A pyrogram of one of the paint flakes mentioned in example b. Upper trace: nitrogen-phosphorus detector. Lower trace: FID. Two peaks (at 13,5 and 47 min respectively) in the NPD-trace were attenuated by a factor of 4 (peaks indicated by rings). Experimental conditions: Gas chromatograph: Perkin-Elmer Σ2. Detection: Exit split with FID and NPD. Column: 3 m × 1/8 inch (inner diameter 0,8 mm) copper tubing. Filling: Porapak Q, 80–100 mesh. Carrier gas: 30 ml N_2/min. Temperatures: injector 200°, detectors 250°. Temperature program: cf diagram. Pyrolysis unit: CDS-pyroprobe 120. Pyrolysis conditions: 10 sec at 750°. The sample is placed into a silica tube which is heated by a platina spiral. The rate of heating is about 75°/msec

c. Power-line Cut by Private Plane

Witnesses observed a small private plane flying low colliding with a high-tension power-line which was severed by the collision. A plane suspected of the accident landed on a ploughed field nearby.

The police took a paint sample from the plane and sent it, together with the damaged parts of the power-line, to the laboratory.

The power-line was about 1.1 cm thick and consisted of six aluminum wires of 3.3 mm diameter. The damaged part is shown in Fig. B. 33. A thin layer of red paint was observed on parts of the wires.

The reference samples had 7 red layers and one grey layer of paint.

The top layer of the reference sample and the thin red layer on the wires had the same color hue. Energy-dispersive X-ray analysis in the SEM and pyrolysis gas chromatography of the two samples gave indistinguishable results.

The conclusion drawn ran as follows:

The red paint traces on the power-line and the red paint sample from the airplane were not distinguishable by the methods used. This shows that the paint samples are of the same type. However, it cannot be decided if the two samples have the same origin.

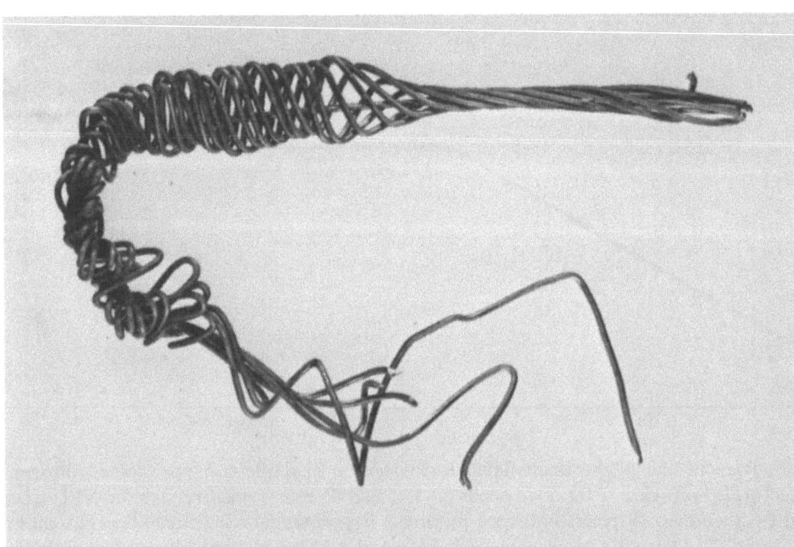

Fig. B. 33. Part of the high-voltage wire on which paint particles from the aircraft were found

9. General references (In Chronological Order)

A) D. A. Crown: The Forensic Examination of Paints and Pigments. Springfield, Ill.: C. C. Thomas 1968
B) N. I. Gaynes: Formulation of Organic Coatings. Princeton, New Jersey: van Nostrand 1967
C) P. Grandou, P. Pastour: Peintures et Vernis. Paris: Hermann 1966
D) A. Kornerup, J. H. Wanscher: Methuen Handbook of Colour, 3rd edit. London: Eyre Methuen 1978. A collection of 1266 standard color samples
E) W. Machu: Handbook of Electropainting Technology. Glasgow: Electrochemical Publications 1978
F) W. M. Morgans: Outlines of Paint Technology. London: Griffin 1968
G) P. Nylén, E. Sunderland: Modern Surface Coatings. London, New York, Sydney: Interscience 1965
H) Oil and Colour Chemists Association: Introduction to Paint Technology. London: Chapman & Hall 1967
I) W. A. Riese: Löserfreie Anstrichstoffe. Hannover: Curt R. Vincentz 1967
J) D. H. Solomon: The Chemistry of Organic Film Formers. New York: Wiley 1967
K) Swedish Standard Commission: Colour Atlas. Stockholm 1978
L) F. Tauchel, W. Helbig: Anstrichstoffe – kriminalistische Bedeutung und Untersuchungsmöglichkeiten. Kriminalistik und forens. Wiss. No. 29, 49 (1977). Review with 35 refs., mostly from Eastern Europe
M) H. Wagner, H. F. Sarx: Lackkunstharze. München: Carl Hanser 1971

10. Specific References

1) R. J. Audette, R. F. E. Percy: A Novel Pyrolysis Technique for Micro Paint Analysis. J. Forens. Sci. *23*, 672 (1978)
2) R. J. Audette, R. F. E. Percy: A Rapid, Systematic, and Comprehensive Classification System for the Identification and Comparison of Motor Vehicle Paint Samples. I: The Nature and Scope of the Classification System. J. Forens. Sci. *24*, 790 (1979)
3) L. Aufroix, S. Pougheon, P. F. Ceccaldi: Identifying Paint Flakes in Evidence I, II, III. Intern. Crim. Police Rev. *21*, I 39, II 67, III 99 (1966)
4) L. W. Bradford, J. Devaney: Scanning Electron Microscopy Applications in Criminalistics. J. Forens. Sci. *15*, 110 (1970)
5) G. Brochard, J. F. Eloy: Identification Analytiques de Différentes Peintures de Carosserie Automobile par Spectrographie de Masse à Sonde Laser. Analusis *5*, 242 (1977)
6) J. L. Brown, J. W. Johnson: Electron Microscopy and X-ray Micro-analysis in Forensic Science. J. Ass. Off. Anal. Chem. *56*, 930 (1973)
7) E. M. Butler: The Examination of Paint With the Electron Microprobe. J. Crim. Law Criminol. Police Sci. *58*, 596 (1967)
8) A. Butterworth: The Laser Microspectral Analyser. J. Forens. Sci. Soc. *14*, 123 (1974)
9) A. Butterworth et al.: The Use of Spark Source Mass Spectrometry in Forensic Science. Intern. Microf. J. Legal Med. *8* (1973)
10) B. Cleverley: Comparison of Plastic Materials and Paint Films Using Infrared Spectroscopy. Med. Sci. Law 7, 148 (1967)
11) D. Deaken: Automotive Body Primers: Their Application in Vehicle Identification. J. Forens. Sci. *20*, 283 (1975)
12) D. M. Ellen: Examination of Very Small Samples, Using Emission Spectrography. J. Forens. Sci. Soc. *5*, 196 (1965)
13) C. Fouweather, R. W. May, J. Porter: The Application of a Standard Color Coding System to Paint in Forensic Science. J. Forens. Sci. *21*, 629 (1976)
14) R. H. Fox, H. I. Schuetzman: The Infrared Identification of Microscopic Samples of Man-Made Fibers. J. Forens. Sci. *13*, 397 (1968)
15) J. A. Gothard: Evaluation of Automobile Paint Flakes as Evidence. J. Forens. Sci. *21*, 636 (1976)
16) H. Hantsche, A. Schöntag: Die Untersuchung von Lacksplittern mit dem Raster-Elektronenmikroskop als wichtiger Beitrag zur deren Identifizierung. Arch. Kriminol. *147*, 92 (1971)
17) W. R. Heilman: Non-destructive Infrared Analysis and X-Ray Diffraction Analyses of Paints and Plastics. J. Forens. Sci. *5*, 338 (1960)

18) C. R. Howden, R. J. Dudley, K. W. Smalldon: The Non-destructive Analysis of Single-Layered Household Paints Using Energy Dispersive X-Ray Fluorescence Spectrometry. J. Forens. Sci. Soc. *17*, 161 (1977)
19) J. C. Hughes, B. B. Wheals, M. J. Whitehouse: Simple Technique for the Pyrolysis – Mass Spectrometry of Polymeric Materials. Analyst *102*, 143 (1977)
20) N. C. Jain, C. R. Fontan, P. L. Kirk: Identification of Paints by Pyrolysis-Gas Chromatography. J. Forens. Sci. Soc. *5*, 102 (1965)
21) S. S. Krishnan: Examination of Paints by Trace Element Analysis. J. Forens. Sci. *21*, 908 (1976)
22) S. S. Labana, M. Wheeler: Characterization of Coatings by Electron Microprobe Analysis. Appl. Polym. Symp. No. 23, 61 (1974)
23) K. Liebscher: Vergleich von Beweismaterial mittels Spurenelementanalyse. Mikrochim. Acta 2. Heft, 272 (1971)
24) E. R. Lippincott, F. E. Welsh, C. E. Weir: Microtechnique for the Infrared Study of Solids. Anal. Chem. *33*, 137 (1961)
25) D. C. Locke, O. H. Riley: Chemical Analysis of Paint Samples Using the Weisz Ring Oven Technique. Studies in Conservation *15*, 94 (1970)
25 a) L. Svensson: unpublished results
26) J. J. Manura, R. Saferstein: Examination of Automobile Paints by Laser Beam Emission Spectroscopy. J. Assoc. Off. Analyt. Chem. *56*, 1227 (1973)
27) R. W. May et al.: A Reproducible Pyrolysis Gas-chromatographic System for the Analysis of Paints and Plastics. Analyst *98*, 364 (1973)
28) R. W. May, J. Porter: An Evaluation of Common Methods of Paint Analysis. J. Forens. Sci. Soc. *15*, 137 (1975)
29) H. R. Mac Queen, G. Judd, S. Ferriss: The Application of Scanning Electron Microscopy in the Forensic Evaluation of Vehicular Paint Samples. J. Forens. Sci. *17*, 659 (1972)
30) K. Mogami: The Analysis of Small Fragments of Paint Film. Rep. Nat. Res. Inst. Police Sci. *31*, 56 (1978)
31) H. Neuninger: Criminalistic Applications of Laser Spectral Microanalysis. Intern. Crim. Police Rev. *27*, 66 (1972)
32) F. W. Paul et al.: Reflection Spectra of Small Paint Samples: A Potential Solution. J. Forens. Sci. *16*, 241 (1971)
33) J. L. Peterson et al.: Crime Laboratory Proficiency Testing Research Program. Washington: National Institute of Law Enforcement and Criminal Justice 1978
34) V. Reeve, T. Keener: Programmed Energy Dispersive X-Ray Analysis of Top Coats of Automotive Paint. J. Forens. Sci. *21*, 883 (1976)
35) P. G. Rodgers et al.: The Classification of Automotive Paint by Diamond Window Infrared Spectrophotometry. Part I: Binders and Pigments. Part II: Automotive Topcoats and Undercoats. Can. Soc. Forens. Sci. J. *9*, 1, 49 (1976)
36) R. Saferstein, J. J. Manura: Pyrolysis Mass Spectrometry – A New Forensic Science Technique. J. Forens. Sci. *22*, 748 (1977)
37) A. Schöntag: Anwendung von Elektronenstrahl-Mikrosonde zur Identifizierung von Lacksplittern. Arch. Kriminol. *138*, 125 (1966)
38) B. W. Schultz, T. P. Perros: Pyrolysis-Gas Chromatographic Analysis of Black Paints. J. Assoc. Off. Analyt. Chem. *58*, 1150 (1975)
39) K. A. Siegesmund, G. M. Hunter: Scanning Electron Microscopy of Selected Crime Laboratory Specimens. Fourth IITRI Symposium, Chicago 1971, pp. 577–584
40) D. Smale: The Examination of Paint Flakes, Glas and Soils for Forensic Purposes, With Special Reference to Electron Probe Microanalysis. J. Forens. Sci. Soc. *13*, 5 (1973)
41) K. W. Smalldon: The Identification of Paint Resins and Other Polymeric Materials from the Infrared Spectra of Their Pyrolysis Products. J. Forens. Sci. Soc. *9*, 135 (1969)
42) K. B. Snow: Comparison of Paints by Neutron Activation Analysis. I Black Paints. Intern. Crim. Police Rev. *24*, 221 (1969)
43) K. B. Snow, W. D. Washington: Comparison of Paints by Neutron Activation Analysis. II Colored Paints. J. Assoc. Off. Analyt. Chem. *54*, 917 (1971)
44) W. D. Stewart, Jr.: Pyrolysis-Gas Chromatographic Analysis of Automobile Paints. J. Forens. Sci. *19*, 121 (1974)

45) W. D. Stewart, Jr.: Pyrolysis-Gas Chromatographic Techniques for the Analysis of Automobile Finishes: Collaborative Study. J. Assoc. Off. Analyt. Chem. *59*, 35 (1976)
46) R. C. Sullivan et al.: The Use of the Laser Microprobe in Forensic Science. J. Forens. Sci. *19*, 486 (1974)
47) F. Tauchel, W. Helbig: Anstrichstoffe – kriminalistische Bedeutung und Untersuchungsmöglichkeiten. Kriminalistik forens. Wsch. *29*, 49 (1977)
48) C. F. Tippett: The Identification of Make, Model and Year of Manufacture of a Car by an Examination of its Paint Flakes. Med. Sci. Law *4*, 22 (1964)
49) C. F. Tippett et al.: The Evidential Value of the Comparison of Paint Flakes from Sources Other than Vehicles. J. Forens. Sci. Soc. *8*, 61 (1968)
50) F. T. Tweed et al.: The Forensic Microanalysis of Paints, Plastics and Other Materials by an Infrared Diamond Cell Technique. Forens. Sci *4*, 211 (1974)
51) V. S. Vasan, W. D. Stewart, Jr., J. B. Wagner Jr.: X-Ray Analysis of Forensic Samples, Using a Scanning Electron Microscope. J. Assoc. Off. Analyt. Chem. *56*, 1206 (1973)
52) J. C. West: Methods for Forensic Comparison and Characterization of Commercial Polymer-Based Materials. J. Polym. Sci. Symp. No 49, 141 (1975)
53) B. B. Wheals, W. Noble: The Pyrolysis Gas Chromatographic Examination of Car Paint Flakes as an Aid to Vehicle Characterization. J. Forens. Sci. *14*, 23 (1974)
54) R. L. Williams: An Evaluation of the SEM with X-Ray Microanalyzer Accessory for Forensic Work. Fourth IITRI Symposium, Chicago 1971, pp. 537–544
55) R. Wilson, G. Judd: The Application of SEM and Energy Dispersive Analysis to the Examination of Forensic Paint Samples. Adv. X-Ray Anal. *16*, 19 (1972)
56) R. Wilson, G. Judd, S. Ferriss: Characterization of Paint Fragments by Combined Topographical and Chemical Electron Optics Techniques. J. Forens. Sci. *19*, 363 (1974)
57) A. H. Witte: 2nd Forens. Sci. Symp. Interpol, St. Cloud, France, 1968; cited from Ref. (31)

11. Tables B. 27–31

Table B. 27. Some historical dates of paint vehicle manufacture (in general, acc. to D. A. Crown [A]

~1600 BC	Distemper paints (Egypt)
Classical times	Natural resins
~130 AD	Linseed oil – based paints
~1900	Tung oil
~1900	Casein
1920	Nitrocellulose lacquers
1925	Coumarone – indene resins
1925	Poly(vinyl acetate) resins
1927	Alkyd resins
1928	Urea – formaldehyde resins
1928	Phenolic – formaldehyde resins
1931	Vinyl chloride resins
1931	Acrylic resins
1932	Polysulfide rubber
1935	Polyisobutyl resins
1937	Ethyl cellulose resins
1939	Melamine – formaldehyde resins
1940	Silicone resins
1940	Unsaturated polyester resins
1940	Urethane resins
1940	Poly(vinyl chloride)resins
1948	Styrene – butadiene vehicles
1956	Methacrylate finishes (automobiles)
1956	Alkyd – melamine resins (automobiles)

Table B.28. A summary of the pigments and dyes used in paints and lacquers acc. to Crown [A]

Color	Number of pigments (inorganic)	Number of dyes (natural and synthetic organic)
White	50	1 (bone)
Yellow	38	40
Orange	11	17
Red	22	96
Green	26	14
Blue	23	21
Purple	7	13
Brown	9	14
Black	15	4
Fluorescent	5	0
Pearlescent	0	1
Metallic	7	0
	213	221

Table B.29. Methods of comparative paint analysis used by 110 American laboratories in a Proficiency Test Program published in 1978 [33]

Methods used	Test no. 1 (automobile paint)	Test no. 2 (building paint)
Visual examination		
Microscopy	98	104
Fluorescence	3	43
Comparison with a reference collection	14	–
Filter observation	1	–
Color densitometry	1	–
Reflectance spectrum	1	1
Inorganic analysis		
Emission spectroscopy	41	39
X-ray fluorescence	22	26
X-ray diffraction	10	7
Elemental analysis	1	–
Atomic absorption	–	1
Neutron activation analysis	–	–
Electron microprobe	–	–
Organic analysis		
Solubility testing	88	100
Infrared analysis	52	57
(Thereof pyrolysis IR)	1	1
Pyrolysis gas chromatography	40	57
Ultraviolet spectrophotometry	14	4
Thin-layer chromatography	3	–
Density measurement	3	8
Spot tests	1	1
Differential thermal analysis	–	–
Pyrolysis mass spectrometry	–	–

Table B. 30. Experimental conditions for pyrolysis gas chromatography used by two teams of researchers

Item	HOCRE [27]	MPFSL [53]
Gas chromatograph	Pye Unicam 104	Varian 1400
Column	5 ft. × 4 mm i. d. (glass)	4 m × 2.2 mm i. d. (stainless steel)
Packing	Porapak Q (50–80 or 80–100 mesh)	15 % Carbowax 20 M on Chromosorb P (80–100 mesh)
Temperature program	100–200 °C at 8° C/min, 25 min at 200° C	70–190° C at 10° C/min
Carrier gas	60 ml N_2/min	30 ml N_2/min
Pyrolyzer	Curie point, 10 s at 610° C	
Detector	FID	

Table B. 31. Some of the experimental conditions of PGC used by 8 American forensic science laboratories in a collaborative study [45]

Laboratory no.	Column dimensions	Solid phase	Pyrolysis temp. (° C)	Sample weight (μg)	pyrolyzer
1	10 ft. × 1/8 inch	15% carbowax 20M	800	100	curie point
2	6 ft. × 1/8 inch	15% dexsil 300	800	50	quartz boat
3	4 ft. × 1/4 inch	3% OV-17	850	≪400	quartz boat
4	12 ft. × 1/8 inch	15% FFAP	800	100	curie point
5	6 ft. × 1/8 inch	3% OV-17	850	<100	quartz boat
6	–	SCOT[a] carbowax 20M	400	<200	reactor type
7	12 ft. × 1/8 inch	10% SE-30	700	50	quartz boat
8	6 ft. × 6 mm	15% carbowax 20M	400	100	Hamilton probe

[a]) Support-coated open tubular column

VI. Glass

1. Introduction

According to Flinders Petrie [12] the earliest known glass-like glaze dates from about 12000 B.C., and the earliest pure glass from about 7000 B.C.; both were found in Egypt. The art of glass making spread slowly westward. Around 1500 A.C., Venice became the center of progress in this field.

Definitions of "glass" have changed with time and with scientific and technical progress. It is certainly an amorphous solid, but opinions differ on the question whether it should be regarded as a rigid solid or as a cooled liquid. A discussion of this problem can be found in Ref. [A].

The chemical composition of glass or glasses can vary enormously. Some authors even include organic polymers in the collective term of glass. In this chapter, however, we shall limit ourselves to a discussion of *inorganic glasses;* polymers are to some extent covered in Chap. B. IV. The components of the majority of glasses are three classes of oxides:

(i) Network-forming oxides, usually silicon dioxide (SiO_2, silica) and to a minor extent boron trioxide (B_2O_3).
(ii) Network-loosening oxides that lower the melting point. These include Na_2O, and K_2O or for special glasses Li_2O.
(iii) Stabilizing oxides, lowering the sensitivity of the glass to hydrolysis. To these belong CaO, MgO and BaO.

Obviously, glasses with special properties (high chemical resistance, low expansion coefficient, high refractive index, color effects etc.) have a large variety of additives or special components. Glass is no stoichiometrically defined chemical compound. Rather, it consists of a SiO_2 or B_2O_3 network interspaced with alkali and alkaline earth oxides. A schematic example is given in Fig. B. 34.

The raw materials used for producing glass melts may include quartz sand (SiO_2), limestone ($CaCO_3$), dolomite ($CaCO_3 \cdot MgCO_3$), soda (Na_2CO_3), sodium sulfate (Na_2SO_4), pottash (K_2CO_3) borax ($Na_2B_4O_7$), nitrates ($NaNO_3$ and KNO_3), feldspar (e. g. $K_2O \cdot Al_2O_3 \cdot 6SiO_2$) and special ingredients such as PbS, $BaSO_4$, ZnO, As_2O_3, NaCl etc. Also used glass objects are sometimes added in the production of cheaper kinds of glass. These raw materials are listed in order to point out that the chemical composition of glass can vary a great deal. From the analytical point of view one can roughly divide the elements found in glass into three groups, viz.

(i) major components (Si, B, C, O, S, Na, K, and Ca),
(ii) minor components (Mg, Al, Ba, Zn, Pb, Fe, Cl, F etc.), and
(iii) trace components due to special additions or to impurities in the major and minor components.

This forms the basis for a discrimination between glass samples by chemical means.

The physical properties of glass vary, depending on the desired function of the product. The density can vary from about 2.1 up to very high values of 8 or more. The "normal" range of window glass is about 2.43–2.52. Higher values are found in crystal glass and optical glasses.

The expansion coefficient is lowest for pure silica (5×10^{-7}/K), higher for chemical glass ware such as Pyrex (32×10^{-7}/K) and reaches 80 to 90×10^{-7}/K for window glass.

The mechanical strength resembles that of porcelain but varies from batch to batch and even from object to object. Values can only be obtained by statistical

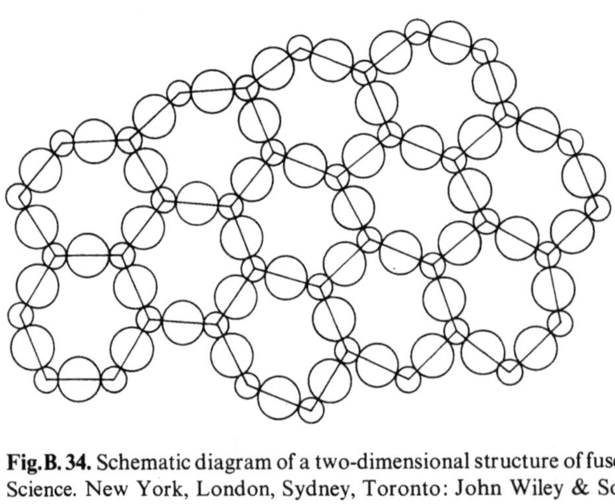

Fig. B. 34. Schematic diagram of a two-dimensional structure of fused silica. From R. H. Doremus: Glass Science. New York, London, Sydney, Toronto: John Wiley & Sons 1973

processes – using a large number of identical objects. Hardness measurements can be carried out by pressing a small diamond pyramide on the glass surface and measuring the "impression" in the glass surface. Specific heat and conductivity can be measured but these properties vary with the absolute temperature of the glass.

The optical properties of glass play a major role in forensic science. One obvious parameter is the color of glass. In cases where glass from bottles, drinking glasses, lamps, church windows and the like are investigated the color is of great value. Most glass objects are, however, either colorless or very lightly tinted, often unintentionally, due to impurities. Optical properties of great importance are the refractive index, and – to a minor extent – the dispersion of a given piece of glass. Normal window glass has a refractive index of around $n_D = 1.52$, whereas glass specimens rich in fluorine can have as low values as $n_D = 1.3$, and glasses with high concentrations of Ba and Pb can reach values of $n_D = 1.9$.

Some physical and chemical properties of glasses are assembled in Table B. 32 (p. 167). The density and the refractive index of glass can be measured with great precision and accuracy, and these properties have been used in forensic science for a long time. More recently, chemical methods for comparing glass samples have come into increased use. The most powerful investigation procedure is a combination of physical and chemical methods.

The evaluation of the results of the forensic investigation of glass has been the subject of several publications and will be briefly discussed below. The forensic importance of glass comparisons and the collection of traces of glass will be treated in Section C. I. 7 of this book.

2. Measurement of Physical Properties

Even though this book is supposed to deal with chemical methods, the determination of the physical properties of glass must be discussed in order to evaluate the total investigation.

As far as window glass is concerned, we have seen above that the range of density (d) is small, from about $d = 2.43$ to 2.52. If a precision of $\Delta d = \pm 0.001$ is achieved window glass can be divided into 90 groups. The situation for the determination of the refractive index (n) is less satisfactory, considering the narrow range of $n_D = 1.51$ to 1.54 usually encountered in window glass. With an accuracy of $\Delta n_D = \pm 0.001$ this gives the possibility of dividing window glass into 30 groups.

If the two properties, d and n, were independent of each other, one would achieve a discrimination of window glass into about 2700 groups which looks quite satisfactory. Unfortunately, d and n are not at all independent variables but show a rather close dependency. This can be clearly seen in Fig. B. 35 a–c where d- and n-measurements from the literature are shown. If this fact is considered, the number of possible combinations of d and n decreases to perhaps a few hundred. Finally, it must be considered that some of these combinations are more frequent than others (cf. Fig. B. 35 a–c). Modern production techniques tend to sharpen this uneven distribution. It is conceivable that a major glass pane delivery to e. g. a large housing development has a pretty uniform distribution of density and refractive index. It

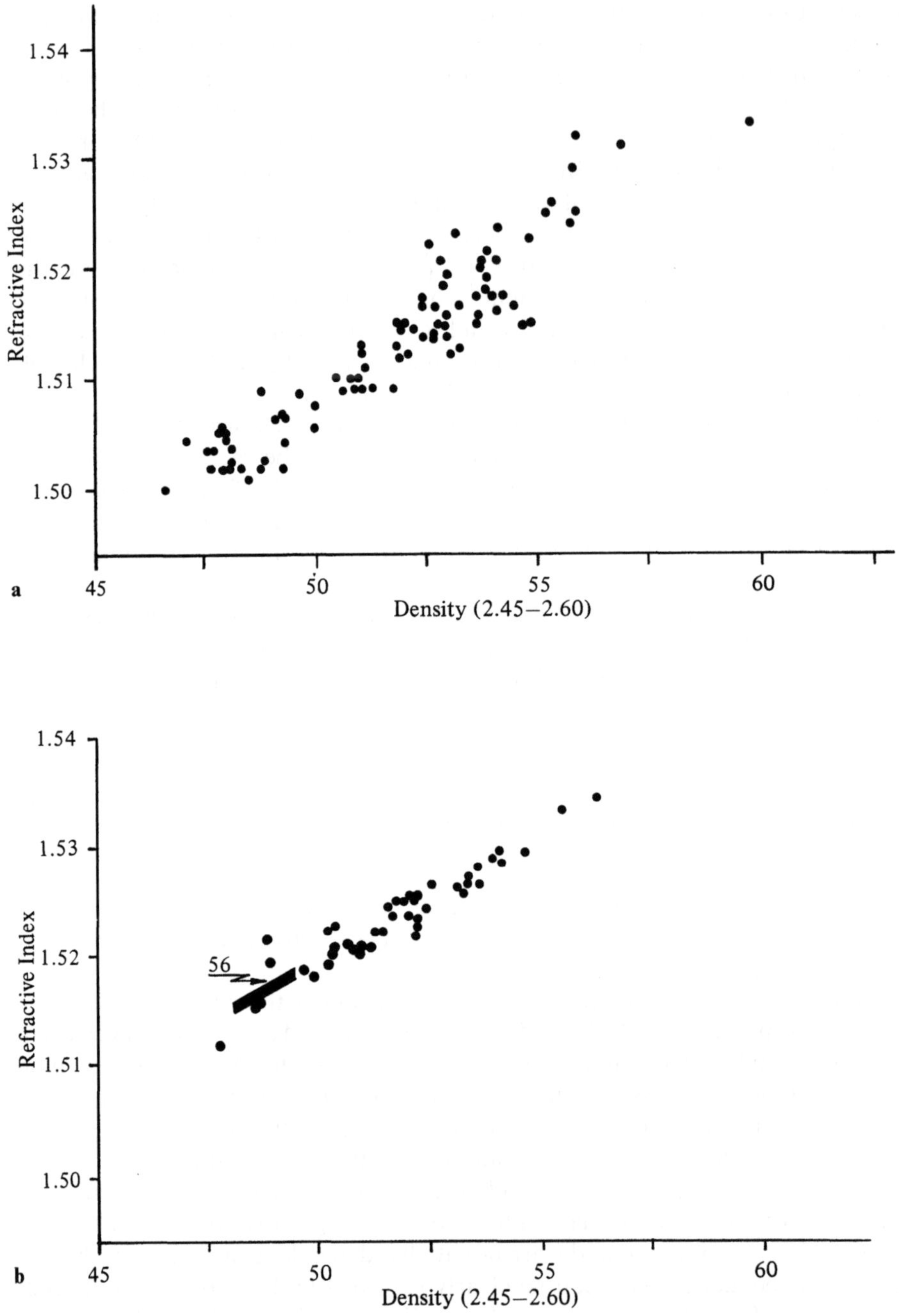

Fig. B. 35 a–c. Diagrams showing the correlations of density and refractive index in three populations of glass samples. All diagrams are replotted on the same scale. The refractive index (1,50–1,54) is plotted on the ordinate. The density is plotted on the abscissa and ranges from 2,45–2,60 g/cm^3 (only the decimals are shown on the figure). **a** Ninety-seven glass samples, according to Gamble [16], **b** One hundred glass samples, according to Dabbs and Pearson [6], **c** One hundred and seventy-five glass samples, according to Cobb [6]

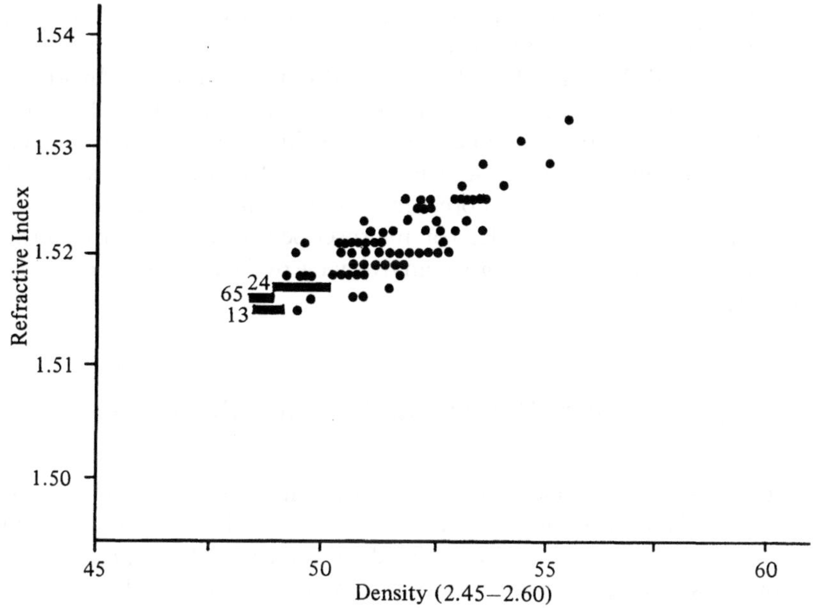

Fig. B. 35 c

would then not be possible to pinpoint a particular pane or even a particular building as a source of a splinter of glass found in the clothing of a suspected burglar.

One possible remedy to this situation is an increased precision and accuracy of the investigative methods. Improved methods for the measurement of the refractive index have, during the last decade, been published e. g. by Evett [10, 11], Heideman [18] and Mc Crone [21] as well as by Ojena and De Forest [22].

Techniques for the measurement of density have been improved, since 1970 e. g. by Beveridge and Semen [2] and by Fong [13, 14].

Dabbs and Pearson [7, 8] have reported on combinations of these properties. The earlier literature on this subject can be found in these recent publications.

There is a limit to the approach of increasing the precision and accuracy of measurements of density and refractive index due to the variations of these properties within the same glass pane. These variations are small [1, 7] but become more significant with further refinements of measurement techniques.

3. Chemical Analysis

If the comparative measurements of physical properties show no difference between the samples, a chemical analysis is indicated. The method(s) used for this task depend on the availability of the necessary equipment. The possibility of chemical analysis of small pieces of glass has existed for quite some time, but the first practical application of this idea to a forensic case seems to be as recent as 1970, when Schmitt and Smith [23] applied neutron activation analysis to the investigation of glass fragments. Two years later, spark source mass spectrometry was introduced for the same purpose [17]. These methods are very sensitive but also quite expensive and are not always available to forensic scientists.

Fuller and Whitehead [15] introduced flameless atomic absorption into glass analysis and Blacklock et al. [3] emission spectroscopy (limited to six elements). A modern emission spectroscopic technique using an inductively coupled Ar plasma was introduced by Catteric and Hickman in 1979 [5]. Howden et al. [19] used energy-dispersive X-ray fluorescence spectrometry. Electron induced X-ray fluorometry in a scanning electron microscope was described by Korda et al. [20]. Table B. 33 (p. 167 and 168), gives an overview of the methods used for forensic glass analysis.

4. Combination of Physical and Chemical Methods

The inclusion of chemical analysis in the comparison methods for glass samples increases the discriminating power of a forensic glass investigation considerably. The logical procedure is a stepwise determination of the physical and chemical properties of the glass specimens to be compared. A rapid and convenient start is the use of a dispersion measurement. The glass samples are placed side by side on a microscope slide, a drop of methyl salicylate is added and a cover glass placed on top. The samples are viewed under a microscope and illuminated with the help of light passed through a monochromator. The wavelength of the illumination is changed until Becke's lines disappear. If they disappear simultaneously for both samples, their diffractive index is the same and density measurements are taken as the next step.

Even here, a comparative method is more practical and rapid than absolute measurements. In an electrically heated glass vessel equipped with an exact reading thermometer and magnetic stirrer, a test tube is mounted, filled with a mixture of about 90% bromoform and 10% bromobenzene. The two glass samples are added to the liquid, and the proportions of the components carefully adjusted so that the samples just stay on the surface. The temperature is slowly raised until one (or both!) samples begin to sink. The event and the temperature are noted. On cooling, the temperature is again noted when the samples begin to rise again.

If the comparative measurements of refractive indices show no difference between the samples, a chemical analysis is indicated. The method(s) used for this task depend on the availability of the necessary equipment.

In the following, the analysis of glass samples with the help of the scanning electron microscope (SEM), equipped with an energy-dispersive X-ray analyzer will be described.

The glass samples to be compared are mounted next to each other on a flat piece of graphite (e. g. from graphite electrodes for emission spectroscopy), working under a binocular microscope at low magnification and using dissecting needles, fine pincers etc. as tools. The mounting is carried out with suitable glue (ordinary office rubber glue is usually quite satisfactory). A sketch is made of the shapes and the position of each glass particle. Care has to be taken to mount the specimens in such a way that one flat surface is parallel to the surface of the carbon rod.

The rod is mounted in the sample holder of the SEM and tilted so that the angle between the normal to the specimen surface and the electron beam as well as the X-ray take-off angle are the same for each sample. The surface area should be the same when measuring the excited X-ray emission from each sample. Also the number of counts are to be approximately the same. Figure B. 36 shows the X-ray analysis of two pieces of glass with the same refractive index (within then limits of error). Instead of measuring the relative intensity of various elemental lines, the ratios of chosen elements are computed (cf. Ref. VI, [8] and [12]). Should the result of these measurements be indistinguishable for both samples, a decision must be reached if further analyses are needed. The SEM method can only account for the main and the minor components of the glass, to a limit of about 0.1–1% concentration. The determination of trace elements constitutes an additional step towards a possible discrimination between two samples. In order to be able to determine traces (less than 0.1%, and down to 1 ppm or less) very sensitive methods must be employed such as NAA, AA, spark source mass spectrometry, electron probe microanalysis or emission spectroscopy (ES). Only AA and ES are generally available in forensic science laboratories. The former method is only applicable to one or a few elements at a time, whereas the latter registers all elements but involves a lot of work for quantitation. In this section a short description of the application of ES to glass analyses follows.

A glass sample (1–2 mg) is soaked in nitric acid, washed with deionized water and dried. It is ground in an agate mortar together with a matrix mixture (graphite powder, lithium fluoride, and internal standard, such as CuO or $PdCl_2$). The

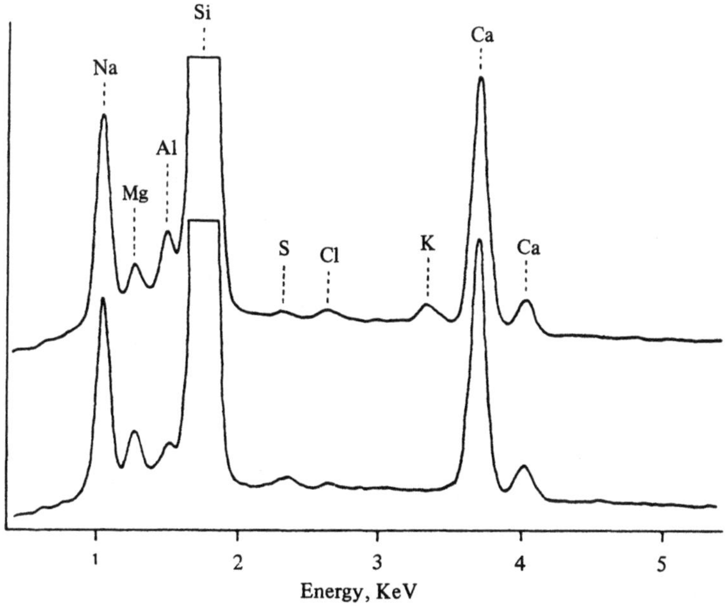

Fig. B. 36. Energy-dispersive X-ray analysis of two glass particles with identical densities and refractive indices (1.5198). Significant differences in peak intensities are seen for Mg, Al and K. From Andrasko and Maehly [1]

mixture is transferred to a high purity graphite electrode tipped by a micro-cup. It is essential to make sure that no pockets of air are formed in this operation lest sputtering would occur in the arcing process.

The sample is burnt to completion in a DC arc, if possible in an O_2/Ar atmosphere. After development of the resulting photographic plate, the position and densities of the desired elemental lines are measured, using a densitometer.

For semiquantitative comparative analysis, the ratios of the densities of the selected elements to that of the internal standard are determined and compared with those of the reference glass sample.

5. Evaluation of the Physical and Chemical Investigation of Glass Comparisons

Several authors have discussed how the analytical data should be interpreted. The earlier papers were concerned with the evaluation and the statistical treatment of refractive index measurements. Some of the more recent examples are quoted here for the reader's reference [2, 10, 11, 18]. Smalldon and Brown have considered the combination of density and refractive index data [24].

The numerical evaluation of the elemental composition of glass samples has been discussed by Blacklog et al. [3], Andrasko and Maehly [1] and by Zetlein [27].

Finally, Stone [25] has compiled physical and chemical data in a short paper published in 1976.

6. Practical Examples

a. Burglary of a News Stand

A burglar was suspected of having broken and entered a news stand. Glass from his trousers and glass from the news stand window were sent in for comparative analysis.

The pieces of glass were large enough for a thickness determination. Both items were 2.90–2.95 mm thick. The glass pieces had the same density and the same refractive index within the limits of error. When placed in the scanning electron microscope, virtually identical energy-dispersive X-ray spectra were obtained for the incriminated glass sample and the reference sample (cf. Fig. B. 37). The conclusion was worded as follows:

"The investigation lends strong support for the assumption that the two specimens of glass investigated stem from the same piece of glass".

b. Hit- and Run Accident

Two pedestrians were hit by an automobile and badly injured. The suspected car was later found at a different place.

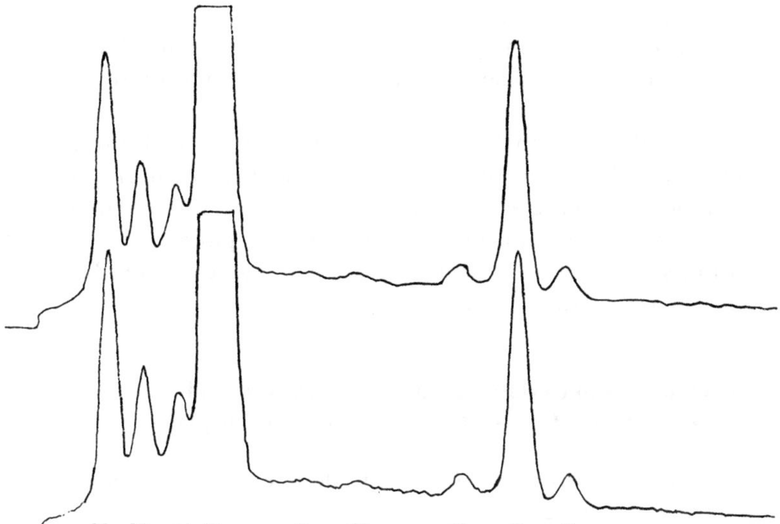

Fig. B. 37. The energy-dispersive X-ray analyses of the crime sample (above) and the reference sample (below). The silicon signal is naturally off-scale. Sodium, magnesium, aluminum, potassium and calcium (2 peaks) gave clear and measurable signals; sulfur and chlorine were barely detectable. The following intensity ratios were obtained:

Ratio	Na/Mg	Na/Al	Mg/Al	Ca/K[a]
Crime sample	1,92	2,33	1,21	11,8
Reference sample	1,92	2,33	1,21	11,6

[a] The higher Ca-peak was measured

The laboratory received the following items for analysis:

A : 1 A jeans jacket form one of the pedestrians (S. O.)
A : 2 A pair of jeans from S. O.
A : 3 A pair of black shoes from S. O.
B : 1 A pair of jeans from the driver E. H.
B : 2 A pair of brown boots from E. H.
C : Pieces of glass from the wind-shield of the car, found on the front seats.

The question was posed if there were any glass particles to be found on A : 1–3 or B : 1 and 2, and if these particles orginated from the same glass object as the particles C.

The following results were obtained:

A : 1 Six small pieces of glass could be picked up with pincers. Vacuum suction yielded a few additional glass particles but these were too small for a meaningful comparative analysis.
A : 2 Vacuum suction yielded two small glass particles.

A : 3 No glass was found.
B : 1 Vacuum suction produced five small glass particles.
B : 2 Two glass particles were picked up from the soles.

The results of the analyses are summarized in Table B. 34 (p. 169). Based on these results the laboratory reported that "The investigation lends support to the assumption that one glass particle from the jacket A : 1 and four glass particles from the trousers B : 1 stem from the same window pane as the reference sample C".

In this way, the glass from the cars' wind-shield could be connected with the victim as well as the suspected driver.

c. Attempt to Link Glass Splinters on Two Pairs of Boots and From a Car Seat With 4 Different Burglaries

The police in a town in Middle Sweden secured glass splinters at 4 scenes of burglaries. These were:

glass specimen no. 1 a: from the door of a supermarket,
glass specimen no. 1 b: on a wooden stick found at the scene,
glass specimen no. 2: from the door of another supermarket,
glass specimen no. 3 a: from the door of a loading bay of a hospital,
glass specimen no. 3 b: from the door to the central hospital kitchen,
glass specimen no. 4: from the door of a radio store.

A man and a woman, riding in a stolen car were stopped and arrested. The following suspect items were secured:

A. A piece of glass from the front seat of the car.
B. A pair of brown lady's boots (worn by the woman).
C. A pair of black men's boots (worn by the man).

These 9 exhibits were sent to the laboratory and the following questions were asked:

I. Can the glass splinter A, and possible glass splinters on the soles of B and C have originated from the glass panes represented by exhibits 1 a, 2, 3 a, 3 b or 4?
II. Can possible glass splinters on exhibit 1 b have the same origin as glass 1 a, A and possible glass splinters on B and C?

The investigation gave the following results:

a. On exhibit B (lady's boots) no traces of glass were found.
b. On exhibit C (men's boots), two small splinters of glass were located.
c. On exhibit 1 b (wooden stick), three small glass splinters were found.
d. All comparison glass specimens (no. 1 a, 2, 3 a, 3 b, and 4) could be distinguished from one another by measuring their density and refractive index respectively.

Exhibit 3 b (hospital kitchen) and glass from exhibit C (men's boots) had the same density and refractive index, respectively, within the limits of error. Energy-dispersive X-ray analysis gave virtually identical X-ray spectra.

The same 3 methods gave identical results (within the limits of error) for exhibit 1 a and glass from exhibit 1 b.

Exhibit A (glass from the car) differed from all other glass specimens with respect to density and refractive index.

Conclusions: The glass found in the stolen car did not have the same origin as any of the other glass particles investigated.

No glass was found on the lady's boots.

There is reason to believe that the glass splinters from the men's boots and the comparison glass 3 B (hospital kitchen door) had a common origin. There is also reason to believe that the glass splinters on the stick 1 b and the comparison glass specimen 1 a had a common origin.

d. Driver Linked to Crashed Car

A car had crashed into a tree in the center of a city in Northern Sweden. The car was severely damaged and the windshield splintered. A man, suffering from some minor injuries and showing clear signs of intoxication was found nearby. The police established that he was the owner of the car. He admitted this but claimed that he had lent his car to an unknown person and denied having driven it himself.

The police sent to the laboratory a pair of blue jeans and a blue sweater (both worn by the car's owner when he was found), together with glass samples from the car's windshield. The questions posed were:

(i) are there traces of glass on the clothes sent in ? and
(ii) if yes, do these pieces of glass originate from the windshield of the crashed car?

In the laboratory, the clothes were vacuum-swept and the filters searched under a microscope. On the filter used for vacuuming the sweater no glass was found, but the jeans yielded several glass particles.

One of these was compared with a reference glass particle from the windshield with respect to their density and their refractive index. In both cases, comparative methods were used where the two particles were observed simultaneously. Finally, both particles were subjected to energy-dispersive X-ray analysis in a scanning electron microscope.

All three measurements gave identical results for both glass specimens within the limits of error.

The conclusions were worded as follows:

In the vaccum sweeping from the blue sweater, no glass particles were found. The vacuum sweeping from the blue jeans contained several glass particles. One of these was investigated further and corresponded to a reference glass specimen from the windshield in the methods used. This finding supports the theory that the glass sample from the trousers originated from the car's windshield.

7. General References (In Chronological Order)

A) G. W. Morey: The Properties of Glass. New York: Reinhold 1954
B) R. Günther, H. Löffler, W. A. Weyl: Glas. In: Ullmans Encyklopädie der technischen Chemie, vol. 8, W. Forest (ed.). München, Berlin: Urban & Schwarzenberg 1957, pp. 133–183
C) F. V. Tooley (ed.): Handbook of Glass Manufacture, vols. I and II. New York: Ogden 1961
D) D. F. Nelson: The Examination of Glass Fragments. In: Methods of Forensic Science, vol. 4, A. S. Curry (ed.). London, New York & Sydney: Interscience 1965, pp. 99–140
E) H. Scholze: Glas, Braunschweig: Vieweg 1965
F) W. A. Weyl, Marboe, E. C.: The Constitution of Glasses, vols. I and II. New York: Wiley 1964–1967
G) R. H. Doremus: Glass Science. New York, London, Sydney, Toronto: Wiley 1973

8. Specific References

1) J. Andrasko, A. C. Maehly: The Discrimination Between Samples of Window Glass by Combining Physical and Chemical Techniques. J. Forens. Sci. *23*, 250 (1978)
2) A. D. Beveridge and C. Semen: Glass Density Measurement Using a Calculating Digital Density Meter. J. Can. Soc. Forens. Sci. *12*, 113 (1979)
3) E. C. Blacklock et al.: The Quantitative Analysis of Glass by Emission Spectrography: A Six Element Survey. Forens. Sci. *7*, 121 (1976)
4) A. Butterworth et al.: The Use of Spark Source Mass Spectrometry in Forensic Science. Int. Mic. J. of Leg. Med. *8* (1973)
5) T. Catterick, D. A. Hickman: Sequential Multi-element Analysis of Small Fragments of Glass by Atomic-emission Spectrometry Using an Inductively Coupled Radiofrequency Argon Plasma Source. Analyst *104*, 516 (1979)
6) P. G. W. Cobb: A Survey of the Variations in the Physical Properties of Glass. J. Forens. Sci. Soc. *8*, 29 (1968)
7) M. D. G. Dabbs, E. F. Pearson: The Variation in Refractive Index and Density Across Two Sheets of Window Glass. J. Forens. Sci. Soc. *10*, 139 (1970)
8) M. D. G. Dabbs, E. F. Pearson: Some Physical Properties of a Large Number of Window Glass Specimens. J. Forens. Sci. *17*, 70 (1972)
9) G. Erdtman, B. Kaysser, H. Petri: Aktivierungsanalytische Untersuchungen an Glassplittern in einem Kriminalfall. Z. Anal. Chem. *274*, 301 (1975)
10) I. W. Evett: The Interpretation of Refractive Index Measurements. Forens. Sci. *9*, 209 (1977)
11) I. W. Evett: The Interpretation of Refractive Index Measurements II. Forens. Sci. Int. *12*, 37 (1978)
12) Sir W. M. Flinders Petri, quoted from Ref. B
13) W. Fong: A Thermal Apparatus for Developing a Sensitive, Stable and Linear Density Gradient. J. Forens. Sci. Soc. *11*, 267 (1971)
14) W. Fong: The Value of Glass as Evidence. J. Forens. Sci. *18*, 398 (1973)
15) C. W. Fuller, J. Whitehead: The Determination of Trace Metals in High-Purity Sodium Calcium Silicate Glass, Sodium Borosilicate Glass, Sodium Carbonate and Calcium Carbonate by Flameless Atomic-Absorption Spectrometry. Anal. Chim. Acta *68*, 407 (1974)
16) L. Gamble, D. Q. Burd, P. L. Kirk: Glass Fragments as Evidence. A Comparative Study of Physical Properties. J. Crim. Law and Criminol. *33*, 416 (1943)
17) B. German, A. W. Scaplehorn: The Use of Spark Source Mass Spectrometry for the Analysis of Glass. J. Forens. Sci. Soc. *12*, 367 (1972)
18) D. H. Heideman: Glass Comparisons Using a Computerized Refractive Index Data Base. J. Forens. Sci. *20*, 103 (1975)
19) C. R. Howden, R. J. Dudley, K. W. Smalldon: The Analysis of Small Glass Fragments Using Energy Dispersive X-Ray Fluorescence Spectrometry. J. Forens. Sci. Soc. *18*, 99 (1978)
20) E. J. Korda, L. H. Pruden, J. P. Williams: Elemental Analysis of Glass and Glass Ceramics With the Scanning Electron Microscope. Amer. Ceramic Soc. Bull. *52*, 279 (1973)
21) W. C. Mc Crone: Collaborative Study of the Microscopical Characterization of Glass Fragments. J. Assoc. Off. Analyt. Chem. *56*, 1223 (1973)

22) S. M. Ojena, P. R. De Forest: Precise Refractive Index Determination by the Immersion Method, Using Phase Contrast Microscopy and the Mettler Hot Stage. J. Forens. Sci. Soc. *12*, 315 (1972)
23) R. A. Schmitt, V. Smith: Identification of Origin of Glass by Neutron Activation Analysis in a Forensic Case. J. Forens. Sci. *15*, 252 (1970)
24) K. W. Smalldon, C. Brown: The Discriminating Power of Density and Refractive Index for Window Glass. J. Forens. Sci. Soc. *13*, 307 (1973)
25) I. C. Stone: The Value of Glass as Evidence. Forens. Sci. Gazette 7, 7 (1976)
26) V. S. Vasan, W. D. Stewart, Jr., J. B. Wagner, Jr.: X-Ray Analysis of Forensic Samples, Using a Scanning Electron Microscope. J. Assoc. Off. Analyt. Chem. *56*, 1206 (1973)
27) M. B. Zetlein: Glass Classification by Elemental Composition: Numerical Evaluation. Forens. Sci. Intern. *13*, 55 (1979)

9. Tables B. 32–34

Table B. 32. Some properties of important commercial glasses

Glass type	Approximate composition (wt-%)								Density (g/cm^3)	Refractive index
	SiO_2	B_2O_3	Al_2O_3	CaO	MgO	PbO	Na_2O	K_2O		
Soda lime	72.6	0.8	1.7	4.6	3.6		15.2		2.47	1.51
"Pyrex" boro-silicate	81	13	2				4		2.23	1.47
Soda boro-silicate	68	24	1				7		2.24	1.48
Alkali lead	77			1		8	9	5	2.86	1.54
Alkaline earth alumino-silicate	64	4.5	10.4	8.9	10.2		1.3	0.7	2.52	1.53

Table B. 33. Methods used for the chemical analysis of glass samples in forensic science

Probable year of introduction	Method	Author(s)	Later references
1970	NAA (neutron activation analysis)	Schmitt and Smith[a]	Erdtman et al.[j], 1975
1972	Spark source MS (mass spectrometry)	German and Scaplehorn[b]	Butterworth et al.[k], 1973 Dabbs et al.[l], 1973 Haney[m], 1977 German et al.[n], 1978
1973	Electron probe microanalysis	Smale[c]	

Table B. 33 (continued)

1973	Wavelength dispersive X-ray analysis using the SEM (scanning electron microscope)	Vasan et al.[d]	
1973	Energy dispersive X-ray analysis using the SEM	Korda et al.[e]	Knuth[o], 1976 Reeve et al.[p], 1976 Howden et al.[q], 1978 Andrasko and Maehly[r], 1978
1974	Flameless AA (absorption analysis)	Fuller and Whitehead[f]	Hughes et al.[s], 1976 Wise and Solsky [t], 1977 Catterick and Wall[u], 1978
1976	ES (emission spectrography)	Blacklock et al.[g]	Andrasko and Maehly[v], 1978 Catterick and Hickman[w], 1979 Zetlein[x], 1979
1978	Phosphorescence analysis	Calloway and Jones[h]	
1978	EPR (electron paramagnetic resonance)	Horncastle[i]	

[a] cf. Ref. [23] [b] cf. Ref. [17]

[c] D. Smale: The Examination of Paint Flakes, Glass and Soils for Forensic Purposes, with Special Reference to Electron Probe Microanalysis. J. Forens. Sci. Soc. *13*, 5 (1973)

[d] cf. Ref. [26] [e] cf. Ref. [20] [f] cf. Ref. [15] [g] cf. Ref. [3]

[h] A. R. Calloway, P. F. Jones: Enhanced Discrimination of Glass Samples by Phosphorescence Analysis. J. Forens. Sci. *23*, 263 (1978)

[i] D. C. J. Horncastle: Discrimination Between Window and Container Glasses by Electron Paramagnetic Resonance. Lab. Pract. *27*, 477 (1978)

[j] cf. Ref. [9] [k] cf. Ref. [4]

[l] M. D. G. Dabbs et al.: The Use of Spark Source Mass Spectrometry for the Analysis of Glass Fragments Encountered in Forensic Applications. J. Forens. Sci. Soc. *13*, 281 (1973)

[m] M. A. Haney: Comparison of Window Glasses by Isotope Dilution Spark Source Mass Spectrometry. J. Forens. Sci. *22*, 534 (1977)

[n] B. German et al.: A Survey of British Container Glass Using Spark Source Mass Spectrometry with Electrical Detection. J. Forens. Sci. Soc. *18*, 113 (1978)

[o] K. Knuth: Glasuntersuchung mittels energiedispersiver Röntgenfluoreszenzspektroskopie in Verbindung mit einem Rasterelektronenmikroskop. Arch. Kriminol. *158*, 87 (1976)

[p] V. Reeve, J. Mathiesen, W. Fong: Elemental Analysis by Energy Dispersive X-Ray: A Significant Factor in the Forensic Analysis of Glass. J. Forens. Sci. *21*, 291 (1976)

[q] cf. Ref. [19] [r] cf. Ref. [1]

[s] J. C. Hughes, T. Catterick, G. Southeard: The Quantitative Analysis of Glass by Atomic Absorption Spectroscopy. Forens. Sci. *8*, 217 (1976)

[t] W. M. Wise, S. D. Solsky: Determination of Aluminium in Glass by Atomic Absorption Spectroscopy. Anal. Lett. *10*, 273 (1977)

[u] T. Catterick, C. D. Wall: Rapid Analysis of Small Glass Fragments by Atomic Absorption Spectroscopy. Talanta *25*, 573 (1978)

[v] cf. Ref. [1] [w] cf. Ref. [5]

[x] cf. Ref. [27]

Table B. 34. Results obtained in the comparative glass analyses of case a

Item investigated	Number of Particles			
	Total	With same X-ray spectrum as C[a]	With same refr. index as C[a]	With same density as C[a]
A : 1, jacket	6	1	1	1
A : 2, trousers	2	–	0	–
A : 3, shoes	0	–	–	–
B : 1, trousers	5	4	4	too small
B : 2, boots	2	–	0	–

[a] Within the limits of experimental error

VII. Soil

by Jan Andrasko, the National Laboratory of Forensic Science, Linköping, Sweden

1. Introduction

There is no universally satisfactory definition of the term "soil". To the ecologist or agriculturist the soil is what plants grow in. The engineer deals with large bodies of material many feet in depth and is interested mainly in the mechanical behaviour of such material he uses for foundations. To the forensic scientist soil may be defined as the surface layer of the land surface of the earth.

Soil is an important type of physical evidence. It is found adhering to the suspect or transferred to or from the scene of crime. If the source of the soil can be identified with that found on the suspect it will support a hypothesis that the suspect had visited the scene.

The forensic examination of soils may be divided into two different types. The first type is very rare and requires an absolute identification of the geological source of its origin, but not necessarily of a particular place. The purpose of the second, more common type of investigation is to identify a particular place but not necessarily the geological source of the soil. Soil found on the shoes or clothing of a suspect is compared with the soil from a scene of crime. Likewise, soils from the tires or the undercarriage of motor vehicles are compared with those from specific locations.

2. The Composition of Soil

a. Introduction

Soil, as we know it today, has been formed over geological epochs. It is a dynamic part of the earth's geomorphic cycle of surface weathering, erosion, deposition, sinking, metamorphosis, mountain building etc. The exposure of the parent rock to the atmosphere, water, volcanic, glacial and other activities resulted in physical and chemical weathering and disintegration of many minerals. Since climatic, biotic, topographic and time factors influenced the chemical and mineralogical composition of soils, these soils represent today a very complex material.

The earth's mantle contributes mineral matter while living organisms introduce organic matter into soils. Thus, the solid phases of soils may be divided into two classes – *mineral* and *organic*. The liquid and gas phases of soils will not be discussed here.

b. Soil Minerals

Soil minerals can be classified into two groups – primary and secondary.

Primary Minerals

Primary minerals were formed at elevated temperatures and are inherited from the parent rocks. These minerals make up the main part of the sand (particles ranging from 1 to 0.05 mm in diameter) and silt (particles ranging from 0.05 to 0.002 mm in diameter) fractions of most soils. Primary minerals may also occur in the clay fraction (<0.002 mm) of some soils. The most abundant primary minerals are quartz (SiO_2) and feldspars ($M_x\,Al_y\,Si_3O_8$). Other important minerals in this group are pyroxenes, amphiboles, olivine etc.

Secondary Minerals

These minerals with particle sizes less than 0.002 mm are called clay and were formed from the parent rock by low-temperature reactions such as weathering. Montmorillonite, illite and kaolinite are classified as the major types of clay minerals.

In addition to the cristalline inorganic minerals, soils may also contain some amorphous materials, predominately oxides of silicon and aluminum. These materials are generally found in soils from areas of volcanic activity.

c. Organic Matter of Soils

The major part of the solid parts of soil is made up by minerals. The mineral content can be as high as 99% in sandy soils but drops to about 92% in well-drained dark-coloured soils. Organic soils (e. g. peat soils) contain much less mineral matter (5–60%), but a typical soil is made up of 95% of minerals and only 5% (by weight) of organic matter. The organic matter of soil, however, is one of the most complex materials existing in nature. It has been derived from the metabolic products and tissues of plants, animals and microorganisms. The organic compounds in soils consist of both the remnants of cellular organisms and of biochemicals adsorbed onto the solid phase of the soil. Some of the classes of organic compounds which occur in soils will be mentioned here.

Carbohydrates

The total carbohydrate level in soils represents 5–20% of the total organic matter. Most of the carbohydrates in soils are presumably present as polysaccharides.

Lipids

Lipids are different classes of compounds insoluble in water – waxes, phospholipids, fatty acids, glycerides etc. The lipid content of most soils is 1–5% of the total organic matter.

Nitrogen Compounds

These compounds are contained in soils as purines, pyrimidines, amino sugars, amino acids, peptides etc.

Organic Phosphorus Compounds

The total organic phosphorus in soils ranges from 300–500 ppm, most of it inositol phosphates.

Organic Sulfur Compounds

The nature of many of the organic sulfur compounds in soils has not yet been determined. Ester sulfates as well as sufur-containing amino acids can be given as an example.

Humus

Humus is a mixture of hydrophilic acidic polymers, heterogeneous with respect to molecular weight. It is usually extracted from soils by alkali and may be fractionated into humic acids (gray, brown and hymatomelanic) and fulvic acid.

The organic matter of soils probably contains most of the naturally occurring organic compounds. Compounds not stated above include e. g. hydrocarbons, ketones, enzymes, inorganic-organic complexes etc.

3. The Forensic Comparison of Soils

a. Introduction

The soil samples are prepared for the general examination by removing gross fragments like twigs, leaves and pebbles by means of a sieve. Such particles often present in control samples, do not adhere to clothing and are not representative for a given soil. For the purpose of investigating the possible identity of two soil samples, a number of different experimental methods have been devised. Most of the methods used to compare soil samples are concerned with the inorganic components of soils. The majority of forensic laboratories employs only two techniques for the characterization of soil evidence: color comparison and the density gradient technique. A short survey of the methods for forensic comparison of soils will be given here.

b. The pH of Soil Samples

The determination of the pH of a soil has been recommended as a preliminary discriminatory test in forensic laboratories [B]. The approximate pH of a soil was determined for a small quantity of the soil sample using indicator papers. Dudley [10] investigated the usefulness of pH measurement as a discriminatory technique for soils. Colorimetric indicators provided a simple and rapid method for determining pH in intervals of 0.5 pH units for sample volumes down to 100 µl. It has been shown that pH variations can be quite large within dry soil of a given color. The reproducibility of pH measurements was found to be good down to a sample size of 50 mg. Since 14 day's storage did not influence the pH values of soil samples, the method was suggested as a useful discriminating technique for use by forensic scientists [10].

c. Sieving Test; Particle Size Distribution

The particle size distribution is a fundamental property suitable for forensic comparisons of soils. Soil scientists characterize a soil in this manner [I]. Forensic scientists need a suitable technique for handling very small samples. Quantitative sieving of soils is a form of analysis which has been carried out for about 25 years [B, E]. A weighed quantity of a soil sample is sieved through a series of sieves – a series of 20, 35, 60, 120, 250 and 250 mesh is a suitable one [B]. Each fraction is weighed and the amount compared with the soils investigated. This method has the advantage that all the material is recoverable. It needs, however, relatively large quantities of soil – upwards of 5 g. Using a Coulter counter, an instrument first designed for blood cell counting [4], the particle size analysis of soils in quantities down to 200 mg has been performed [11, 12]. The instrument is capable of measuring the particle size distribution with as little as 30 mg of sample. According to one study reported [11], the sample is sieved through a 63 µm sieve and the particle size distribution of the silt fraction (passing the sieve) and of the sand fraction (retained on the sieve) is determined on a Coulter counter after mixing with an electrolyte and a non-ionic surfactant. Good reproducibility for duplicate samples from the same source has been obtained. The discriminating power for the sand and silt fractions has been determined [12].

d. Microscopic and Mineralogical Examination

Microscopic examination of soil is usually a lengthy procedure demanding a great deal of skill and experience from the operator.

Soil scientists have developed procedures for the quantitative petrographic examination of soil minerals [5]. For forensic purposes, a method for the rapid quantitative mineralogical classification of soils has been devised [18]. Using small samples, soil particles are identified with the help of a polarizing microscope using the 100–150 mesh fraction. The optical properties of commonly occurring soil particles have been described.

Quartz is almost universally present in soils. The microscopic examination of soils includes the examination of size and shape of quartz grains. Two grain-shape parameters have been defined [27]: roundness (= the lack of sharpness of corners) and sphericity (=the resemblance to a sphere). Fitzpatrick and Thornton [15] described the models of grain-shape determinations and evaluated the usefulness of Krinsley and Margolis' [22] method of grain texture-environment interpretation in forensic characterization of sand. The examination of grain surfaces by scanning electron microscopy has been reported [15].

The microscopic examination may also reveal the presence of foreign particles which have a typically local distribution. The presence of materials derived from local sources (cement dust, coal, brick dust etc.) may have a considerable evidential value.

e. Color of Soil and Ash

In most forensic laboratories, the routine comparison of soils includes color matching. This comparison is subjective – various fractions of the soils examinated are placed in heaps on white paper and inspected [7]. Soil specimens are usually dried in an oven (105–110° C) before examination. If the soil samples seem to be very similar, a further examination may be carried out by comparing the colors of ashed samples. The recommended ashing temperature is 500–550° C [B] or 800–900° C [E].

Dudley [7] used a standard color system (The Munsell Soil Color Charts) for evaluating the discrimination power in the comparison of dried ashed soil colors. The author also suggested comparing moist soil colors although ashing was shown to give a superior discrimination.

f. Density Gradient Distribution

This technique is probably in use in most forensic laboratories. The method involves observing the distribution of soil particles and their aggregates in a density gradient and was introduced into forensic science by Goin and Kirk [16]. The density gradient is constructed from mixtures of bromoform and bromobenzene in varying proportions in narrow glass tubes. A series of the same volumes of each liquid mixture with a steadily decreasing density is placed in each tube. The density of the top layer is $\sim 1.50\,g/cm^3$ and that of the bottom layer is $\sim 2.90\,g/cm^3$. A simple 8– to 10–step gradient is usually sufficient to obtain a usuable resolution for soil samples. A minimum of 10 mg of dried, sieved soil is carefully added to the gradient tubes and the resulting distribution patterns are photographed for obtaining a visual record. To make the patterns strictly comparable, the columns should be made in a strictly identical manner. The best discriminations between density distribution patterns are obtained from the finer fractions of soils (passing a 100 mesh sieve).

The density gradient technique is a well established procedure and provides considerable and pertinent information. Soils from different geographical locations often show clearly different distribution patterns while soils from the same locations show similar patterns. In fact, soils taken only a few meters apart have been found to be different in this respect.

However, some criticism of the discrimination power of the density gradient technique is justified. The clay minerals have essentially the same particle density as the "sandy" soils and quartz also has a density close to the same value. The densities of some of the principal mineral constituents of soils are listed in Table B. 35 (p. 181).

A critical re-evaluation of the density gradient columns has been the subject of a recent paper [13]. Density gradient columns alone were not always found sufficient for identifying the soils for forensic purposes. The method of comparing the distribution patterns of the soils usually gives good and reproducible results, even though it is more subjective in character.

The classical density gradient method [16] is a time – consuming process if a sensitive discrimination is to be achieved. Mc Crone and Hudson [23] developed an agitation technique for fast preparation of density gradients, but the method had a limited reproducibility. A flow system which prepares a series of identical gradients in one run has been devised [24]. The construction of an automated, minicomputer-controlled system for the preparation of a reproducible density gradient, having a large number of steps, has been recently described [19].

Milligram quantities of soil distributed in the density gradient column may be recovered as intact fractions. The column is frozen in dry ice, the bottom cut off and successive fractions extruded and collected using a glass rod [17]. In this way, the soil fractions may be subjected to additional examinations.

g. Instrumental Methods of Chemical and Mineralogical Analysis

A large amount of sophisticated instrumental techniques can be (and have been) applied to the characterization of soils. Most of these have been used for the analysis of lunar samples [F]. The methods were reviewed by Thornton [H]. Some of the instrumental methods of analysis have been applied to the forensic comparison of soils. The differences in elemental composition of soils were investigated by emission spectroscopy [A], atomic absorption spectrophotometry [20], neutron activation analysis [20] and electron microprobe analysis [25]. It seems, however, that the elemental composition of many soils does not vary very much and that the differences observed for soil samples from one and the same location are of the same order of magnitude as the precision of the analysis.

Other instrumental techniques applied to the forensic examination of soils are those that determine the mineral composition of soils. Methods like X-ray diffraction [A], differential thermal analysis [C], thermoluminescence [21] and cathodoluminescence [8] have been advocated for the comparison of soils. Although inorganic infrared spectroscopy exhibits some advantages and additional information (chemical classification, structural clues, proton locations) [14] compared with X-ray diffraction, this method has not been tested in forensic characterization of soils.

h. Examination of Organic Constituents of Soils

Organic matter represents a minor component of the solid phase of the majority of soils and there have been very few studies dealing with the forensic comparison of soils on the basis of their organic composition alone.

The total organic matter content of a soil may be determined by means of the Walkley-Black wet oxidation method [D]. The loss of soil weight on ashing is due to the combustion of organic matter, and also to the loss of combined water and carbonates. Any total characterization of the organic fraction of soils would undoubtedly be impossible but the characterization of some components may be very useful.

A colorimetric method for the determination of soil *saccharide* content and its use in discriminating between soils has been examined [9]. The variation of the saccharide content between dry soils with the same color was found to be quite considerable. No correlation between the saccharide content and the pH of soils has been found, but there was some correlation between the color of dry soils and the saccharide content.

The levels of *enzyme activity* in soils have been studied [26, H]. The enzymes considered were phosphatase, urease, aryl sulfatase, invertase and trypsin. The type and activity of enzymes present in soils vary greatly since the microorganisms in soils show great variations. However, comparisons of enzyme activity with the organic matter content failed to establish a correlation between these two factors [H]. The enzyme patterns of soils collected from sites close to each other were often found to be distinguishable.

Polycyclic aromatic hydrocarbons (PAH) represent a group of compounds that were found in soils. It has been suggested [3] that these substances may be attributed to atmospheric pollution. PAH from many sources are accumulated in soils. The main source of PAH in soils are presumably the air-borne particles formed in various combustion processes (combustion of oils, of coal, of refuse, automobile exhaust etc.). Thus, the amount and the composition of PAH in a soil sample should vary with the geographical location. The analysis of PAH in soils was suggested for forensic purposes by Wheals [29] and a good discrimination between soils taken from different places has been achieved [1]. The PAH content of the soils reflects the geographical position of the sample rather than the type of soil. The main drawback of this method is a risk of contamination of samples taken from suspects. Furthermore, the discrimination capacity for soils from the countryside is poorer, due to relatively low levels of PAH.

From the forensic point of view, it would appear desirable to be able to characterize the *humic components* of soils. The residence time of humus is very long and its composition is not apt to change much from one week to another. It is supposed that microbial activity plays an important role in humus formation [G]. It has further been postulated that two humus samples have never exactly the same composition [6]. The subject of soil humic materials is, however, complex and many questions related to the chemistry of humic materials are not yet answered. Two attempts to analyze and characterize soil humic materials for forensic purposes have been made. Electrophoretic separation of humic compounds was not successful [H]. Column electrofocusing, slab electrofocusing in acrylamide gels, and disc acrylamide electrophoresis resulted either in an apparent fractionation of the humic material or the separation into discrete fractions was not observed.

A limited study of pyrolysis gas chromatography (PGC) analysis of acetone extracts from soils has been reported [28]. Differences between PGC chromatograms from different soils have been observed. These presumably reflect differences in the humic composition of the soils, since PGC was later shown to be a suitable method for the discrimination between different types of humus [2]. The discrimination offered by this technique is probably limited to clearly different humus types. However, soil samples found to be different by this method would presumably be distinguishable by other techniques used in the forensic examination of soils.

4. Practical Example

The police arrested a person who was suspected of having smuggled illicit drugs into a prison in southern Sweden. He had spent the night in a hotel room. When this room was searched, lumps of clay and earth were found on the floor. Footsteps were discovered, leading through a field to the prison area. Two samples of soil [2, 3] were secured from this field and sent to the laboratory for comparison with the soil samples from the hotel room [1].

Samples 1 and 2 consisted of a brownish mixture of sand and clay whereas sample 3 differed in color and grain size. All three samples were dried and sieved. The fractions passing the sieve DIN 40 (<0.15 mm diameter) were placed in the bottom of dry glass tubes (4 mm internal diameter $\times$ 32 cm). Mixtures of bromoform and bromobenzene were then added very carefully layer by layer according to Goin and Kirk [16], cf. Table B. 36 (p. 181). The soil particles were then allowed to rise, helped along by gentle tapping of the tubes. The result is shown in Fig. B. 38.

The conclusion was worded as follows:

"Based on the results of the investigation, it cannot be excluded that sample 1 (from the hotel room) has come from the field in question.

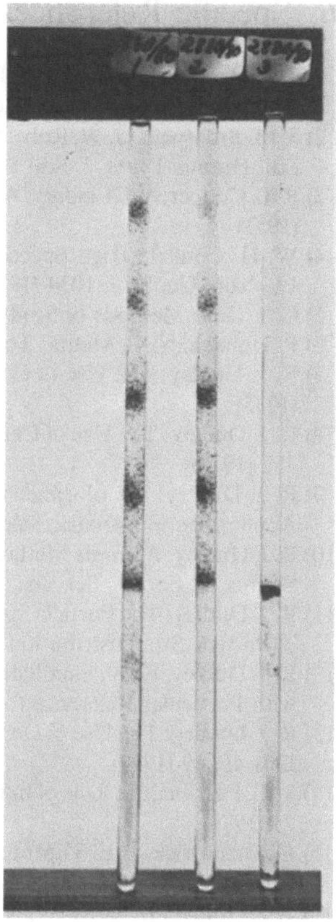

Fig. B. 38. The distribution of soil particles from the three samples sent in. *1* sample from the hotel room, *2* sample from the field, *3* another sample from the field

5. General References

A) C. E. O'Hara, J. W. Osterburg: An Introduction to Criminalistics. New York: Macmillan 1952

B) L. C. Nickolls: The Scientific Investigation of Crime. London: Butterworth and Co. Ltd. 1956.

C) P. A. Sabine: Petrographical and mineralogical techniques in forensic science. In: Soil. M. W. Neil. F. L. Warren (eds.). British Acad. of Forens. Sci. Teaching Symposium No. 1. London: Sweet & Maxwell 1962.

D) H. S. Jacobs, R. M. Reed: Soils Laboratory Exercise Source Book. Madison, Wis.: Amer. Soc. of Agronomy 1964.

E) H. J. Walls: Forensic Science. London: Sweet & Maxwell 1968.

F) Lunar Sample Edition, Science. Vol. 167, No. 3918 (1970)

G) G. T. Felbeck, Jr.: Chemical and biological characterization of humic matter. In: Soil Biochemistry, Vol. 2 A. D. Mc Laren, J. J. Skujins (eds.). New York: Dekker 1971

H) J. I. Thornton: A Forensic Biochemical Characterization of Soils. Thesis. School of criminology, University of California. Berkeley 1973.

I) B. W. Avery, C. L. Bascomb: Soil Survey Laboratory Methods. Technical Monograph No. 6. Soil Survey of England and Wales 1974.

6. Specific References

1) J. Andrasko: An Analysis of Polycyclic Aromatic Hydrocarbons in Soils and Its Application to Forensic Science. Intern. Microfilm J. Leg. Med. *13*, 4, 135 (1978)
2) J. M. Bracewell, G. W. Robertson: A Pyrolysis-Gas Chromatography Method for Discrimination of Soil Humus Types. J. Soil Sci. *27*, 196 (1976)
3) R. L. Cooper, A. J. Lindsey: Atmospheric Pollution by Polycyclic Hydrocarbons. Chem. Ind. *44*, 1177 (1953)
4) W. H. Coulter: High Speed Automatic Blood Cell Counter and Cell Size Analyzer. Proc. Natl. Electron. Conf. *12*, 1034 (1956)
5) C. I. Dell: Methods of Study of Sand and Silt from Soils. Can. Mineralogist *6*, 363 (1959)
6) P. Dubach, N. C. Mehta: The Chemistry of Soil Humic Substances. Soils Fertilizers *27*, 293 (1963)
7) R. J. Dudley: The Use of Colour in the Discrimination Between Soils. J. Forens. Sci. Soc. *15*, 209 (1975)
8) R. J. Dudley: The Use of Cathodoluminescence in the Identification of Soil Minerals. J. Soil Sci. *27*, 487 (1976)
9) R. J. Dudley: A Colorimetric Method for the Determination of Soil Saccharide Content and Its Application in Forensic Science. Med. Sci. Law *16*, 226 (1976)
10) R. J. Dudley: A simple Method for Determining the pH of Small Soil Samples and Its Use in Forensic Science. J. Forens. Sci. Soc. *16*, 21 (1976)
11) R. J. Dudley: The Particle Size Analysis of Soils and Its Use in Forensic Science – the Determination of Particle Size Distribution within the Silt and Sand Fractions. J. Forens. Sci. Soc. *16*, 219 (1977)
12) R. J. Dudley, K. W. Smalldon: The Objective Comparison of the Particle Size Distributions in Soils with Particular Reference to the Sand Fraction. Med. Sci. Law *18*, 278 (1978)
13) R. J. Dudley: The Use of Density Gradient Columns in the Forensic Comparison of Soils. Med. Sci. Law *19*, 39 (1979)
14) V. C. Farmer: The Role of Infrared Spectroscopy in a Soil Research Institute, Europ. Spectr. News *25*, 25 (1979)
15) F. Fitzpatrick, J. E. Thornton: Forensic Science Characterization of Sand. J. Forens. Sci. *20*, 460 (1975)
16) L. J. Goin, P. L. Kirk: Application of Microchemical Techniques: Identity of Soil Samples. J. Crim. Law and Criminol. *38*, 267 (1947)
17) G. Goldman, J. I. Thornton: Recovery of Soil Fractions from a Density Gradient Column. J. Forens. Sci. Soc. *16*, 171 (1976)
18) W. J. Graves: A Mineralogical Soil Classification Technique for the Forensic Scientist. J. Forens. Sci. *24*, 323 (1979)
19) D. N. Harbin et al.: A Liquid Gradient Density Screening System for Soil Sample Identification Studies Using a Minicomputer. J. Forens. Sci. *24*, 511 (1979)
20) C. M. Hoffman, R. L. Brunelle, K. B. Snow: Forensic Comparisons of Soils by Neutron Activation and Atomic Absorption Analysis. J. Crim. Law Criminol. and Police Sci. *60*, 395 (1969)
21) J. D. Ingham, D. D. Lawson: Thermoluminescence: Potential Applications in Forensic Science. J. Forens. Sci. *18*, 217 (1973)
22) D. Krinsley, S. Margolis: A Study of Sand Grain Textures with the Scanning Electron Microscope. Trans. N. Y. Acad. Sci., Series II *31*, 457 (1969)
23) W. C. Mc Crone, W. Hudson: The Analytical Use of Density Gradient Separations. J. Forens. Sci. *14*, 370 (1969)
24) H. D. Nute: An Improved Density Gradient System for Forensic Science Soil Studies. J. Forens. Sci. *20*, 668 (1975)
25) D. Smale: The Examination of Paint Flakes, Glass and Soils for Forensic Purposes, with Specific Reference to Electron Probe Microanalysis. J. Forens. Sci. Soc. *13*, 5 (1973)
26) J. I. Thornton, A. D. Mc Laren: Enzymatic Characterization of Soil Evidence. J. Forens. Sci. *20*, 674 (1975)
27) H. Wadell: Volume, Shape and Roundness of Rock Particles. J. Geology *43*, 250 (1935)
28) B. B. Wheals, W. Noble: Forensic Application of Pyrolysis Gas Chromatography. Chromatographia *5*, 553 (1972)
29) B. B. Wheals: Forensic Aspects of High-Pressure Liquid Chromatography. J. Chromatogr. *122*, 85 (1976)

7. Tables B. 35, 36

Table B. 35. The densities of some soil minerals

	Density (g/cm^3)		Density (g/cm^3)
Clay minerals	2.6–2.7	Labradorite	2.68–2.71
Quarz	2.6–2.7	Albite	2.61–2.64
Chlorite	2.6–2.96	Olivine	3.2–3.5
Orthoclase	2.54–2.57	Biotite	2.8–3.2
Oligoclase	2.65	Limonite	3.6–4.0

Table B. 36. Solutions used for the density gradient employed in the soil particle suspension test. The data are taken from the original paper by Goin and Kirk [16].

Layer no.	Density	Bromoform (ml)	Bromobenzene (ml)
1	2.890	10.0	0.00
2	2.800	10.0	0.69
3	2.700	10.0	1.58
4	2.600	10.0	2.63
5	2.500	10.0	3.90
6	2.300	10.0	7.36
7	2.000	10.0	15.06
8	1.499	0.0	10.00

7 Tables B 35, 36

VIII. Firearm Discharge Residues

1. Introduction

Modern ammunition consists, as a rule, of a cartridge case of copper, brass, aluminum etc. containing the primer, the propellant charge and the bullet (cf. Fig. B. 39). When the cartridge is to be fired, the gun mechanisms' firing pin strikes the primer (Fig. B 40). The primer ignites the propellant charge, and the bullet is driven out of the weapon through the barrel and nozzle. The cartridge case is usually ejected backwards/sideways.

During the firing sequence, particles form the primer are ejected both forward (through the nozzle) and backward/sideways (through the breech of the firearm). Residues of not completely burnt propellant are also ejected, mostly, forward, following the bullet.

These events permit the chemical analysis of objects that are in the way of the forward and backward blasts respectively for residues of primer and/or propellant charge. Such analyses are of great value for establishing the shooting distance (distance between gun muzzle and target) and for determining if a certain person has fired a gun. If correctly carried out and interpreted, such analyses can discriminate between suicide and murder (or manslaughter) or can be instrumental in the reconstruction of a crime or a shooting accident.

The analytical procedures depend on whether the organic residues from the propellant (unburnt smokeless powders and/or their reaction products) or the inorganic residues from the primers are to be determined.

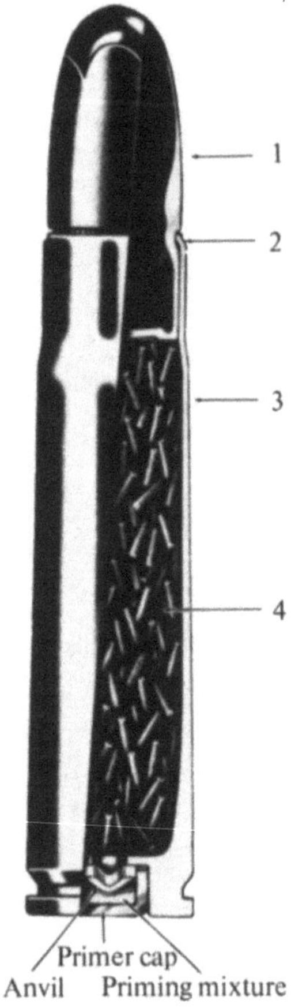

Fig. B. 39. Schematic view of an ammunition round. *1* Bullet, *2* Cannelure, *3* Cartridge case, *4* Powder charge. According to FC Barnes: Cartridges of the World. Chicago.: J. T. Amber 1965

There are a number of physical methods, such as X-ray or infrared techniques, available for investigating the size and pattern of firearm discharge residues. For these, textbooks on criminalistics have to be consulted. In this book, chemical methods alone are considered.

Modern propellants for handgun ammunition are usually based on nitrocellulose (single-base powder) but can also contain appreciable amounts (up to 40%) of nitroglycerine (in double-base powders). Stabilizers (diphenylamine and related compounds) and plasticizers (dibutyl phthalate, centralites, camphor etc.) are added in smaller amounts. The physical properties of the major components of smokeless powders and their analyses are described in Sect. B.VIII. 2.

Mach et al. [20, 21] have recently described the qualitative composition of 33 commercially manufactured smokeless powders and their analysis by gas chromatography combined with chemical ionization mass spectrometry (GC-CIMS).

Of these 33 powders, 27 contained nitroglycerine (double-base powders). Twenty-one of these double-base products contained also diphenylamine, in 12 powders, 2,4-dinitrotoluene was found, and 11 preparations contained dibutyl

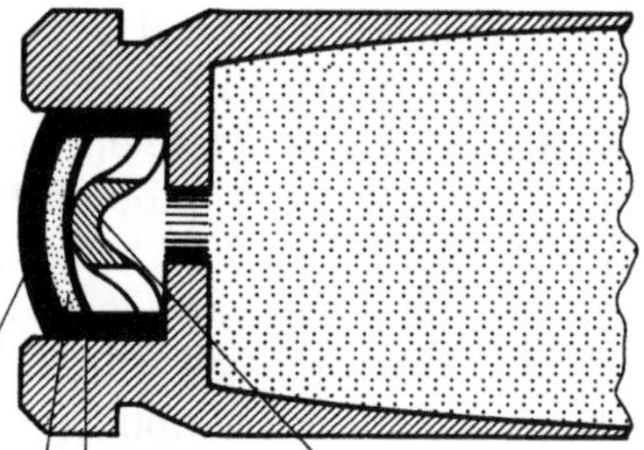

Fig. B.40. Details of a center fire cartridge primer. According to WJ Howe: Professional Gunsmithing. Kingsport, Tennesse: Kingsport Press 1946

phthalate. In another 11 double-base powders, centralite (1,3-diethyl-1,3-diphenylurea) was detected.

The composition of primers has undergone some changes since the beginning of this century. Components such as mercury fulminate and lead azide have been replaced by safer and yet efficient compounds. Most primers contain nowadays compounds of Pb, Sb and Ba, and the detection methods are therefore aimed at these elements. However, it must be kept in mind that some of these three elements may be missing in certain types of ammunition.

Some typical primer compositions are listed in Table B.37 (p. 199).

2. The Chemical Determination of Propellant Residues

Modern instrumental methods for the analysis of the constituents of the propellant charge are discussed in Sect. B.II.4 in this volume. The residues from a fired cartridge contain traces of unburnt powder and of reaction products of the charge. In practice, the analysis is directed towards the nitrite and nitrate groups present in these residues.

One of the classical reagents for these groups is diphenylamine dissolved in strong sulfuric acid. The reagent gives a blue color with NO_3 and NO_2 but is not specific for these ions alone; other oxidizing agents can also give a positive reaction. The test has nonetheless been used for the detection of propellant residues, both on the target material and on the so-called shooting hand. It seems that the first application of the reaction to forensic problems was suggested in 1911 by Wellenstein and Kober [41]. Other authors have developed the technique. A superior method was published by Walker in 1937 [38]. The same author wrote a thorough review article on the state of the art in 1940 [39]. His test became known as the "Walker-test".

Walker used gelatinized paper treated with "C" acid (2-naphthylamine-4,8-disulfonic acid). The following is a verbatim passage from Ref. [39], p. 513:

"Ordinary glossy photographic paper is completely desensitized by the usual photographic hypo bath, washed thoroughly and dried. It is then immersed in a warm 5% solution of "C" acid for ten minutes. The treated paper is allowed to dry. A pad of cotton cloth or a towel is laid upon the table, on top of which a piece of the prepared paper is placed face up. This must be of sufficient size to accommodate all of the powder residue. On top of this, face down, is laid the fabric containing the bullet hole. Next are placed a thin layer of dry toweling, a thin layer of toweling slightly moistened with 20% acetic acid, and a final thin layer of dry toweling. The whole pack is then pressed with a warm electric iron for from five to ten minutes.

The prepared photographic paper when removed is found to have impressed upon it a number of dark red spots which correspond to the position of the partially burned powder grains around the bullet hole. This test is sensitive to black and smokeless powder residue, and is insensitive to all other usual chemicals except nitrites. The test is specific for nitrites and highly indicative of powder residue if consideration is given to the spatial distribution of the particles around the bullet hole. A permanent graphic representation of the powder residue pattern is produced, without in any way destroying or significantly altering the fabric of the bullet hole. Blood stains do not react nor greatly interfere."

In 1974, Fewari [6] suggested yet another reagent, 2-(N-benzyl-anilinomethyl)-iminazoline, also known as antazoline, for the visualization of nitrate groups from propellant powder residues. This test has not found widespread use.

A related problem is the determination of propellant residues on the so-called shooting hand. For the removal of the residues from the surface of the hand, casting with paraffin was suggested. This procedure was apparently introduced by Teodoro Gonzales from the Mexico City police laboratory while he was working in the United States in 1933 [5]. The cast was then treated with diphenylamine to reveal nitro groups. The technique became known under many names such as the "Gonzales Test", "Paraffin Test", "Dermal Nitrate Test" or "Diphenylamine Test".

The test has since been critically evaluated and has been shown to be unreliable by many authors (e.g. [5, 3]). At the Interpol Conference in Paris 1968 it was officially stated that the test no longer should be used.

3. The Chemical Determination of Primer Residues

a. Introduction

The determination of gunshot residues has in recent years been concentrated on the metallic elements present in the residues. These include constituents of the primer (especially Pb, Sb, and Ba), and – as far as the target is investigated – to a lesser degree even metal traces which are abrased from the bullet during its passage through the barrel. In the latter case, Sb may be observed even with ammunition the primer of which is lacking this element. If the bullet is jacketed, Cu may also be found on the target. If the barrel is contaminated from previous shots, these residues are partly carried along by the bullet as well.

The analytical procedures consist in principle of two steps, i.e. the collection of the sample, and the analysis itself. A considerable amount of work has been laid down on both of these problems. Several monographs and review articles describe the state of the art. Such reviews were published in 1967 by Sellier [A], in 1975 by

Midkiff [E] and in 1977 again by Sellier [C]. With very few exceptions, the original work was done either in the United States or in Germany – and often in parallel.

The main methods used are listed in Table B. 38 (p. 200), together with the (probable) first publication and one of the more recent ones containing references to previous work.

The choice between these methods depends on the purpose of the analysis and on the availability and cost of the equipment needed. For investigations of the shooting hand the requirements on selectivity, reliability, and high signal-to-noise ratio are high, while there is no, or little need for quantitation. Methods for the determination of firearm discharge residues targets, on the other hand, can often dispense with some selectivity but should allow for quantitative measurements when the shooting distance has to be evaluated.

b. Investigation of the Shooting Hand

When a handgun is fired, a measurable amount of particles from the primer is usually deposited on the shooting hand. Long weapons (rifles, shotguns) leave generally very small deposits on the hand of the shooter. For the analyst, and before all for the police officer in charge, it is of importance to know where on the hand the major part of this deposit is located. This problem has been studied by Seamster et al. [32], by Sellier [33, 34] and several others. In general, the "upper side" of the hand, especially the index finger and the thumb, as well as the web between them are most densely covered (Fig. B. 41). The total amount of the primer elements deposited depends on the type of weapon and ammunition, the number of shots fired, the time elapsed between shooting and collection etc. A single round fired can typically yield a recoverable deposit of the order of a few micrograms of primer metal compounds in the form of molten globules with diameters of about 10–30 μm. Several workers have published pictures of such particles taken in the SEM (e. g. [1, 24, 26]), cf. also Fig. B. 42.

A variety of methods have been used for removing these particles for subsequent chemical analysis. The old paraffin method is not suitable for this purpose, due to impurities in the paraffin and the large bulk of the resulting matrix. Pillay et al. [28] have developed a substitute technique, the "film-lift" method. The film consists of

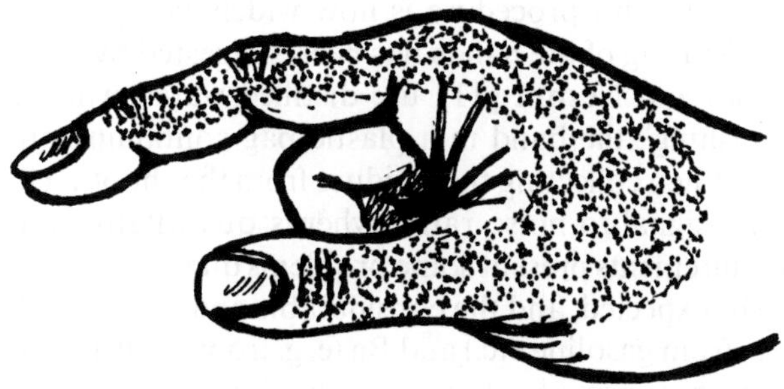

Fig. B. 41. The general region of the hand where gunshot residues from short arms are primarily deposited

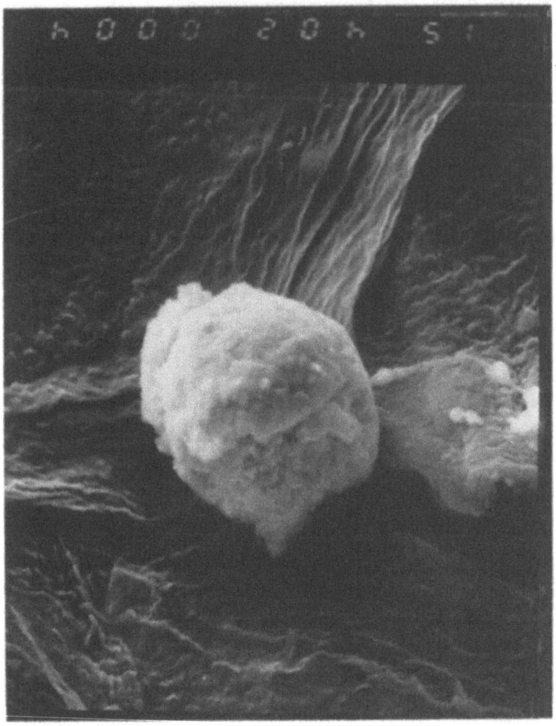

Fig. B.42. A gunshot particle recovered from the right hand of the victim in case a. The diameter of the particle is about 20 μm

cellulose acetate and coconut oil dissolved in acetone. Each of these components must be of the highest purity or must be especially purified.

Adhesive tapes of different makes have been used in criminalistics for quite some time for removing and collecting microtraces of evidence. The main pioneer in this field is the Swiss criminalist E. P. Martin (e. g. [22]). He applied the tape lifting technique to the collection of gunshot residues in 1972 [23]. The method was later taken up by several other groups of workers ([1, 7, 24, 26, 32], and others).

A different approach is based on the removal of the residues from the hand by rinsing or swabbing with water or dilute acid. Harrison and Gilroy [10] seem to have been the first to have used cotton swabs, moistened with acid for collecting primer residues. Their technique was later modified by other authors (e. g. [7, 9, 13, 33, 35]), This procedure is now widely accepted and routinely used.

Rinsing of the hands has been suggested by a few authors; Kilty [15] as well as Jones and Nesbitt [11] use distilled water, whereas Krishnan [17] recommends "washing" the hand in a plastic bag containing nitric acid.

When collecting the residues from the shooting hand it is important to achieve a high signal-to-noise ratio, wheras quantitative removal is not necessary. These requirements present some problems due to the low concentrations of Pb, Sb and Ba to be expected, and the considerable and very varied background concentrations of Pb (from gasoline etc.) and Ba (e. g. from white paper) present on many hands. If the person suspected to have fired the shot is a car mechanic, factory worker or the like, the background interference is serious. It is not advisable to increase it by extensive washing of the whole hand, and even swabbing removes such of the background "dirt". Naturally, both hands have to be treated in the same way in order to find differences in the metal content. Methods that remove the particles from the primers mechanically are to be preferred, and among these, tapes are optimal, provided that they are free of detectable amounts of Pb, Sb and Ba.

At present, the most reliable and elegant method is the combined investigation of the morphology and the chemical composition of gunshot residue particles in the scanning electron microscope (SEM). This technique was developed almost simultaneously by researchers in the United States, the United Kingdom, Israel and Sweden [1, 12, 18, 24, 26, 37, 40, 42–44].

Only those particles that show the characteristics of gunshot residue are analyzed, and only those containing the elements from the primer used are considered as positive evidence. The background from the hand surface is virtually eliminated. False positive results have not been reported to date when this method was used. On the other hand, false negative results can be expected to occur since the possibility exists that the necessarily time-limited search of the tapes may not detect single, widely spaced particles.

For a detailed description of the technique, the reader is referred to recent original articles [1, 12, 18, 24, 26, 37, 40, 42–44] and Sect. C.I. 9 of this book.

An important detail is the correct collection of the evidence and the appropriate way for transporting it to the laboratory. This will be discussed in Sect. C.I. 9.

If no SEM is available, swabbing with nitric acid, confined to the pertinent parts of both hands, is probably the best choice for collecting the residue [7, 8, 35]. The swabs can then be analyzed by neutron activation analysis (NAA) for Sb and Ba, by AA for Pb, Sb and Ba, or by ES (cf. Table B. 38 (p. 200). If the swabbing technique is used, it is advisable to read the thorough and critical paper by Cornelis and Timperman [4] in order to avoid the pitfalls of this procedure.

The use of photoluminescence for the detection of gunshot residues was described by Nesbitt et al. [27].

c. Investigation of the Target

In connection with the investigation of gunshot residues on the target area the following questions can arise among others:

(i) which of the bullet holes (especially in a body) is the entrance hole or the exit hole, respectively?

(ii) what was the shooting distance (distance between muzzle and target)?

When "shot" has been used, both of these questions can usually be answered by physical examination in combination with test firings.

When a bullet has been fired, the first problem can be solved by inspection of the powder residues on the target in visible light, in infrared light, or by X-ray examination. If the shot had been fired point blank or at very close range (1–2 cm), powder residues could be invisible or absent. In such cases it is essential to examine each garment separately and carefully from both outside and inside – and to secure the wound and its immediate surroundings at autopsy. These materials must then be analyzed instead of performing standard tests for shooting distance determination.

The investigation of shooting distances occurs more frequently and is more complicated since quantitative determinations are needed in this case. The kind of elements to be determined can, on the other hand, be limited to those known to occur in the ammunition used. Even if most primers contain Pb, Sb and Ba, this is not always the case – and certain types of primers contain additional elements [1]. If the primer composition of the ammunition used is not known, it is advisable to perform a qualitative exploratory analysis, e. g. by examining the residues left inside of a fired cartridge case.

Any correct analytical procedure for the determination of the shooting distance must consider the effect of the bullet, the propellant charge, and the primer.

The bullet can leave particles from its surface around the edge of the entrance hole. This can contain traces of the bullet itself (Pb, Sb, Cu), of the propellant charge (organic material containing nitrites and nitrates), of the primer (Pb, Sb, Ba and possibly Hg), and residues left in the barrel from previously fired ammunition. The ring formed by these deposits has been called the "dirt-ring". Since the bullet usually rotates around its axis, the contaminants are gradually lost in flight, but this diminuation cannot be used as a reliable measure of firing distance even if this approach has been tried [14, 31].

Incompletely burnt particles of the propellant charge leave the muzzle at high speed but are rapidly losing speed during their flight. They form a halo around the bullet hole. The density of their deposit decreases with firing distance. At very close range the density can again approach zero (cf. above). Finally, we have to consider the "smoke", consisting of very small particles (less than about 30 μm) of primer residue and propellant residue. Since the air resistance of these particles is relatively higher than that of the unburnt little pieces of propellant, their density decreases even more rapidly with firing distance.

The expansion zone around the bullet hole is of minor significance with respect to firing distance examination. Due to the relatively high concentration of residues in the "dirt-ring", and due to this ring's uncertain relation to the firing distance, it should not be removed together with the halo of residue deposits around the bullet hole for the determination of the shooting distance. It must be kept in mind that gunshot residues can penetrate the upper layer (s) of a victim's clothing. A knitted sweater, e. g. can act as a sort of sieve, so that residues are deposited also on underlying clothes, and even on its backside.

If only qualitative results are needed, such as in establishing the residue pattern, this consideration is of minor importance. For such studies, a variety of "lifting-techniques" has been used. These include paper for colorimetric pattern development [36], tape for distribution studies by NAA and autoradiography [30], and qualitative analysis by taping followed by X-ray microprobe and X-ray fluorescence measurements [2, 19]. Also, extraction with nitric acid [16, 29] has been used for subsequent analysis by atomic absorption (AA) and NAA.

As soon as quantitative results are needed, the area to be investigated has to be cut out. The procedure involves firing a series of reference shots, analyzing the resulting halos of gunshot residue, and comparing these with the analysis of the questioned impact halo. In order to obtain reliable results, as many parameters as possible must be kept constant. The shots must be fired with the questioned weapon (or at least with exactly the same type of weapon), the identical ammunition (not only the same type, but from the same lot as the questioned rounds), the target material should be chosen from the questioned garment, and – most important – the identical areas from the target material must be secured and analyzed.

It is further necessary to punch blanks from the target material in order to be able to subtract the background values from the experimental values.

The method permits the estimation of shooting distances up to about 1 m with a precision of about $\pm$ 0.1 m, depending on the weapon and ammunition used. This distance is sufficient for discriminating between suicide and murder or homicide respectively but it is not suitable for the reconstruction of certain shooting cases where the distances exceed those of the arm's length.

The finding of lead alone around a bullet hole is not conclusive. Messler and Armstrong [25] have shown that metal from the bullet can severely contaminate the edges of a hole. This is another reason for discarding the target region adjacent to the hole before analysis.

4. Practical Examples

a. Gunshot Residues on the Shooting Hand

The Stockholm police was called to investigate the facts behind the discovery of a body, killed by a pistol shot. Next to the body, a pistol (Walther, model PP, calibre 7.65 mm) was found. One round (Norma 7.65 mm) had been fired through the head of the victim.

The technical squad taped the hands of the victim and sent the tapes to the National Laboratory of Forensic Science. Correct ways for sending such tapes are discussed in Chap. C.I. On arrival at the laboratory, the tapes were vacuum coated with carbon (to avoid charging by the electron beam) and searched for gunshot particles (GSP) using a scanning electron microscope (SEM). A suitable magnification for this electron-optical search is about 600. On the tape from the right hand several objects with a morphology typical of GSP were found. Two of them were checked at higher magnification (Fig. B.42) and subjected to energy-dispersive X-ray analysis. The result of one of these analyses is shown in Fig. B.43. Since Pb, Sb and Ba were found in the same particle of typical morphology, the presence of GSP on the right hand was considered proven.

The case could be classified as suicide, and a search for a presumptive murderer could be dispensed with.

Time needed for a conclusive result:

1. Taping, packaging, mailing: 2 hours
2. Mail: 1 day
3. Analysis: coating: 1 hour
   search: 0.5 hours
   analysis: 0.2 hours
4. Reporting: telex: 0.5 hours

b. Shooting Distance

The police in a city in northern Sweden had to investigate a double killing. The suspect (S) had killed a man (M) and a woman (W) during an argument. In order to reconstruct the crime it was necessary to establish the shooting distance of some of the shots fired. Also, it was desirable to establish the presence of gunshot residues on the shooter's hands. The latter investigation gave no positive result, probably due to the time elapsed (about 20 hours) between firing the shot and taping the hand.

The first shot was fired straight ahead at M, entered his chest and killed him almost instantly. Three shots were fired at W, the third of which caused the mortal wound according to the autopsy findings.

The weapon, a revolver Smith & Wesson, model 28-2, caliber .357 Magnum, was sent to the National Police for fingerprinting, then to the National Laboratory of Forensic Science (NLFS). The laboratory received further:

2) 10 sharp rounds .357 Magnum from the suspect's pocket,
3) a dark blue skirt from M,
4) a brassière from W,
5) a blouse from W,
6) tape from the suspect's hands.

One bullet hole in the shirt from M and two out of a total of three in the blouse from W could be identified as entrance holes by inspection in daylight and by infrared photography, a method for strengthening the optical contrast between gunpowder particles and the target.

Using the revolver and the ammunition sent in, reference shots were fired at pieces of cloth (the target material could not be used in this case, due to extensive blood spots (M) and small areas (W)). The shooting distances were 0.1 – 0.3 – 0.5 – 0.7 – 0.9 and 1.1 meters.

The analytical method used at the NLPS is the following:

1. Rings around the bullet holes (the experimental and the crime samples) are punched out with an inner diameter of 2.0 cm and an outer diameter of 4.0 cm.

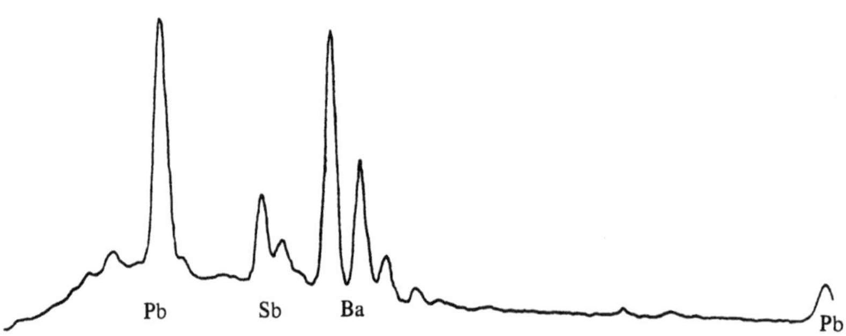

Fig. B. 43. The result of the energy-dispersive X-ray analysis of the gunshot particle shown in Fig. B. 42. Two peaks for Pb, a twin peak for Sb, and a triplet for Ba are seen

2. Each ring is cut into pieces, placed in a porcelain crucible and heated in an electric furnace to 500–600° C until they are ashed completely (1–5 hours approx.).
3. The residues are dissolved in 10 drops (about 0.2 ml) of conc. HNO_3 pro analysi. The solution is transferred to a 10-ml volumetric flask with deionized water.
4. Aliquots of this solution are analyzed by graphite-rod atomic absorption spectrometry for Pb, Sb, and Ba in parallel with solutions containing known concentrations of these metals.
5. A calibration curve is established by plotting log (Pb, Sb, Ba) versus shooting distance. The distance of the unknown samples is read from this curve.

In the particular case described above the shooting distance in the case of the shirt of M was found to have been less than 0.2 m and in the case of the blouse of W between 0.4 and 0.5 m.

c. Suspect Freed by the Determination of Shooting Distance And Gunshot Particles on the Victim's Hand

Four men had a drinking party in a town in Lapland. The following day one of the men (J. S.) found the dead body of his comrade (S. H.) outdoors with a shotgun lying several meters away. He was arrested on suspicion of having killed the victim. The suspect told the police that S.H. had talked about suicide and even had dictated a farewell letter to J. S. which S. H. had then signed.

The crucial question was if S. H. could have killed himself with the shotgun, and if the firearm could have been thrown backwards several meters. Among the 13 different exhibits sent to the laboratory were the following:

1. A large section of skin and underlying tissue (stomach) around the hole caused by the shot. This material was sent in by the forensic pathologist.
2. A single-barreled shotgun of the Russian make Baikal. Both the barrel and the stock had reduced lengths (sawn off).
3. Pieces of adhesive tape, used to tape off the webs of both hands of the victim S. H.
4. Pieces of tape from both hands of the suspect J. S.

The shooting distance was not determined by chemical analysis in this case since shot can disturb such analyses. Instead, stomach tissue from pigs was used to compare the damage caused by the shotgun at various distances using the same type of ammunition as in the actual case. Figures B. 44–46 shows some of the results. The gun was thereafter placed loosely in a supporting stand with the pipe pointing down 45° from a height of about one meter. The gun was fired thrice and thrown backwards 4.5 m, 8 m, and 20 m respectively.

On the tapes from the suspect's hands no primer residues were found. This had to be expected with regard to the time difference between the shooting and the taping.

On the tapes from both hands of the victim, primer residues were found in the scanning electron microscope and were shown to contain Pb, Ba, and Sb. Some of the particles also contained As which may have orginated from the shot pellets.

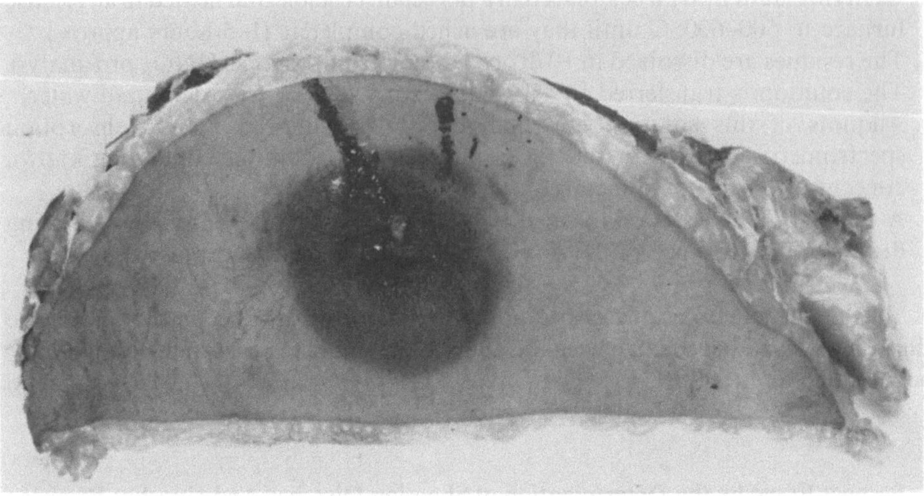

Fig. B. 44. The shot wound into the stomach of the killed S. H.

The conclusions drawn at the laboratory were:

a. The shooting distance was less than 10 cm.
b. It is possible to fire the gun with both hands and cause a wound in the stomach.
c. The gun can easily be thrown backwards by the recoil of the shot.
d. Primer residue particles were found on the hands of the victim.

The case was pronounced to be a suicide and the suspect was released.

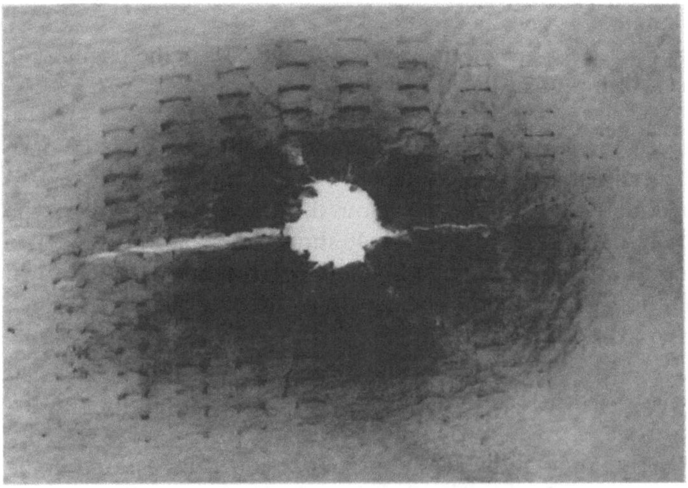

Fig. B. 45. A reference shot through a pig's skin at a distance of 0 cm

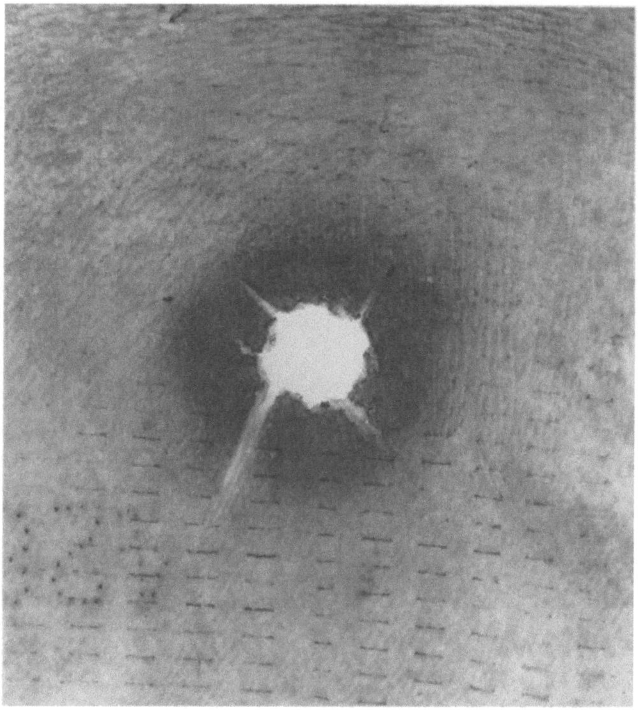

Fig. B. 46. A reference shot through a pig's skin at a distance of 10 cm

d. Shooting Incident in a City Near Stockholm

For political reasons a fight broke out between several antagonists who had met in a shopping center. One of the men involved was illegally in the possession of a pistol. During the fight, this man was grabbed from behind and in order to free himself he fired several shots backwards between his legs. Two shots hit the opponent through the left lower leg.

Since the assailant, the victim, and witnesses gave different versions of the shooting it was necessary to determine the shooting distance.

The police sent the following exhibits to the laboratory:

1. A pistol, FN, 6,35 mm.
2. Two rounds of ammunition from the pistol's magazine.
3. A pair of grey trousers worn by the victim.

The questions posed were:

a) What was the shooting distance for the two shots that hit the victim?
b) Has the pistol been used in previous crimes?

The trousers' left leg is shown in Fog. B. 47 a, b. The bullet holes marked 1 were entrance holes, those marked 2 exit holes, as judged by the direction of the broken

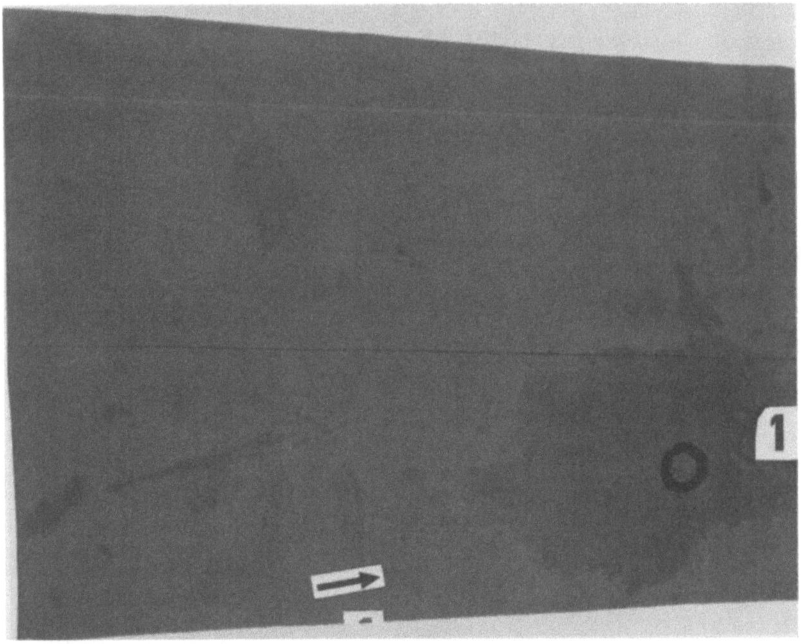

a

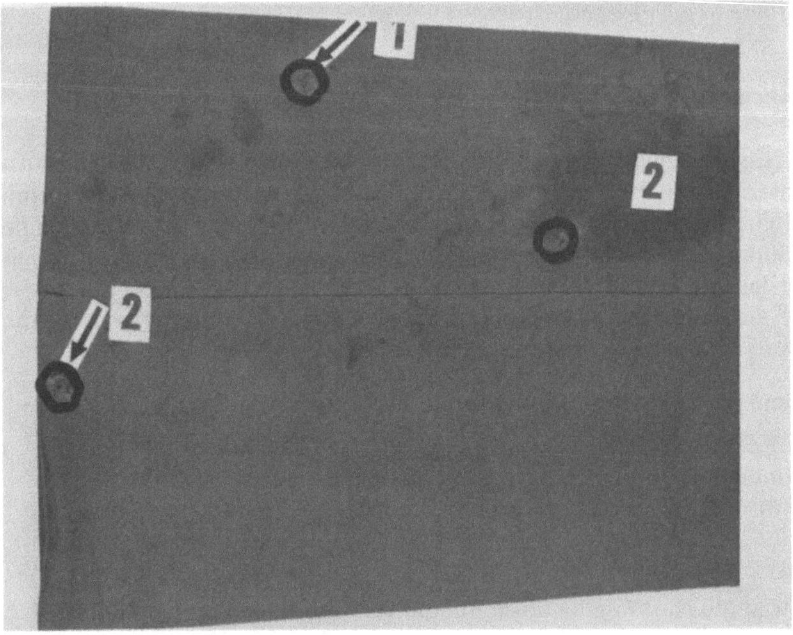

b

Fig. B. 47. a One of the bullet entrance holes (marked 1) in the left trouser leg, **b** The other entrance hole (marked 1) as well as the two exit holes (marked 2) in the left trouser leg. (Case d)

textil fibers. Visual observation showed no blackening by gunshot residues, nor did the observation with the help of an infrared sensitive TV instrument.

Rings around the two entrance holes were punched out. Using the pistol and the two rounds shown in Fig. B.48, shots were fired at other cut-out pieces of the trousers from 40 and 80 cm distance respectively. Rings were punched out also in these cases. After ashing, the concentrations of Sb, Ba and Pb were determined by atomic absorption spectroscopy.

The shooting distance to the upper hole was found to have been less than 40 cm, that to the lower hole 1 40–80 cm.

The bullets and cartridge cases used in these test shootings were compared with bullets and cases of the same type used in previous crimes, but no correlation was found.

Based on the statements of witnesses and on the results of the laboratory investigation, the shooter was sentenced to one year and six months prison for assault and battery as well as for illegal possession of a firearm.

5. General References

General works and review articles on firearms and ammunition containing information on the determination of firearm discharge residues.

A) K. Sellier: Schußentfernungsbestimmung. Arbeitsmethoden der medizinischen und naturwissenschaftlichen Kriminalistik vol. 7. Lübeck: Max Schmidt-Römhild 1967

B) K. Sellier: Schußwaffen und Schußwirkungen. Ballistik, Medizin und Kriminalistik. Arbeitsmethoden der medizinischen und naturwissenschaftlichen Kriminalistik, vol. 8. Lübeck: Max Schmidt-Römhild 1969

Fig. B.48. The pistol, the magazine and the two rounds of ammunition sent in. (Case d)

C) K. Sellier: Schußwaffen und Schußwirkungen II. Forensische Ballistik, Wundballistik. Arbeitsmethoden der medizinischen und naturwissenschaftlichen Kriminalistik, vol. 15. Lübeck: Max Schmidt-Römhild 1977
D) Ch. G. Wilber: Ballistic Science for the Law Enforcement Officer. Springfield Ill.: Charles C. Thomas 1977
E) Midkiff: Detection of Firearms Discharge Residues on the Hands – A State of the Art Report. Intern. Crim. Police Rev. *34*, 170 (1979)

6. Specific References

1) J. Andrasko, A. C. Maehly: Detection of Gunshot Residues on Hands by Scanning Electron Microscopy. J. Forens. Sci. *22*, 279 (1977)
2) H. Bürger, H. Neuninger: Röntgenfluoreszenz- und emissionsspektralanalytische Schußentfernungsbestimmung. Arch. Kriminol. *145*, 11 (1970)
3) J. L. Clement et al.: On the Validity of the Gonzales Test. Intern. Crim. Police Rev. *31*, 238 (1976)
4) R. Cornelis, J. Timperman: Gunfiring Detection Method based on Sb, Ba, Pb and Hg Deposits on Hands. Evaluation of the Credibility of the Test. Med. Sci. Law *14*, 98 (1974)
5) M. E. Cowan, P. L. Purdon: A Study of the "Paraffin Test". J. Forens. Sci. *12*, 19 (1967)
6) S. N. Tewari: A New Micro-chemical Reagent for the Detection of Gun Powder Residues in Shooting Cases. Intern. Crim. Police Rev. *28*, 220 (1974)
7) J. A. Goleb, C. R. Midkiff, Jr.: Firearms Discharge Residue Sample Collection Techniques. J. Forens. Sci. *20*, 701 (1975)
8) C. M. Hoffman: A Simplified Method of Collecting Gunshot Residue for Examination by Neutron Activation Analysis. Identification News *18*, 7 (1968)
9) C. M. Hoffman: Neutron Activation Analysis for the Detection of Firearm Discharge Residues Collected on Cotton Swabs. J. Assoc. Offic. Analyt. Chem. *56*, 1388 (1973)
10) H. C. Harrison, R. Gilroy: Firearms Discharge Residues. J. Forens. Sci. *4*, 184 (1959)
11) P. F. Jones, R. S. Nesbitt: A Photoluminescence Technique for Detection of Gunshot Residue. J. Forens. Sci. *20*, 231 (1975)
12) R. H. Keeley: Some Applications of Electron Probe Instruments in Forensic Science. Proc. Anal. Div. Chem. Soc. *13*, 178 (1976)
13) B. Keisch, R. C. Callahan: Potential Uses of Lead Isotope Ratios in Gunshot Cases. J. Assoc. Offic. Analyt. Chem. *61*, 520 (1978)
14) H. Kijewski, J. Lange: Möglichkeiten und Grenzen einer Schußentfernungsbestimmung aus dem Schmutzring. Arch. Kriminol. *154*, 179 (1974)
15) J. W. Kilty: Activity After Shooting and Its Effect on the Retention of Primer Residue. J. Forens. Sci. *20*, 219 (1975)
16) S. S. Krishnan: Trace Element Analysis by Atomic Absorption Spectrometry and Neutron Activation Analysis in the Investigation of Shooting Cases. Can. Soc. Forens. Sci. J. *6*, 55 (1973)
17) S. S, Krishnan: Detection of Gunshot Residue on the Hands by Neutron Activation and Atomic Absorption Analysis. J. Forens. Sci. *19*, 789 (1974)
18) S. S. Krishnan: Detection of Gunshot Residues on the Hands by Trace Element Analysis. J. Forens. Sci. *22*, 304 (1977)
19) E. Liebhardt, G. Hauck, W. Spann: Röntgenfluoreszenzanalytische Bleibestimmung bei bekannten Schußentfernungen. Arch. Kriminol. *144*, 92 (1969)
20) M. H. Mach, A. Pallos, P. F. Jones: Feasibility of Gunshot Residue Detection Via Its Organic Constituents. Part I: Analysis of Smokeless Powders by Combined Gas Chromatography-Chemical Ionization Mass Spectrometry. J. Forens. Sci. *23*, 433 (1978)
21) M. H. Mach, A. Pallos, P. F. Jones: Feasibility of Gunshot Residue Detection Via Its Organic Constituents. Part II: A Gas Chromatography-Mass Spectrometry Method. J. Forens. Sci. *23*, 446 (1978)
22) E. P. Martin: Proof in Criminal Cases by Means of Microscopical Evidence. The Microscope *17*, 111 (1969)
23) E. P. Martin, H. Schwander, W. B. Stern: A Bullet in the Back of the Neck. Suicide or Murder? Intern. Crim. Police Rev. *27*, 9 (1972)
24) V. R. Matricardi, J. W. Kilty: Detection of Gunshot Residue Particles from the Hands of a Shooter. J. Forens. Sci. *22*, 725 (1977)

25) H. R. Messler, W. R. Armstrong: Bullet Residue as Distinguished from Powder Pattern. J. Forens. Sci. *23*, 687 (1978)
26) R. S. Nesbitt, J. E. Wessel, P. F. Jones: Detection of Gunshot Residue by Use of the Scanning Electron Microscope. J. Forens. Sci. *21*, 595 (1976)
27) R. S. Nesbitt et al.: Evaluation of a Photoluminescence Technique for the Detection of Gunshot Residue. J. Forens. Sci. *22*, 288 (1977)
28) K. K. S. Pillay, W. A. Jester, H. A. Fox: New Method for the Collection and Analysis of Gunshot Residues as Forensic Evidence. J. Forens. Sci. *19*, 768 (1974)
29) E. Rudzitis, M. Kopina, M. Wahlgren: Optimization of Firearm Residue Detection by Neutron Activation Analysis. J. Forens. Sci. *18*, 93 (1973)
30) J. Schmitz: Erfahrungen mit der Methode der Autoradiografie zur Bestimmung der Schußentfernung. Kriminalistik *27*, 207 (1973)
31) A. Schöntag: Vorschlag einer neuen Methode: Bestimmung der Schußentfernung mittels des „Schmauchringes". Arch. Kriminol. *120*, 62 (1957)
32) A. Seamster et al.: Studies of the Spatial Distribution of Firearms Discharge Residues. J. Forens. Sci. *21*, 868 (1976)
33) K. Sellier, W. Geller: Nachweis von Schmauchrückständen an der Schußhand. Deut. Z. ger. Med. *60*, 22 (1967)
34) K. Sellier: Bemerkungen zur Schußhanduntersuchung. Kriminalistik *26*, 269 (1972)
35) I. C. Stone Jr., C. S. Petty: Examination of Gunshot Residues. J. Forens. Sci. *19*, 784 (1974)
36) H. Suchenwirth: Ein einfaches spezifisches Abdruckverfahren zum Erfassen und Beurteilen von Schmauchbildern. Arch. Kriminol. *150*, 152 (1972)
37) M. Tassa, N. Zeldes: Characterization of Gunshot Residue Particles by Localizations of Their Chemical Constituents. Scanning Electron Microscopy II, 175 (1979)
38) J. T. Walker: Chemistry and Legal Medicine. New Eng. J. Med. *216*, 1024 (1937)
39) J. T. Walker: Bullet Holes and Chemical Residues in Shooting Cases. J. Crim. Law and Criminology *31*, 497 (1940)
40) J. S. Wallace, R. H. Keeley: A Method for Preparing Firearms Residue Samples for Scanning Electron Microscopy. Scanning Electron Microscopy II, 179 (1979)
41) A. Wellenstein, H. Kober: Über den Nachweis von Pulvereinsprengungen bei Verwendung moderner Handfeuerwaffen mit rauchlosem Pulver. Z. Unters. Nahrungsmittel *21*, 544 (1911)
42) G. M. Wolten et al.: Particle Analysis for the Detection of Gunshot Residue. I: Scanning Electron Microscopy/Energy Dispersive X-Ray Characterization of Hand Deposits from Firing. J. Forens. Sci. *24*, 409 (1979)
43) G. M. Wolten et al.: Particle Analysis for the Detection of Gunshot Residue. II: Occupational and Environmental Particles. J. Forens. Sci. *24*, 423 (1979)
44) G. M. Wolten, R. S. Nesbitt, A. R. Calloway: Particle Analysis for the Detection of Gunshot Residue. III: The Case Record. J. Forens. Sci. *24*, 864 (1979)

7. Tables B. 37, 38

Table B. 37. Typical compositions of primer charges

Compound or material	Concentration in per cent		
Make no.	1	2	3
Lead trinitroresorcinate	38	40	25–55
Barium nitrate	39	45	25–45
Lead dioxide	5	5	5–10
Calcium silicate	11	7	3–15
Antimony sulfide	5	–	0–10
Tetrazene	2	3	0.5–5
Glass powder	–	–	0–5

Table B.38. Analytical methods used for the detection and determination of metallic elements from firearm discharge residues

Probable year of introduction	Method	First author(s)	Later author(s)
1957	ES (emission spectroscopy)	Schöntag[a]	Sellier[n]
1959	Color reactions	Harrison and Gilroy[b]	–
1963	NAA (neutron activation analysis)	1) Baumgärter et al.[c] 2) Ruch et al.[d]	Cornelis and Timperman[o]
1966	Polarography	Ziegler and Fehse[e]	–
1969	X-ray fluorescence	Liebhardt et al.[f]	–
1971	AA (atomic absorption)	Krishnan et al.[g]	Kijewski and Lange[p]
1972	X-ray spectroscopy	Martin et al.[h]	(cf. combination of X-ray spectroscopy and SEM)
1973	Autoradiography	Schmitz[i]	–
1975	Photoluminescence	Jones and Nesbitt[j]	–
1976	Combination of SEM (scanning electron microscopy) and X-ray spectroscopy	1) Nesbitt et al.[k] 2) Keeley[l]	1) Matricardi[q] 2) Andrasko and Maehly[r]
1978	Lead isotope ratios	Keisch and Callahan[m]	–

[a] A. Schöntag: Neue Methode: Bestimmung der Schußentfernung durch spektrographische Spurenanalyse der „Schmauchelemente“ Antimon, Blei oder Barium. Arch. Kriminol. *120*, 4 (1957)
[b] cf. Ref. [10]
[c] F. Baumgärtner, H. Stärk, A. Schöntag.: Aktivierungsanalytische Antimonbestimmung im Nanogrammbereich zur Ermittlung von Schußentfernungen. Z. Anal. Chem. *197*, 424 (1963)
[d] R. R. Ruch et al.: Neutron Activation Analysis in Scientific Crime Detection – Some Recent Developments. J. Forens. Sci. *9*, 119 (1964)
[e] M. Ziegler, B. Fehse: Schußentfernungsbestimmung mit dem Tastpolarographen. Arch. Kriminol. *137*, 71 (1966)
[f] cf. Ref. [19]
[g] S. S. Krishnan, K. A. Gillespie, E. J. Anderson: Rapid Detection of Firearm Discharge Residues by Atomic Absorption and Neutron Activation Analysis. J. Forens. Sci. *16*, 144 (1971)
[h] cf. Ref. [23] [i] cf. Ref. [30] [j] cf. Ref. [11] [k] cf. Ref. [26] [l] cf. Ref. [12] [m] cf. Ref. [13] [n] cf. Ref. [34] [o] cf. Ref. [4]
[p] H. Kijewski, J. Lange: Neue Aspekte in der Schußentfernungsbestimmung durch Anwendung der flammenlosen Atom-Absorptions-Spektrophotometrie. Z. Rechtsmed. *74*, 9 (1974)
[q] cf. Ref. [24] [r] cf. Ref. [1]

IX. Fire Investigation

1. Introduction

The forensic investigation of causes for fires is usually called "arson investigation", implying that arson has been proven or suspected from the start. This terminology is somewhat misleading since fires can start essentially in three ways:

(i) pure accident,
(ii) negligence, and
(iii) deliberate arson.

There exist, however, no very sharp borderlines between these categories. Consider e. g. self-ignition of rags soaked in lineseed oil: accident (the person responsible was ignorant of the danger) or negligence (the person took a chance)? In another case, a fuse was shorted in a minor industry: negligence (the perpetrator hoped for the best), or arson (he hoped for the worst)?

The task to investigate causes of fires falls on different agencies in different countries or regions. In the authors' country (Sweden), the police investigates all major fires, and if the cause cannot be established locally, relevant evidence is sent to the National Laboratory of Forensic Science. The laboratory investigates primarily electrical, chemical, and mechanical causes for fires.

Among electrical causes, the following can be named: lightning, sparking due to static electricity, short circuits, faulty construction of electrical equipment, faulty installations and wiring, shorted fuses, overloading of circuits, damage to insulations (by water or indeed by fire) faulty use of equipment etc.

Mechanical causes for fires can be friction of moving parts such as ball bearings, transmission belts, car tires (when one of a pair of twin tires got a flat), fans etc. Also leaks in pipes or hoses carrying liquid or gaseous fuel, leaking valves, faulty burners, malfunctioning manometers etc. can of course cause fires.

There are two major categories of chemical fire causes, i.e. self-ignition and deliberate ignition using fire accelerants or inorganic ignition devices.

In principle, arson, negligence, or accidents leading to fires can be caused by electrical, mechanical or chemical means. It is certainly a simplification to equate arson with the use of fire accelerants.

The investigation of mechanical and electrical causes for fires lies outside of the scope of this book. An exception is the problem of deciding if a spark-over damage has been caused by an initial short circuit (primary event) or was due to fire damage of the insulation (secondary event). It has been claimed that, in the case of copper wires, the state of oxidation (i.e. the composition of the copper oxide) can present a clue for deciding between primary and secondary events. However, this theory has not been adequately tested experimentally to-date.

2. Sampling And Analysis of Fire Accelerants

It must be realized that incendiary devices not only include liquid accelerants (even though these are most frequently used) but also a large number of solid inflammables such as candles and a variety of inorganic mixtures of chemicals.

The investigation of arson cases where candles were used as a starting device was discussed by Specht and Fischer [14] some years ago. They extracted the suspect material with hexane and investigated the extraction residue with the help of melting point determinations, X-ray diffraction and infrared spectrophotometry.

Igniting mixtures are legio and cannot be listed in detail here. A discussion of some of them can be found in the general literature at the end of this chapter [B, F]. In general, they consist of an oxidizing agent (nitrates, chlorates, perchlorates, chromates, permanganates, peroxides etc.) and a combustible compound (sugar, starch, wood, plastics, carbon, phosphorus, magnesium, aluminum etc.).

A prerequisite for a successful investigation of chemical causes for fires is obviously a proper evidence collection. This will be discussed in Chap. C. I. 10 Once the evidence has arrived at the laboratory, proper sampling techniques must be employed in order to be able to isolate and identify suspected fire accelerants. The purpose of any accelerant sampling technique must be isolating the maximum amount of accelerant in a pure state and at a high concentration. Special techniques needed for investigating spontaneous combustion will be mentioned in Sect. IX. 3.

The samples sent to the laboratory from a scene of a fire can consist of an enormous variety of materials. Among the frequent ones the following can be mentioned: wooden boards, soil, parts of furniture, rugs, paper, clothing, sawdust etc. Usually, a mixture of these materials is received, and this mixture is charred and often wet or humid. Any sampling method should be capable to be used in all cases and should give reliable results. It must be emphasized at this point that many of the objects sent in may from the start contain or form inflammable compounds in small amounts (turpentine, methanol, lacquers, paint components, printing ink components etc.). This fact must be kept in mind when interpreting the analytical results of the investigation. In some cases, it will be necessary to treat reference material in the same way as the questioned sample in order to exclude interference from naturally occurring compounds. The volatile products obtained by steam distillation of various materials occurring in buildings (wood, tar, floor tiles, plastic polymers etc.) were investigated by Clodfelter and Hueske [4] who found that these products can easily be distinguished from fire accelerants when using gas chromatographic analysis.

There is nowadays no sharp boundary between sampling methods on the one hand and analytical methods on the other since both procedures can be combined in one and the same technique. Nevertheless, an effort has been made to list sampling techniques in Table B. 39 (p. 211 and 212), and analytical techniques in Table B. 40. (p. 212–214).

The choice of the method(s) depends on a number of factors, among them

(i) the kind of accelerant used,
(ii) the amounts of accelerant available for analysis,
(iii) the availability of instrumentation and of trained specialists,
(iv) the time element,
(v) the legal importance of the arson case.

The *kind of accelerant* or igniting device used can often be determined with fair certainty. Hints in the right direction can come from observations made by the fire fighting personnel, witnesses, police intelligence, and – before all – from the crime scene search carried out after the fire. But there are obviously cases where such indications are completely absent.

The logical way for starting an investigation of this kind is a stepwise procedure where the material sent to the laboratory is first analyzed with respect to relatively volatile compounds, followed by attempts to isolate less volatile or non-volatile components.

If the *amount of accelerant* is relatively large, of the order of a few grams, a number of conventional chemical methods can be used for characterization. These methods include e. g. distillation profile or boiling point determination, measurements of density, refractive index, flash point determination, measurement of the iodine number, infrared and ultraviolet spectrophotometry and elemental analysis. The choice of method becomes much more limited when only milligram amounts or less of an inflammable material are available. It might then still be possible to determine the refractive index or the ultraviolet spectrum but normally only gas chromatography, mass spectrometry, and – in some cases – spectrofluorimetry, X-ray fluorescence and electron excited energy-dispersive X-ray analysis would give useful analytical results.

The *availability of instrumentation* and of operators with experience in such instrumentation is obviously of great importance in the choice of methodology. It may be pointed out in this connection that it is no good to have instruments available at the laboratory if operators have insufficient training and infrequent opportunity to use them. Uncritical use of an advanced method can be worse than no use at all.

The *time element* can be of importance from the scientific as well as the legal point of view. It is obviously necessary to process fire debris suspected of containing volatile compounds as soon as possible. Even well packed items have a tendency to leak volatiles rather rapidly. This fact impinges on the sequence of case work to be taken up. Cases of great legal importance may have to wait. Methods allowing for rapid screening of debris containing volatiles have special merit on occasions where legal and scientific demands clash.

The *legal importance* of arson cases can dictate the choice of analytical methods. In the relatively rare instances where a suspect has been apprehended, in cases where great financial losses occurred, or when fires broke out in buildings of important military or national interest, a more thorough investigation will often be warranted. Time-consuming detailed analyses of minor cases could well impede the thorough work on important fires.

At the authors' laboratory the usual procedure for analyzing fire debris with respect to volatile accelerants can be summarized as follows:

1. olfactory test (smell),
2. testing of the headspace by detector tubes,
3. steam distillation,
4. isolation of the accelerant,
5. test for inflammability,
6. gas chromatography.

In some cases, other methods are employed such as head-space gas chromatography, distillation and flash point determination (if the amount of material is large enough), and mass spectrometry in conjunction with gas chromatography. In addition, extraction of the steam distilled material with ether is used in cases of suspected presence of accelerants with low volatility (heavy oils, waxes, candles).

An olfactory test can give a valuable hint of what to expect of the investigation to follow. It is useful to know which individuals in the forensic laboratory have a particularly good discriminating sense of smell.

There are commercially available testing devices incorporating groupspecific color tests, for, say, aromatic hydrocarbons or alcohols. The frequency of various types of accelerants occurring in Sweden is shown in Table B. 41 (p. 214). It is seen that alcohol is a popular compound for incendiary purposes – in contrast to reports from the United States where gasoline seems to dominate completely [10].

We use a gas and vapor sensing device from Draeger-Werke in Lübeck, West Germany. It consists of a single hand pump to which glass tubes filled with group-specific reaction mixtures are attached. There are a number of other devices commercially available, e. g. the J-W Aromatic Hydrocarbon Indicator (Johnson-Williams Ltd., Palo Alto, California).

Even in cases where neither specific smell nor positive reactions to vapor sensing devices are registered, the material is subjected to steam distillation. Steam is generated from distilled water and led into a brass container holding the material. This container (15 × 15 cm wide and 28 cm high) is heated in order to prevent condensation of steam. The distillate is cooled and collected in such a way that liquids of lower density than water are collected on top of a column of liquid cooled water. In the case of the suspected presence of water-soluble liquids (alcohols, ketones etc.), the very first drops of distillate are collected for analysis.

If the amount of accelerant is large enough to be visible on top of the distilled water, a minute quantity (a few μl) is drawn up into a glass capillary and held into a flame for testing its flammability.

Finally, gas chromatography is carried out on the isolated liquid. We use a 6 ft. column of 4% OV-101 on Varopor and an oven temperature programmed from 40 to 250 °C at 20 °C/min. The chromatograms obtained are compared to a reference collection of chromatograms for a fair number of accelerants. For more sophisticated comparisons between various brands and possibly samples from individual containers it is necessary to use capillary gas chromatography [2, 3].

An alternative screening method used on reception of the samples from a fire scene is the direct injection of the head space above the sample into a gas chromatograph [6, 16]. This technique is easiest to perform if the containers are piercable by a syringe needle. Some remarks on suitable container or packing materials will be found in Chap. C. I.10 of this book.

Mass spectrometry in conjunction with gas chromatography can be useful in certain cases, especially when only one or a few peaks are observed in the chromatogram.

If sufficient amounts of accelerant have been isolated, its flash point, density, refractive index, and distillation profile can be determined.

A novel and very sophisticated method for direct sampling and analysis of accelerants from sent-in debris has been developed by Twibell and Home [15]. They glue active carbon powder to the tip of a so-called Curie point-wire and expose the wire to the head space vapors. After a while (a few minutes to 1.5 hours) the wire is withdrawn and heated for 10 seconds in a Curie-point pyrolyser with attached gas chromatograph.

A review article on arson analysis has recently been published by Bland [1]. In the special case of leaded gasoline, tests for the presence of lead, bromine and chlorine can be carried out. In order to increase the octane number, tetraethyl lead or tetramethyl lead is added to most automobile gasolines. These gasolines also contain ethylene dibromide or ethylene dichloride which serve to transform atomic lead into volatile lead halogenides, thus preventing lead deposits on the cylinder walls. Fire debris can naturally contain lead (e. g. from paint or additions to certain polymers), and tests for Pb, Cl and Br must therefore be carried out on the distillate.

The actual analysis can then be carried out by emission spectroscopy [8], paper chromatography [9], X-ray fluorescence [12], or energy-dispersive X-ray spectroscopy [5]. In cases where no volatile gasoline can be detected it can be worth-while to analyze soot [7, 13] if it is deposited on clean surfaces (glass, tiles, metal doors etc.).

3. Spontaneous Combustion

Spontaneous combustion in the restricted sense is defined as self-accelerating autoxidation of a combustible material. For practical purposes of fire investigation, exothermic processes other than oxidation are also included in this section. We have therefore to consider the following processes:

(i) exothermic autoxidation,
(ii) reaction of reducing agents with water,
(iii) exothermic decomposition,
(iv) exothermic polymerization,
(v) dangerous, metastable, mixtures of chemicals.

The phenomenon of exothermic autoxidation is the most common one encountered in practice. A number of unsaturated oils can react with air in this way if they are dispersed on materials with a large surface such as cotton waste, cellulose wadding, sawdust or rags. Examples of such oils include linseed oil (containing esters of linoleic, linolenic and oleic acids) as well as oils from poppy seed, sunflower seed, hemp seed, walnut, perilla, seal, whale, and herring. In addition to a large surface and free access to air, poor heat conductivity is a prerequisite for a self-accelerating autoxidation to occur. The kinetics and temperature rise of this process depend on the kind and amount of the oil and the porous carrier, the presence of solvents (these have a cooling effect due to the heat of evaporation) and other compounds, the temperature and the air turbulence.

Experience has shown that boiled linseed oil, soaked up into rags, can start a fire two to eight hours after exposure to air. Balls of cotton waste, weighing a few grams and containing 30% of boiled linseed oil can ignite spontaneously if enclosed in a small space (laundry bag, pocket etc.).

In the laboratory, two types of experiments are carried out for investigating suspected spontaneous combustion, i. e. chemical analysis of the material sent in, and practical tests using the same starting materials for reconstruction of the process.

Identification of the oil can include the determination of the iodine number, the infrared spectrum and chromatography (gas or liquid). If siccatives are present it may be worth-while to determine the content of metals since many siccatives contain salts of lead, cobalt, manganese or zirkonium.

Reconstruction experiments can give useful information on the possibility of spontaneous combustion. However, it is not always possible to obtain conclusive results since a large number of factors play an important and sometimes decisive role for the process to take place. Some of these factors are: the nature of the unsaturated compound, the nature of the matrix on which the compound is distributed, the concentration of the compound, the absolute amounts of the reactive species involved, the type and amount of solvents present (if any), the temperature, the ventilation of air through and around the sample, reaction retarding or accelerating impurities etc.

In general, it can be stated that a large temperature rise in the reconstructed material is quite good positive evidence of a spontaneous combustion theory, whereas a small or zero rise in temperature does not always exclude such a theory.

One way for measuring an unsaturated compound's ability for spontaneous combustion has been introduced by Mackey [11]. The suspected oils is soaked into cheese-cloth (weight ratio 1 : 1), the cloth is tightly rolled and placed in an air bath at 100 °C. The temperature of the cloth is measured by thermometer at known intervals or by a thermocouple which may be connected to a recorder. If the presence of volatile solvents is suspected, the test is repeated after evaporation of the solvent in a stream of air.

4. Practical Examples

a. Attempted Arson Through the Letter Slit in an Apartment Door

An unknown person had thrown a burning, illustrated travel-catalogue through the letter slit into an apartment. The fire was put out by pouring water over it. A little while later, a second, similar catalogue, not afire, was pushed through the door. Both items were sent to the laboratory for analysis for fire accelerators: 1) partially burnt catalogue, (2) undamaged catalogue.

No. 2 smelled like alcohol whereas no. 1 did not have any typical smell. On testing with reagent tubes (Draeger-Werke, Lübeck, cf. Sect. B. IX. 2) a positive reaction for alcohols was obtained from both samples. Each sample was now distilled with overheated steam and the first few drops isolated. Upon continued distillation, a total of 50 µl of a yellow inflammable liquid collected on top of the distillation tube in both cases.

The first drops of distillate were examined by gas chromatography and mass spectrometry and were shown to contain ethanol. No additives present in denatured spirits could be found. The yellow liquid gave a gas chromatogram of unknown character. This led to the suspicion that the liquid could be a component of the printing ink of the catalogue. Another, identical catalogue was therefore purchased and distilled in the same-way. It yielded 50 µl of a yellow liquid which gave a gas chromatogram identical with those obtained above.

This case shows clearly the importance of a critical attitude towards "positive results" in the analysis for fire accelerants.

b. Fire Caused by Carelessness

A fire was suspected of being caused by the house owner who allegedly started a wood-fire in his furnace with the help of Diesel oil. Since he denied this allegation, the following material was sent to the laboratory for analysis:

1. A pair of rubber boots.
2. Crumpled newspapers.
3. Diesel oil bought at the nearest gas station (reference sample).

Materials 1 and 2 smelled of a petroleum product. A test for hydrocarbons with specific reagent tubes (Draeger, cf. above) was positive. Distillation with overheated steam yielded for sample no. 1 about 2 ml, and for sample no. 2 about 37 ml of pale-yellow inflammable liquids. Microdistillation gave similar boiling ranges as for Diesel oil (or heating oil). The flamepoint for the liquid from sample 2 was found to be 72 °C (Setaflash flamepoint apparatus). The gas chromatograms of both liquids were closely similar but lacked some of the low boiling components of gas chromatograms from genuine Diesel oil.

The reference material no. 3 had the same distillation range, flame – point and gas chromatogram as the corresponding values obtained for Diesel oil in the laboratory –as expected.

c. Arson Through the Entrance Door

A successful attempt to start a fire was apparently brought about by pouring an inflammable liquid through the letter slit onto a doormat of cocos fiber. The fire was rapidly put out.

The doormat was sent to the laboratory. Preliminary tests for hydrocarbons were positive. The mat was steam distilled and 45 ml of an inflammable liquid were isolated.

Gas chromatography yielded a curve resembling a chromatogram for gasoline but lacking some peaks for low-boiling compounds. Microdistillation gave the same boiling sequence as for gasoline. Part of the liquid was burnt, and the soot investigated by energy-dispersive X-ray analysis using a scanning electron microscope. Lead as well as bromine were shown to be present.

The conclusion was that leaded gasoline had been used as an accelerant in this fire.

d. Comparison of Accelerants from Fire Debris And a Liquid in a Container

At the scene of a fire the police collected fire debris from a bed (exhibit 1) and from the floor under the bed (exhibits 2 and 3). Closeby in the same room they found a plastic container with about 1 liter of liquid (exhibit 4). The questions posed to the laboratory were:

A. Do exhibits 1–3 contain a fire accelerator?
B. Are the contents of exhibit 4 inflammable?
C. If questions A and B are answered in the affirmative, are the liquids from exhibits 1–3 identical with the liquid in exhibit 4?

Exhibits 1–4 all smelled of a petroleum product and gave positive reactions with Draeger vapor test tubes for hydrocarbons. On steam distillation, exhibits 1–3 yielded 10.5, 6.0 and 7.5 ml respectively of an inflammable liquid. Exhibit 4 had a volume of about 1 l. The results of the investigation of the 4 liquids are collected in Table B. 42 (p. 214), cf. Fig. B. 49 a, b.

The conclusions were formulated as follows:

"There is strong evidence for the fact that the liquids from exhibits 1–3 each contain a mixture of low-boiling components, type kerosene and a higher boiling component, type Diesel oil. There is also strong evidence that the liquid from exhibit 4 consists of a low-boiling component of type ligroin and a high-boiling component of type Diesel oil. The liquids 1–3 are thus of the same kind whereas liquid 4 had a different composition".

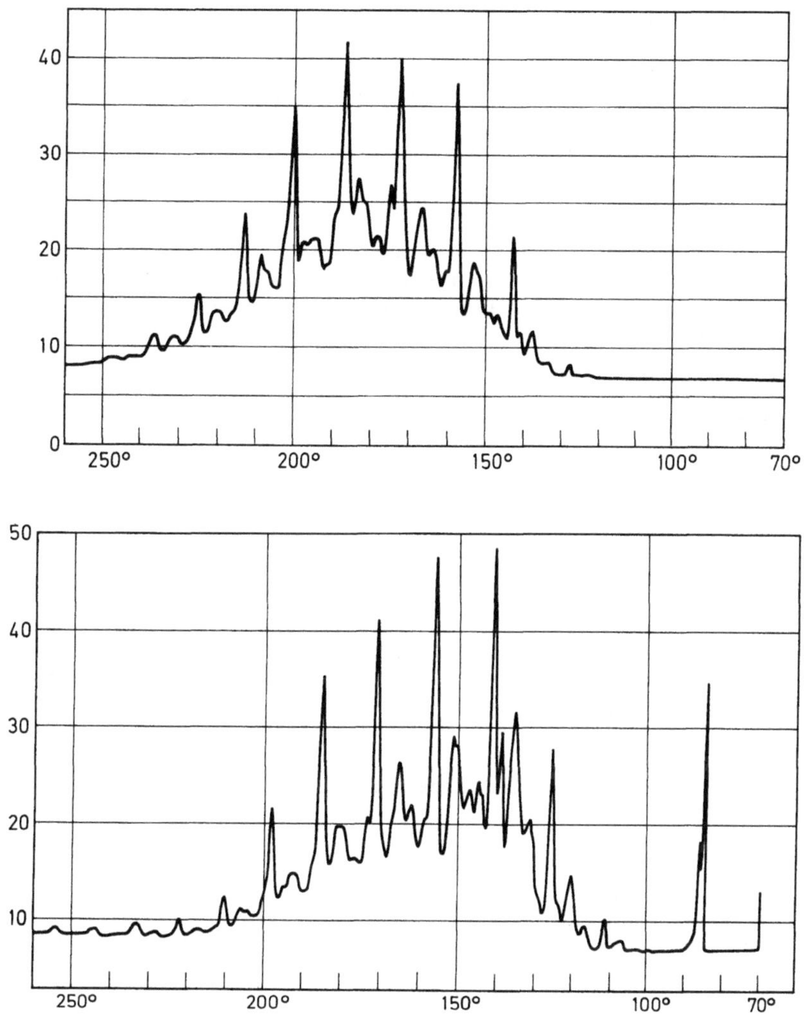

Fig. B. 49 a, b. Gas chromatograms obtained from liquids sent in in case d. **a** Chromatogram of the water-insoluble fraction of the distillate from exhibit 3, showing a mixture of a liquid of type *kerosene* and a liquid of type diesel oil. The liquids isolated from exhibits 1 and 2 gave chromatograms which did not differ significantly from the one shown here, **b** Chromatogram of the liquid isolated from the plastic container (exhibit 4). The chromatogram shows a mixture of a liquid of type *ligroin* and a liquid of typ diesel oil

5. General References (In Chronological Order)

A) K. Nabert, G. Schön: Sicherheitstechnische Kennzahlen brennbarer Gase und Dämpfe. 2nd edit. Berlin: Deutscher Eichverlag GmbH 1963
B) National Fire Protection Association: National Fire Codes. Vol. 1. Flammable Liquids. Vol. 2. Gases. Vol. 3. Combustible Solids, Dusts & Explosives. Boston: 1966–1967
C) R. D. Fitch, E. A. Porter: Accidental or Incendiary. Springfield: Ch. C. Thomas 1968

D) P. L. Kirk: Fire Investigation. New York, London, Sydney, Toronto: John Wiley & Sons Inc. 1969
E) National Fire Protection Association: Fire Protection Handbook. 13th edit. G. H. Tryon, G. P. McKinnon (eds.): Boston 1969
F) M. Frei-Sulzer, A. Suter: Leitfaden zur Bearbeitung von Brandsachen und Explosionen. Hamburg: Kriminalistik Verlag 1971
G) C. J. Hilado: Flammability Test Methods Handbook. Westport Conn.: Technomic Publishing Co., Inc. 1973
H) C. J. Hilado: Flammability Handbook for Plastics. 2nd edit. Westport Conn.: Technomic Publishing Co., Inc. 1974
I) N.-A. Ek, A. Maehly, E. Rudh: Arson Investigation. In: Forensic Medicine, Vol. 1., C. G. Tedeschi, W. G. Eckert, L. G. Tedeschi (eds.). Philadelphia, London and Toronto: W. B. Saunders 1977
J) J. R. Carroll: Physical and Technical Aspects of Fire and Arson Investigation. Springfield: Ch. C. Thomas 1979

6. Specific References

1) H. H. Bland: Petrol, Paraffin and Arson. J. Forens. Sci. Soc. *19*, 81 (1979)
2) P. M. Cain: Comparison of Kerosenes Using Capillary Column Gas Liquid Chromatography. J. Forens. Sci. Soc. *15*, 301 (1975)
3) W. J. Chisum, T. R. Elzerman: Identification of Arson Accelerants by Gas Chromatographic Patterns Produced by a Digital Log Electrometer. J. Forens. Sci. *17*, 280 (1972)
4) R. W. Clodfelter, E. E. Hueske: A Comparison of Decomposition Products from Selected Burned Materials with Common Arson Accelerants. J. Forens. Sci. *22*, 116 (1977)
5) E. L. Covey: Application of Energy Dispersive X-ray Spectroscopy in Fire Investigation. J. Forens. Sci. *22*, 325 (1977)
6) B. V. Ettling, M. F. Adams: The Study of Accelerant Residues in Fire Remains. J. Forens. Sci. *13*, 76 (1968)
7) G. Hellmiss, B. Wagner: Der Nachweis von verbranntem Benzin (Ottokraftstoff) bei der Untersuchung von Brandasservaten auf dem Wege der Röntgenfluoreszenzspektrometrie. Arch. Krimonol. *158*, 18 (1976)
8) W. Katte, W. Specht: Die chemische Identifizierung von Benzinrückständen in Brandresten. Arch. Kriminol. *115*, 116 (1955)
9) W. Katte: Zur Aufklärung von Brandstiftungen: Vorschlag eines papierchromatographischen Nachweises von Bleibenzin-Spuren in Brandresten und Brandruss. Arch. Kriminol. *120*, 106 (1957)
10) Q. Y. Kwan, C. C. Denault: Needs in Arson Investigation: A Summary of Arson Investigators. A. T. R. -76 (7918-05)-1, July (1976)
11) W. Mackey: Apparat zur Bestimmung der Gefahr der Selbstenzündlichkeit von Oelen auf Baumwolle. J. Soc. Chem. Ind., 15 (1896)
12) K. D. Pohl: Die Röntgenfluoreszenzanalyse bei der Untersuchung von Brandrückständen auf Zündmittel. Arch. Kriminol. *151*, 96 (1973)
13) A. Schöntag: Spektral- und Röntgenanalyse zur Aufklärung von Brandursachen. Arch. Kriminol. *122*, 151 (1958)
14) W. Specht, K. Fischer: Ist der analytische Nachweis von Kerzenbrandstiftungen gesichert? Neuer Beitrag zur Bewertung von Brandmittelrückständen. Arch. Kriminol. *122*, 18 (1958)
15) J. D. Twibell, J. M. Home: Novel Method for Direct Analysis of Hydrocarbons in Crime Investigation and Air Pollution Studies. Nature *268*, 711 (1977)
16) I. H. L. Yip, E. G. Clair: A Rapid Analysis of Accelerants in Fire Debris. Can. Soc. Forens. Sci. J. *9*, 75 (1976)

7. Tables B. 39–42

Table B. 39. Sampling techniques for the detection and/or isolation of volatile fire accelerants from fire debris, chronologically arranged from 1947 on

Year	Technique used	Advantages and disadvantages	First and later author(s)
1947	Distillation under reduced pressure	Loss of most volatile compounds	Farrel[a] Bennet (1958)[b] Hrynchuk et al. (1977)[c]
1955	Steam distillation NaCl added to the distillate	Low yields for high-boiling compounds	Katte and Specht[d] Parker et al. (1962)[e] Stone et al. (1978)[f]
1955	Steam distillation with ethylene glycol	Yields for high-boiling compounds improved	Brackett[g] Wilson (1977)[h]
1957	Extraction with ether after distillation	Non-volatile compounds extracted but also impurities from the debris	Specht[i]
1957	Flushing by air, then extraction with CCl_4	Incomplete recovery of less volatile compounds and extraction of impurities	Adams[j]
1960	Detection of aromatic hydrocarbons with a commercial detector	Alcohols and aliphatic hydrocarbons not detected	Burd[k]
1963	Extraction with CH_2Cl_2, evaporation, dissolution in hexane and chromatography on Al_2O_3 column	Loss of very volatile compounds, chromatography losses	Ettling[l]

[a] L. G. Farrell: Reduced Pressure Distillation Apparatus. J. Crim. Law Criminol. Police Sci. *38*, 438 (1947)

[b] G. D. Bennett: The Arson Investigator and Technical Aids. J. Crim. Law Criminol. Police Sci. *49*, 172 (1958)

[c] R. Hrynchuk, R. Cameron, P. G. Rodgers: Vacuum Distillation for the Recovery of Fire Accelerants from Charred Debris. J. Can. Soc. Forens. Sci. *10*, 41 (1977)

[d] cf. Ref. [8]

[e] B. P. Parker, P. Rajeswaren, P. L. Kirk: Identification of Fire Accelerants by Vapor Phase Chromatography. Microchemical J. *6*, 31 (1962)

[f] I. C. Stone et al.: Accelerant Detection in Fire Residues. J. Forens. Sci. *23*, 78 (1978)

[g] J. W. Brackett, Jr.: Separation of Flammable Material of Petroleum Origin from Evidence Submitted in Cases Involving Fires and Suspected Arson. J. Crim. Law Criminol. Police Sci. *46*, 554 (1955)

[h] D. Wilson: A Unified Scheme for the Analysis of Light Petroleum Products Used as Fire Accelerants. Forens. Sci. *10*, 243 (1977)

[i] W. Specht: Der Nachweis von Zünd- und Brandmitteln. Arch. Kriminol. *119*, 88 (1957)

[j] D. L. Adams: The Extraction and Identification of Small Amounts of Accelerants from Arson Evidence. J. Crim. Law Criminol. Police Sci. *47*, 593 (1957)

[k] D. Q. Burd: Detection of Traces of Combustible Fluids in Arson Cases. J. Crim. Law Criminol. Police Sci. *51*, 263 (1960)

[l] B. V. Ettling: Determination of Hydrocarbons in Fire Remains. J. Forens. Sci. *8*, 261 (1963)

Table B. 39 (continued)

1968	Direct sampling of head-space vapors and injection into gas chromatograph	Incomplete sampling (run at room temperature)	Ettling and Adams[m]
1976	Three-step procedure: (i) head-space sampling at room temperature (ii) head-space sampling after heating to 130° C (iii) steam distillation	Complicated but efficient method	Yip and Clair[n]
1977	Four-step procedure: (i) qualitative analysis by commercial detector tubes (ii) steam distillation (iii) vacuum distillation (iv) extraction with n-pentane	Complicated but efficient method. Losses may be expected due to distribution of accelerants between three fractions	Wilson[h]
1977	Absorption of vapors on active coal on a pyrolysis wire, followed by gas chromatography	Qualitative sophisticated method without physical isolation of the accelerant	Twibell and Home[o]

[m] cf. Ref. [6] [n] cf. Ref. [16] [o] cf. Ref. [15]

Table B. 40. Methods of qualitative analysis of volatile fire accelerants in chronological order from 1955 on

Year	Method	Remarks	Author(s)
1955	Emission spectroscopy for detecting Pb in a liquid	Restricted to leaded gasoline	Katte and Specht[a]
1957	1) Classical procedures; determination of: flash point, density, refractive index, iodine number and C,H-analysis 2) Analysis for $PbEt_4$, methylanilin and tricresylphosphate 3) Ultraviolet spectrophotometry	These procedures require relatively large amounts of physically separated accelerants	Specht[b]
1957	IR spectrophotometry	Limited selectivity	Adams[c]

[a] cf. Ref. [8]
[b] W. Specht: Der Nachweis von Zünd- und Brandmitteln. Arch. Kriminol. *119*, 88 (1957)
[c] D. L. Adams: The Extraction and Identification of Small Amounts of Accelerants from Arson Evidence. J. Crim. Law Criminol. Police Sci. *47*, 593 (1957)

Table B. 40 (continued)

1957	Paper chromatography for Pb, visualized as rhodizonate	Restricted to leaded gasoline	Katte[d]
1958	1) Emission spectroscopy for detecting Pb in soot 2) Ultraviolet spectrophotometry of oils	The paper describes some actual cases of fire	Schöntag[e]
1960	Thin-layer chromatography of the dyes added to commercial brands of gasoline	The method is of limited use in fire investigations	Häusser[f]
1960	Gas chromatography	At present still the method of choice for routine cases	Lucas[g] as well as Cadman and Johns[h]
1968	Head-space gas chromatography	cf. table B. 39 (p. 211)	Ettling and Adams[i]
1972	Capillary gas chromatography	High discriminating power	Chisum and Elzerman[j]
1973	X-ray fluorescence for analysis of leaded gasoline and lubricating oils (Pb, Br, Zn, Ba & As) directly on the fire debris	Limited to certain types of fire accelerants	Pohl[k]
1976	X-ray fluorescence for analysis of soot	Limited to leaded gasoline	Hellmiss and Wagner[l]
1976	Heated head-space gas chromatography	cf. table B. 39 (p. 211)	Yip and Clair[m]
1977	Absorption on active carbon plus gas chromatography	cf. table B. 39 (p. 211)	Twibell and Home[n]
1977	Gas chromatography – mass spectrometry	Carried out on simulated arson residues	Mach[o]
1977	Energy-dispersive X-ray spectroscopy (EDX spectrometer)	Limited to leaded gasoline	Covey[p]

[d] cf. Ref. [9] [e] cf. Ref. [13]

[f] H. Häusser: Identifizierung von gefärbten Kraftstoffen und Kraftstoffgemischen mit Hilfe der Dünnschicht-Chromatographie nach Stahl. Arch. Kriminol. *125*, 72 (1960)

[g] D. M. Lucas: The Identification of Petroleum Products in Forensic Science by Gas Chromatography. J. Forens. Sci. *5*, 236 (1960)

[h] W. J. Cadman, T. Johns: Application of the Gas Chromatograph in the Laboratory of Criminalistics. J. Forens. Sci. *5*, 369 (1960)

[i] cf. Ref. [6] [j] cf. Ref. [3] [k] cf. Ref. [12] [l] cf. Ref. [7] [m] cf. Ref. [16] [n] cf. Ref. [15]

[o] M. H. Mach: Gas Chromatography – Mass Spectrometry of Simulated Arson Residue Using Gasoline as an Accelerant. J. Forens. Sci. *22*, 348 (1977)

[p] cf. Ref. [5]

Table B. 40 (continued)

1978	Combination of methods, including infrared spectrophotometry, energy-dispersive X-ray analysis, nuclear magnetic resonance, gas chromatography and mass spectrometry	Comprehensive collection of analytical tools	Stone et al.[q]

[q] I. C. Stone et al.: Accelerant Detection in Fire Residues. J. Forens. Sci. *23*, 78 (1978)

Table B. 41. Fire accelerators found in fire debris at the National Laboratory of Forensic Science in Sweden 1968–1979

Accelerator	Number of cases	Per cent
Gasoline	142	36.9
Kerosene	121	31.4
Diesel oil or fuel oil	52	13.5
Ethanol	38	9.9
Other liquids	32	8.3
Total number of cases	385	100.0

Table B. 42. Analytical results obtained for the liquids from exhibits 1–4

Property	Liquid from exhibit			
	1	2	3	4
Flash point (°C)	59	62	70	48
Distillation process	As for a mixture of kerosene and Diesel oil			[a]
Gas chromatogram	As for a mixture of kerosene and Diesel oil (cf. Fig. B. 49 a)			[b]

[a] as for a mixture of ligroin and Diesel oil
[b] as for a mixture of ligroin and Diesel oil (cf. Fig. B. 49 b)

X. Questioned Documents

1. Introduction

The investigation of questioned documents is mainly carried out by optical methods. These methods are well described in the literature (cf. general references at the end of this chapter) and will not be referred to in any systematic manner.

There are instances, however, of investigations that require chemical analyses, usually in the microscale, even in the field of questioned documents. The most frequent problem is the analysis of ballpoint ink on a document. In some cases, printing inks must be analyzed, and once in a while, typewriter ribbon inks have to be investigated for discriminating purposes.

Sometimes, the documents themselves, i.e. in practice the paper, has to be analyzed in order to find possible discrepancies. This problem can arise when bank bills, shares, bonds etc. are suspected to be false – but also company stationary or similar papers can be suspected of counterfeiting. In paper analysis, physical methods predominate, but chemical tests can be performed.

2. Historical Notes

The oldest known writing surface material is papyrus, made from strips cut out of the bush of the same name. At a Paris museum, a piece of papyrus believed to be from 3500 B.C. is kept. Later, parchment (animal skin, treated with lime) became another significant writing surface. One of the most famous parchment documents are probably the "dead sea scrolls", discovered 1947 in Israel and containing parts of the Old Testament, written around the 1st century B.C. Large amounts of parchment (pergament) were produced in ancient Greece in Pergamon.

Paper in the modern sense is a product made from cellulose. According to tradition it was invented in 105 A.C. in China by one Ts'ai Lun. He mixed pieces of old rags, hemp fibres and mulberry fibres with water and macerated the mixture by mechanical stamping. The resulting suspension (pulp) was sieved by scooping it in wooden frames with bamboo netting. The felt was pressed and dried.

The manufacturing secret spred slowly westward and "arrived" in Europe in the 13th century. Until the middle of the 19th century, European paper makers only used lump material. A shredding machine was invented in Holland around 1800, and is in principle still used today ("hollander").

Some technical improvements included "sieving" with metal meshes (and later woven ones), pressing the raw paper mats between layers of felt, and treatment with glue (and later rosin) for reducing the capillary forces in the raw fibres. In the middle of the 19th century, the amount of lump and rag material became insufficient for the growing need for paper. German technicians found several ways for using wood fibres as an additive. The amount of wood pulp used for paper manufacture grew steadily, and by 1960, paper production could have the following distribution of raw materials:

wood pulp	55%
reused paper	35%
straw	4%
lump material	3%
other cellulose materials	3%
	100%

Modern paper has thus a relatively short history, about one century. World production of paper in 1960 was over 70 million tons per year and is doubling in about 20–25 years. In industrial countries, about 50–200 kg of paper are used per capita and year.

From about 1950 on, some kinds of paper were made from synthetic fibres and even from non-fibrous synthetic polymers. Also, paperlike materials are manufactured from glass fibres and metal threads. There is no sharp terminological boundary between such products and cellulose-based paper. So far, however, the document examiner does not normally encounter these kinds of "paper".

3. Modern Paper Manufacture

The raw material for paper making is essentially cellulose. Some of the starting material consists of almost pure cellulose, and needs therefore very little enrichment. The cellulose content of cotton or linen (92–97%), hemp (80%), ramie (85%) and sisal (77%) is thus very satisfactory. Wood, however, contains only 58–67% of cellulose, and cereal straws even less (36–57%). These materials must undergo a preliminary treatment for removing unwanted components. Table B.43 (p. 233) gives the composition of the most common kinds of wood.

The process of eliminating non-wanted constituents of woods consists mainly of a disruption of bonds between cellulose and some of the "impurities", and a transformation of these components into water-soluble forms. The cheapest but least efficient way to do this is mechanical disruption followed by washing. The product (mechanical or groundwood pulp) still contains much lignin, is weak and non-permanent and is mainly used for newsprint paper. In semichemical methods, chemical "cooking" is followed by mechanical treatment in the presence of the hot chemical. The most common, most costly, but also most efficient methods are chemical treatment of wood by various means. There is an acid process and several alkaline processes.

The acid process (*sulfite method*) consists of a treatment with calcium hydrosulfite (prepared from burning of S or FeS and absorption of the SO_2 formed on lime) for 7–9 hours at 145–160° C (quick cook) or for 30 hours at 125–135° C (slow cook). The lignin becomes water-soluble, and polysaccharides are hydrolyzed to soluble sugars.

The most common alkaline techniques are:

(i) *The lime process.* This is a mild procedure, used with materials with high cellulose content (rags) or for low quality, cheap pulp production. The raw material is heated with 5–18 per cent of lime, $Ca(OH)_2$, for 8–17 hours at 25–55 at pressure.

(ii) *The soda process.* Treatment with NaOH or Na_2CO_3 hydrolyzes the lignins and polysaccharides and saponifies fats and waxes. The resulting pulp is used for printing paper making.

(iii) *The soda-chlorine process.* Gaseous chlorine reacts with lignin, replacing H by Cl and forming HCl. The acid is neutralized by adding alkali. It is possible to work at atmospheric pressure at 100° C and to obtain relatively pure and white cellulose.

(iv) *The sulfate (or sulfide) process.* This is the most common alkaline technique. The wood is treated by cooking it for 2 to 3 hours at 8 at in a solution containing about 46% NaOH, 25% Na_2CO_3, 22% Na_2S, and 7% Na_2SO_4. The sulfate is reduced to sulfide by the cellulose carbon, and fresh Na_2SO_4 has to be added. This has given the name to the sulfate process. The resulting pulp is mechanically very strong ("kraft-pulp"). The pulp (stock) obtained has usually a strong grey, yellow or brown color which has to be removed for the manufacture of high-quality paper. This is achieved by bleaching. It can be done by treatment with elemental chlorine followed by neutralization by alkali (which also solubilizes a number of impurities such as lignins) but the most common bleaching agent is a solution of calcium hypochlorite.

All chemically treated pulp has to be washed thoroughly with large amounts of water.

The raw material described above (stock or half-stuff) has to undergo a series of treatments in order to obtain the desired properties of the finished paper product.

The finest paper consists only of lump material, medium quality papers contain cellulose from wood in various amounts whereas low grade paper can contain from 10% up to 80% wood fibres.

Figure B. 50 gives a schematic overview of the paper manufacturing process. The first step consists of mechanical beating, which is usually done in a so-called hollander (see above), a container equipped with a rotating cylinder. The bottom of the container and the cylinder (roll) contain brass bars which chop and cut the cellulose fibres. By controlling the sharpness and working distance of these bars, the treatment leads to fibrillation, cutting, or hydration (by beating) of the fibres. Usually, a combination of these processes is used in order to produce fibres with the optimal properties for their subsequent use. The physicochemical processes brought about by beating are not clear, but it seems that the fibrillation, destroying part of the cell walls, leads to an uptake of water and to bonding through hydrogen bonds.

After beating, the pulp has acquired the ability to form aggregates, to get "sticky". But the resulting material has also a high capillary power. Except for blotting paper and the like, this property is undesirable for the finished product. The remedy is the treatment of the stock with rosins or glues, a process called "sizing". Sizing prevents the spreading of writing ink on the paper, increases its resistance to moisture, consolidates the sheet and facilitates the uptake by the stock of other additives (loading and coloring).

In 1806, Illig found that rosin, a wood resin, is well suited for sizing. Rosin is still the most used material for this purpose. Usually, it is treated with alkali to give a water-soluble soap consisting mainly of sodium abietate. On subsequent treatment with alum or aluminum sulfate, the aluminum salt is formed on the fiber. Other sizing agents are starches, animal gelatine, synthetic polymers, aluminum hydroxide etc.

Finally, it is necessary to give the paper stock more body, to load it with inorganic products. The purpose of loading is to smooth the surface, thus improving the printability of the paper, to improve the opacity, and to increase the whiteness of the finished product.

As loading agents, or "fillers", a large variety of minerals are used. The more important ones are listed in Fig. B.50. Kaolins (aluminum silicates) are probably the most widely used fillers. For quality papers of great whiteness, barium sulfate or – best but most expensive – titanium dioxide is used. The choice of fillers is made according to the desired properties of the end product and cannot be discussed in this short survey. Fillers may constitute about 5–30% of the weight of the finished product.

The document examiner very often meets with the special kinds of paper used for money bills and documents of high value (shares, bonds, passports etc). This paper is almost invariably made from cellulose fibres (lump, rags) in the "kraft" process. It has high physical strength and no (or very week) fluorescence. Usually, it has watermarks of high sophistication. A genuine watermark is produced by placing a three-dimensional mask (usually of metal) with the appropriate design on top of the mesh, carring the pulp suspension. This mask can be attached to rotating cylinders, or – in hand-made paper – to the wire mesh attached to a hand-held frame.

The watermark production is thus part of the machinery producing sheets of paper from the pulp. The first paper machine came into use in 1803, and modern

Major steps in paper manufacture

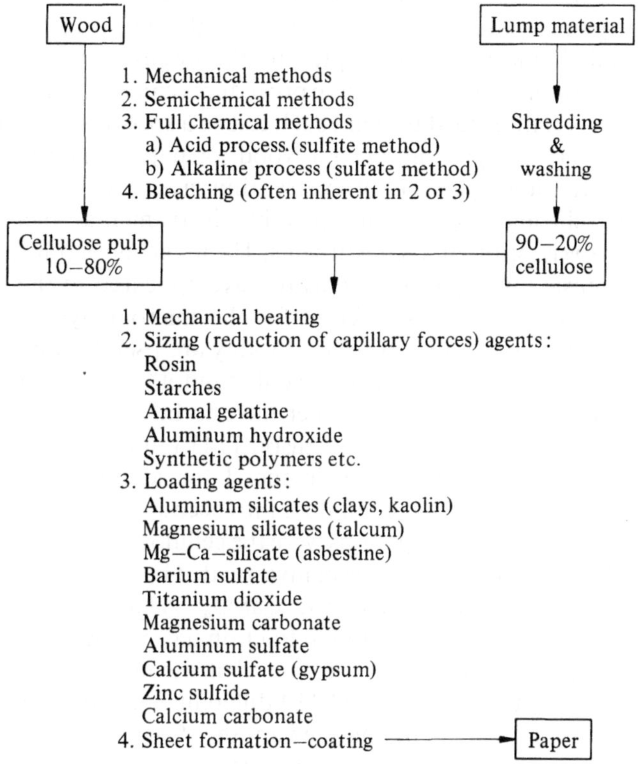

Fig. B.50

machinery can be as long as a city block. A pulp suspension of around 0.8% enters the machine, and after drying by draining, suction, pressing, and heat, a sheet of paper leaves the machine. A number of additional treatments such as coating, dying, glazing etc. can be incorporated.

4. Paper Analysis

It is evident from the discussion above that paper is quite a complex chemical mixture containing, except for cellulose, a fair number of organic and inorganic additives. These additives ahould in principle lend themselves to comparative analyses. There are, however, a number of difficulties. One of these concerns the variations in composition from batch to batch as compared to those from one type or manufacture of paper to another. Extensive background information based on a large number of analyses will be needed to evaluate those variations. In particular cases this information may be available from the paper mill which produced the paper for the document – if this can be ascertained. The other difficulty is the need for micro – or ultramicroanalytical methods due to the limited amount of paper available. It is usually not legally permissible to sacrifice sizable parts of a questioned document for destructive analysis.

In principle, both organic and inorganic components of papers can be analyzed, but the need for micromethods restricts the choice to a few simple color tests as far as organic matter is concerned. Most of these tests require a great deal of experience and should not be attempted by the forensic chemist who would use them only occasionally. Details of these procedures can be found in the literature (e. g. [J]), but it is recommended to remit such test to a specialist working either in the laboratory of a large paper mill or in a wood (or paper) research institute.

Infrared spectrophotometry, thin-layer-, high pressure liquid-, and pyrolysis-gas-chromatography are possible instrumental methods for the comparative analysis of organic constituents. However, no application of these techniques to forensic investigations seems to have appeared in the literature.

The situation is different for the microanalysis of inorganic components in paper. The methods published today are pseudo non-destructive, i. e. the paper is not destroyed, but the instruments cannot be used on large areas of paper. Thus, in practice, small pieces of paper have to be removed for analysis.

In 1960, Garey and Swanson [14] published a method of paper analysis by X-ray diffraction. It was intended for identifying fillers and had to take the diffraction patterns of cellulose into account. The method should be well suited for comparative investigations were the cellulose "background" so to say "cancels out". The following minerals were investigated by Garey and Swanson: kaolin, talc, TiO_2 (anatase and rutile), bentonite, $CaSO_4 \cdot 2H_2O$, $2CaSO_4 \cdot H_2O$, and $BaSO_4$. To the authors' knowledge, X-ray diffraction has not been used in forensic paper analysis.

Guinn et al. [18 a] introduced neutron activation analysis (NAA) for paper analysis; they investigated 119 different kinds of paper.

Brunelle et al. [6] analyzed another 600 paper samples of 26 main types from 11 American manufacturers. The measurement precision for the following 11 elements was determined: Mn, Na, Sb, Sa, Sc, Ta, Ba, La, Cr, Cs, and Au. Also, a further 11 elements were detected. Schlesinger and Settle [24] analyzed 120 different white bond papers for the following elements: Ti, Al, Ca, Na, Mn, Cl, Cr, Sb, Zn and Ta.

The NAA data obtained by Guinn et al. [18 a] were later analyzed by Duewer and Kowalski [12] using a pattern recognition method. In 1978, Blanchard and Harrison [2] applied NAA to fine paper identification.

In 1975, two groups of investigators [1, 22] published a micromethod for inorganic paper analysis for forensic purposes by scanning electron microscopy (SEM) combined with energy-dispersive X-ray analysis. All elements with $Z > 10$ can, in principle, be determined in a semi-quantitative manner. The filler particles have diameters in the range of 0.1–5 μm [1] and their topography can be studied by SEM. The specimens are punched out or cut out with a new razor blade and mounted e. g. with a transfer adhesive [3M No 465, Ref. 22] or Duco cement diluted with acetone to a graphite rod on the holder of the SEM. Sometimes, the analysis of an ashed sample (6 mm diameter) can be advantageous [1].

Barnard et al. [1] list the analytical results for 54 specimens of paper with regard to Na, (Mg), Al, Si, (P), S, Cl, (Ag), K, Ca, (I), Ti, (La), (Cr), Fe, and (Zn), where the elements within brackets were only found in 1–4 samples. The same group of workers as in [1] but for one author published another paper two years later [23]. There, compositional variations among identically labeled sheets of paper were examined.

Totty et al. [29] employed X-ray microprobe analysis for examining coated papers used in photocopying processes. Kondo and Tanaka [18] identified paper by X-ray fluorescence analysis. Finally, Bober and Berkshire [3] used attenuated total reflectance infrared spectrophotometry for determining clay coatings on paper.

In summary, it seems that instrumental methods for the analysis of organic constituents of paper for forensic purposes have not been published to date. Inorganic components, on the other hand, have been studied by X-ray diffraction (XD), neutron activation analysis (NAA), and scanning electron microscopy combined with X-ray analysis (SEM-X). The choice between these methods depends in part on the relative availability of the instruments required. It should be noted, however, that XD, while useful for evaluation of particular minerals, can only detect crystalline components. The increased use of SEM-X equipment in forensic laboratories makes this technique the method of choice. In some cases XD is an excellent complement, however.

A discussion of the role of paper in forensic science has been published by Grant [16], and a review article by Martin appeared in 1962 [L].

5. Ink Analysis

a. Introduction

Whereas the chemical analysis of paper for forensic purposes has not found widespread use, the situation is very different for the forensic analysis of inks. The question arises frequently if a given text on a document has been written with one and the same ink, or with different types of inks. Also, the age of the writing can be questioned. In printed documents, printing inks need to be analyzed in order to find compositional differences that might prove counterfeiting. Furthermore, stamp pad inks might need to be analyzed in cases of suspected forgeries with rubber or metal stamp marks.

Witte (O) lists the following types of inks that can be met with in document work:

1. Indian inks.
2. Logwood inks.
3. Iron gallotannate inks.
4. Fountain pen inks.
5. Dyestuff inks.
6. Water-resistant inks.
7. Quick-drying alkaline inks.
8. Ballpoint inks.
9. Stamp pad inks.
10. Copying inks (hectograph inks).
11. Typwriter ribbon inks.
12. Printing inks.

Readers are referred to his article for the less frequently occurring types of inks. In the context of this book we will confine the discussion to four kinds of inks: ballpoint ink, fluid inks (roughly types 1–7), typewriter ribbon inks, and printing inks. Only the chemical analysis of ballpoint ink, however, can be covered in some detail.

b. Ballpoint Inks

The development of ballpoint inks closely followed the improvements of the ballpoint pen (BPP). The early history of this type of pen is not clear. The first US patent regarding BPP's has Nr 600, 299/1895 (cit. from Ref. [D]), and a German patent (DRD 126, 302/1900) mentions that the idea in itself was known by then (cit. from Ref. [K]). However, the BPP did not come into practical use until in the 1940's when it was developed in the US for military use.

The main parts of the BPP are the *shaft*, carrying the cartridge, the *cartridge* itself with a carefully machined *point* holding the ball – and the *ball* itself (cf. Fig. B.51 a, b). The ball has usually a diameter of 0.6–1.0 mm and must be machined to very

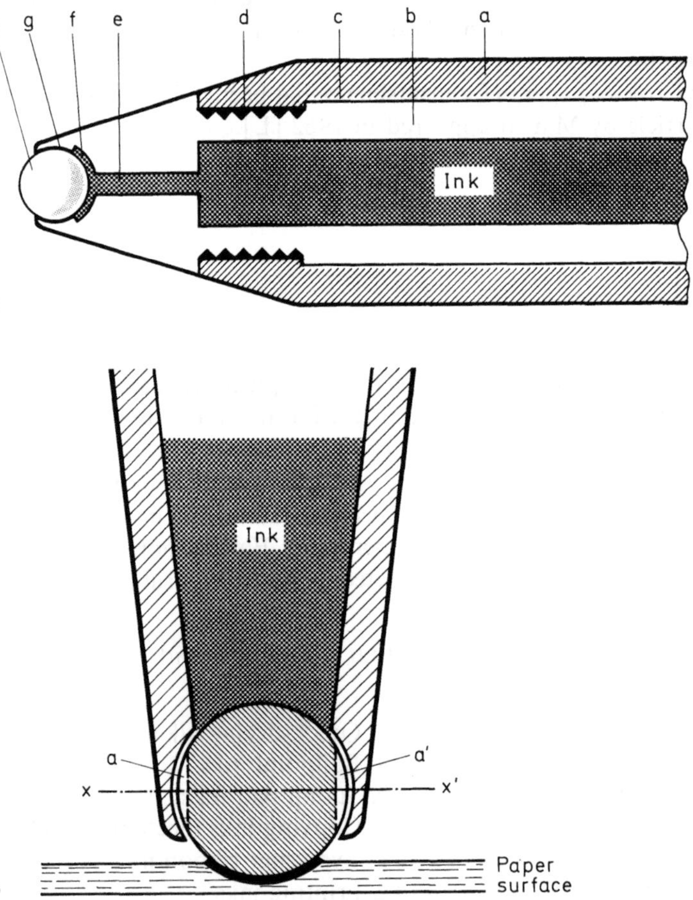

a

b

Fig. B. 51. a A schematic view of a ball-point pen. *a* holder for refill-cartridge, *b* cartridge, *c* empty space, allowing the plastic cartridge to bend slightly, *d* thread, *e* jet-tube, connecting the ink reservoir to the tip of the pen, *f* distribution channels (3–6 in all) assuring an even flow of the ink, *g* bearing, holding the ball in place, *h* the ball, diameter 0,6–1,0 mm, **b** A simplified view of the tip of a ball-point pen: x–x′ axis of ball, rolling towards the observer, a and a′ parts of the ball which are not in contact with the ink. Note that there are two frictions involved, viz the external friction (ball-paper) and the internal friction (ball-bearing). Both figures are reproduced from a publication by Mally [19]

small tolerance indeed (down to 1 μm) since the "gap" between the point (bearing) and the ball is of the order of 10 μm. The speed of rotation of the ball can exceed 3000 rpm. In the beginning, chromium plated iron balls were used, but modern pens contain balls made of hardened steel, in some cases of ceramic materials, and in high-quality pens of semiprecious stones or sapphire.

The ball bearing has to be made with similar tolerances as the ball itself. The small gap between bearing and ball precludes the use of inks consisting of suspensions of carbon black or other pigments. Instead, true solutions of dyes must be used. The demands for these inks are very high and include

1. A correct viscosity, so that no ink leaks out until the pen is used, and no excess ink leaves the pen on writing.
2. A correct adhesion, so that the ball is always covered with an even layer of ink.
3. A correct vapor pressure of the solvent, so that the ink does not clog the pen but dries quickly on the writing surface.
4. Non-hygroscopicity.
5. Stability to light, oxidation and polymerization.
6. Absence of corrosive properties.
7. High intensity of color.
8. Good adherence to paper without smudging.
9. Absence of "striking" (spreading) on the paper.
10. Absence of impurities that may clog the pen.
11. High resistance to solvents and bleaching agents.

During the first 10 years or so, oil-soluble dyestuffs dissolved in oleins, such as castor oil, were used. In 1952, poly (ethylene glycols) and to some extent glycols, glycol ethers and benzyl alcohol replaced the olein, and the choice of dyes became considerably larger. Eventually, single dyes were replaced by mixtures of dyes in order to improve the properties of the ballpoint inks. This fact is of great importance to the forensic chemist who is now able to discriminate between various ballpoint inks, even those which do not differ in color to the naked eye or even when regarded through suitable color filters.

The film of the writing is very thin, of the order of 0.1 μl cm^{-2}, and the concentration of the dyes must be high (15–45%) in order to achieve the necessary color intensity. One gram of ink (1–3 cartridges) can produce a straight line of 3–12 km in length.

Thorough review articles on the BPP appeared already in 1954 and in 1956 [15, 19] but they are still worth consulting.

Analytical methods for comparing ballpoint inks are legio and some of them will be discussed further on. A much more difficult problem is the determination of the age of a ballpoint-written text. In contrast to conventional writing inks where oxidation and other chemical changes can be observed for weeks, months or years (cf. the general reference works), ballpoint inks usually do not undergo readily measurable changes. During the first minutes or hours, a certain amount of smudging may be observed when pressing a clean sheet of paper on the written text, but this method is inexact and of little use.

If the date of first manufacture of a certain type of ink is known, an "earliest date" of the writing can be established. For this purpose, a large ink collection together with detailed data on production dates is needed. To our knowledge, there exists only one such collection – the one at the Bureau of Alcohol, Tobacco and Firearms (ATF) in Washington D. C., USA. The collection consisted of about 3500 different inks in 1978 [9 and 11] and includes inter alia information on the date of their first marketing.

In order to conduct an efficient chemical analysis of ballpoint inks, a knowledge of its composition is essential. It has been mentioned above that poly (ethylene glycols) and related compounds have replaced the oils of early BPPs as solvents. It has also been noted that usually mixtures of dyes are used to improve the properties of the inks. The types of dyes include aniline dyes (methyl violet, rhodamine, victoria blue, nigrosin etc.), alcohol soluble dyes (zapone true dyes) and, more recently phthalocyanines (astra blue base, blue base KG etc). It is difficult to obtain detailed information on the composition of the inks since it is more often covered by manufacturing secrets then by published patents. A list of 55 dyestuffs used in 1965 together with some of their chemical properties has been published by Nakamura and Shimoda [21].

Essentially, analysis will have to be directed in three ways, i. e. covering the solvents, the organic components of the dyes, and the traces of metals (from the phthalocyanines, inorganic acids, preservatives etc.). All of these approaches have been published, but the vast majority of papers concerns the comparative, qualitative analysis of the dyestuffs themselves.

The early attempts of chemical analysis were carried out on the document itself, using spot test techniques. Obviously such procedures are unsatisfactory from the legal point of view since they generally lead to irrevisible changes which in some cases can destroy parts of the document investigated. A somewhat later publication on this methodology is that by Crown et al. [10] who investigated 210 ballpoint pens.

Around the year 1954, chromatographic methods were introduced which still required the destruction of a document but on such a limited scale that most courts would not object. The development of these methods was rather rapid, and some of the improvements introduced during the following 25 years are listed in Table B.44 (p. 234 and 235). Brunelle and Pro [7] discussed a systematic approach to ink identification.

At present, it seems to be most advantageous to examine ballpoint writing optically first. This can be done by direct visual comparison, by observation through suitable filters, by photography through filters, in UV- or IR-light, or by observation with various TV techniques. When the examinations give inconclusive or insufficient results, chemical analysis is first carried out by thin-layer chromatography (TLC). Finally, high pressure liquid chromatography (HPLC) may be used. HPLC has a greater potential than TLC since it allows for an analysis of the vehicles used in the ballpoint ink and since it can discover ink components only visible in the UV or IR regions. Finally, it can perform semiquantitative analyses; true quantitation is hard to achieve due to the difficulty to extract small portions of writing in a reproducible manner. At present, the quantity of ink needed for HPLC is an order of magnitude larger than that for TLC, but with the recent introduction of micro-HPLC equipment this drawback may disappear.

c. Liquid Inks

The use of liquid writing inks has been largely replaced by ballpoint inks. In some parts of the world the former are still in use, and there are of course many older documents that have to be examined at a later date.

The review article by Witte [O] covers the principle methods of analysis of the fluid inks. Since then, Thompson [28] had introduced electrophoresis for the examination of fluid inks. In 1972 Tewari and Bhatt [26] reexamined thin-layer chromatography, and in 1977 Mehrotra and Sidhana [20] analyzed metal ions from coloring matter, inorganic acids, and preservatives by laser emission spectroscopy. Sen and Gosh [25] have reported a method for dating iron-base writing inks.

d. Typewriter Ribbon Inks

As Brunelle et al. [8] have stated, analysis of typewriter ink can provide additional corroborative evidence that the questioned document is authentic or prove that different typewriter ribbon inks were used. Brown and Kirk [5] were apparently the first to analyze typewriter ribbon inks. They used paper chromatography and paper electrophoresis. Thin-layer chromatography was used by Tholl [27].

Brunelle et al. [8] extracted over 150 samples of ribbons (from 7 manufactures) with pyridine and spotted the extracts on thin-layer plates. The ink from one typewritten letter or number suffices for an analysis. In 1978, English published a similar method [13].

e. Printing Inks

A general introduction to the composition of paints, dyes and lacquers is found in Chap. B.III. Only a short discussion pertinent to printing inks will follow here. The properties of printing inks are adjusted to the printing method (typographic, lithographic and intaglio techniques respectively) but these inks have a number of characteristics in common.

In general, printing inks contain pigments suspended in a vehicle (varnish) and blended with a solvent. The pigment gives color and contrast and must

(i) consist of a finely divided powder,
(ii) have good coloring and "hiding" properties,
(iii) form a homogeneous paste with the vehicle,
(iv) be soft so as not to harm the printing plates,
(v) be resistant to light, heat, and chemicals,
(vi) be non-bleeding and,
(vii) have good storage properties.

Both inorganic and organic coloring matter is used in printing inks. Microanalysis, however, is only practical for inorganic pigments, especially if non-destructive methods are to be used. An exception are certain lakes and metal complexes of organic dyes which lend themselves to inorganic analysis by virtue of their content of Al, Cu and a few other metals. The vehicles used include oils, resins and synthetic resins and are hot amenable to non-destructive microanalytical investigations.

Frequently used pigments in printing inks are among others (cit. from [A]:

White pigments: TiO_2, ZnO, ZnS and lithopone (26–30% ZnS, and $BaSO_4$ to give 100%).
Extenders (used for diluting pigments): $BaSO_4$, Al_2O_3, $CaCO_3$, and $Al_2O_3 \cdot 2\ SiO_2 \cdot 2\ H_2O$.
Blue pigments: Ultramarine (approx. $Na_4Al_3Si_3S_2O_{12}$) and Bronze Blue ($Fe_4(Fe(CN)_6)_3 \cdot 10\ H_2O$)
Red-orange-yellow pigments: Lead chromes (mixtures of $PbCrO_4$ and $PbSO_4$) to give shades from red to yellow;
molybdated scarlet chrome, mixtures of Na_2CrO_4, $Pb(NO_3)_2$ and Na_2MoO_4;
Zink chrome ($ZnCrO_4 \cdot K_2CrO_4 \cdot ZnO$);
titanium nickel yellow (obtained by calcination of NiO and TiO_2;
iron oxides.
Green pigments: Guignet's Green ($Cr_2O_3 \cdot 2\ H_2O$).
Black pigments: Carbon black of different varieties.

The chemical analysis of printing inks has not been a frequent problem in forensic science. The examiner of questioned documents has usually the task of comparing a suspected counterfeit document with a genuine reference counterpart. Valuable documents such as bank bills, shares, bonds, bank checks, passports, identification cards etc. are usually protected from counterfeiting by using special quality paper (cf. above) and by employing a combination of printing methods and printing techniques. The counterfeiter usually has neither the technical skill nor the financial means to use these sophisticated techniques. He will try to duplicate the genuine document by cheaper and simpler means. A trained document examiner can then relatively easily determine the printing technique used and show the difference in quality and detail between the genuine and the counterfeit document.

Chemical analysis can nevertheless be useful in special cases. One of these would be the need of comparing inks on documents where the printing method does not differ or where the difference is not certain. Another case is the need to identify the composition of an ink for tracing its origin or for helping to elucidate the printing method(s) used.

Forensic investigations of printing ink have, to the author's knowledge, not been published in the literature. In our own laboratory we have used electron excited X-ray fluorescence in a scanning electron microscope for ink analysis and related investigations on printed documents in a few cases [31]. Other suitable methods may be X-ray fluorescence, electron microprobe analysis and neutron activation analysis.

f. Other Writing Materials

Hilton [17] described a method for differentiating between colored pencils, and Verma et al. [30] developed a thin-layer system for analyzing fiber-tip pen inks.

6. Practical Examples

a. Which Pens Were Used for Writing a Signature?

A postal money order was double-signed (cf. Fig. B. 52). The question posed by the police were:

1. Which one of two suspected persons had written the signature (handwriting specimens from the two individuals were sent in)?
2. Which of the two pens (sent in), if any, had been used for writing the signature?

The first question could not be answered with certainty due to the non-spontaneous character of the writing.

In order to answere the second question, the following procedure was carried out:

1. On ocular inspection one saw that the bottom signature had been written with a dark blue ball-pen ink whereas the color of the ink used for writing the later (top) signature was lighter blue.
2. Small lines were drawn on the document with both pens sent in (A and B respectively).
3. The document was now irradiated in short infrared light and examined with an IR-sensitive TV-camera. The light blue ink showed IR-luminescence (cf. Fig. B. 53) whereas neither the darker blue (bottom) signature nor the traces from pens A and B showed any luminescence.

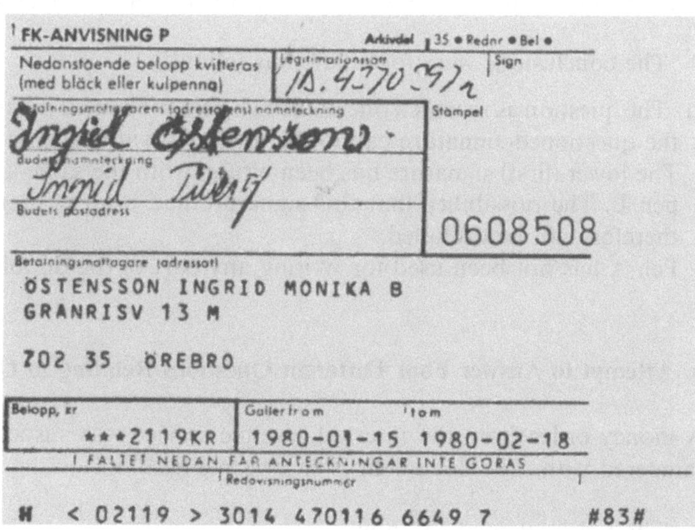

FK-ANVISNING P

Nedanstående belopp kvitteras (med bläck eller kulpenna)

Stämpel

0608508

Betalningsmottagare (adressat)

ÖSTENSSON INGRID MONIKA B

GRANRISV 13 M

702 35 ÖREBRO

Belopp, kr ***2119KR

Gäller fr o m 1980-01-15 t o m 1980-02-18

I FÄLTET NEDAN FÅR ANTECKNINGAR INTE GÖRAS

Redovisningsnummer

H < 02119 > 3014 470116 6649 7 #83#

Fig. B. 52. The postal money order of case a. The recipient's signature is double-written. The second name (I. T.) is the messenger's.

Fig. B. 53. Part of the money order mentioned in case a. The document was illuminated by short-wave IR light and observed by a TV-camera sensitive to long-wave IR. The picture on the TV-monitor is shown in the photograph

4. In order to compare the dark blue ink with those of pens A and B, small amounts of writing were removed from the document and analyzed by thin-layer chromatography. The result is shown in Fig. B. 54. It can be seen that the composition of the ink from the bottom signature and from pen B are indistinguishable whereas the ink from pen A has a different composition.

The conclusions were formulated as follows:

"1. The question as to which one of the two persons named by the police had written the questioned signature cannot be answered with certainty.
2. The lower (first) signature has been written with the same kind of ink as that in pen B. The possibility that this signature had been written with this pen can therefore not be excluded.
3. Pen A has not been used for writing any part of the double signature."

b. Attempt to Answer Four Different Questions Relating to One document

A money order from the internal revenue service was suspected of having been tampered with. The document is shown in Fig. B. 55.

The police wanted to know:

1. If one of three suspects had written the signature of the recipient (J. J.). Handwriting specimens from all of them were sent in.

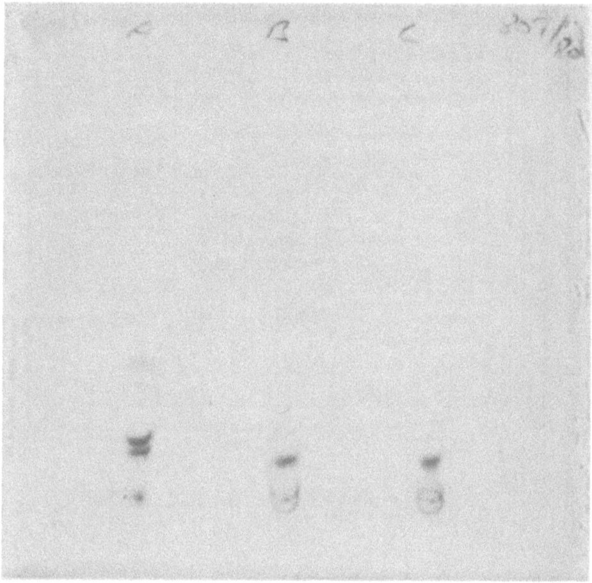

Fig. B. 54. Small parts of the questioned writing were cut from the document and placed in a small "pocket" of a narrow, tile-shaped filter paper strip. The paper was extracted with N, N-dimethylformamide into a small (1 ml) procelain crucible. The extract was evaporated to a volume of a few microliters and transferred to the thin-layer plate using a Microcap capillary (10 µl). Thin-layer plates of 10 × 10 cm were cut from Merck "DC-Plastikfolien Kieselgel 0,25 mm, ohne Fluoreszenzindikator" and the chromatogram was developed with ethyl acetate: 2− propanol: water = 6 : 3 : 2. When the solvent front had risen about 4 cm above the origin of the applied spots (15 minutes or so) the plate was removed and allowed to dry in a hood. *A* Ink from pen A, *B* Ink from pen B, *C* Ink from the "bottom" signature

2. If the signature J. J. and the signature of the messenger (U. K.) had been written with the same pen.
3. If any of the two signatures had been written with the felt-tip pen sent in.
4. If any changes had been made in the typed numbers (Kronor 9584) (096 hundra kronor) and (belopp < 09584).

The handwriting examination was only partly conclusive in that two of the suspects could be excluded whereas the identity of the handwriting of the third suspect with the questioned one could not be decisively established.

The colors of the ink in both signatures were indistinguishable by optical inspection using various color filters. Small parts of the writing from the two signatures were cut out and subjected to thin-layer chromatography. The inks gave identical chromatograms as shown in Fig. B. 56.

The felt-tip pen could not have been used for any of the questioned writing nor for changing any of the typed numbers.

The original typing on the document fluoresces in infrared light. The numbers "9", however do not fluoresce. Under the first two "9" 's, there were fluorescent traces of original zeros.

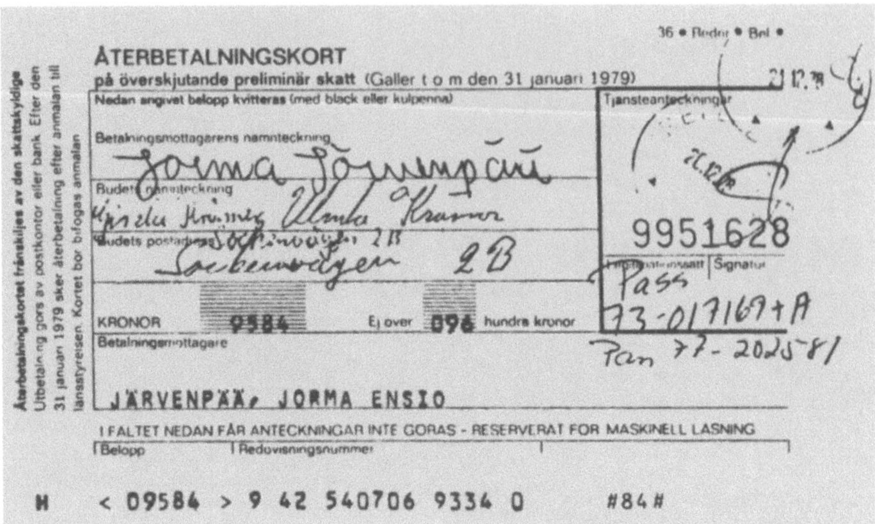
ÅTERBETALNINGSKORT

på överskjutande preliminär skatt (Galler t o m den 31 januari 1979)

Nedan angivet belopp kvitteras (med black eller kulpenna)

Betalningsmottagarens namnteckning

Budets namnteckning

Budets postadress

KRONOR 9584 Ej over 096 hundra kronor

Betalningsmottagare

JÄRVENPÄÄ, JORMA ENSIO

I FALTET NEDAN FÅR ANTECKNINGAR INTE GORAS - RESERVERAT FOR MASKINELL LASNING

Belopp | Redovisningsnummer

H < 09584 > 9 42 540706 9334 0 #84#

Tjansteanteckningar

9951628

Signatur

Fig. B. 55. The money order from the internal revenue service (restitution of tax). The questioned numbers appear on the shaded rectangles and at the left bottom line under "belopp" (case b)

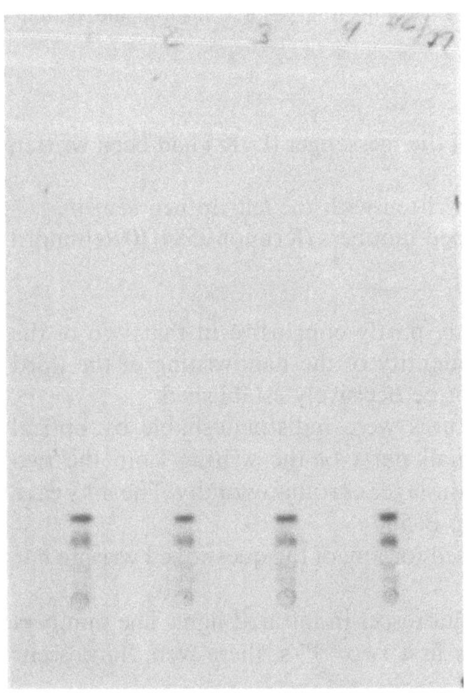

Fig. B. 56. Thin layer chromatograms of parts of both questioned signatures. *1* and *3*: from signature (J J), *2* and *4*: from signature (U K). The experimental conditions are the same as described in the legend of Fig. B. 54 (case b)

Conclusions:

1. It is not very probable that the suspects J. J. or U. K. have written the recipient's signature.
2. The question whether the suspect K. J. had written this signature cannot be answered.
3. The felt-tip pen has not been used for any writing on the document.
4. Both signatures (the recipient's and the messenger's) have been written with the same type of ink.
5. The original numbers "584", "006" and "00584" have been changed to "9584", "096", and "09584" respectively.

c. Three Kinds of Ballpoint Ink on One Check

In a case involving 24 different documents and an extensive handwriting examination the police wanted to know if the same ballpoint ink had been used for writing the whole text on a bank check.

Ocular inspection with the use of color filters indicated that three types of ballpoint ink had been used: One for the signature L. E. B. on the front, another for the remaining text on the front, and a third for the signature H. H. on the back of the check. Thin-layer chromatography confirmed this finding as shown in Fig. B. 57.

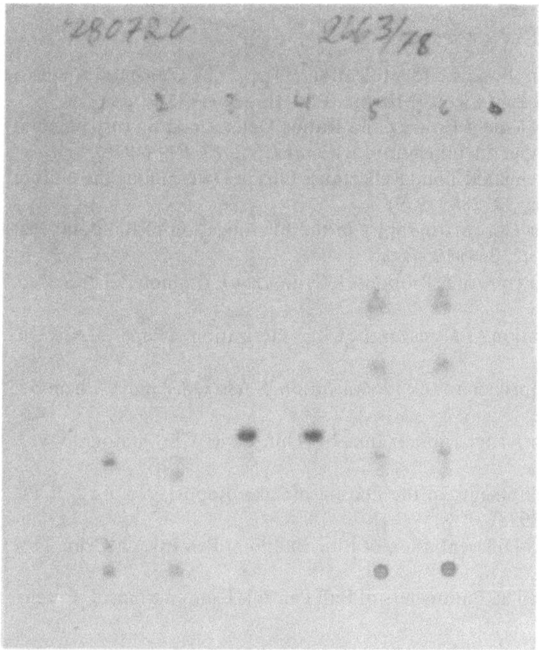

Fig. B. 57. Thin layer chromatograms of parts of the three questioned signatures and text. *1* from signature L E B (front), *2* from signature L E B (front), *3* from other text (front), *4* from other text (front), *5* from signature H H (back), *6* from signature H H (back), (case b)

7. General References and Review Articles

A) E. A. Apps: Ink Technology for Printers and Students. London: Leonard Hill 1963

B) J. P. Casey: Pulp and Paper. Chemistry and Chemical Technology, 2nd edit., vols. 1–3. New York and London: Interscience Publ. 1960–1961

C) R. H. Clapperton: Modern Papermaking, 3rd edit. Oxford: Blackwell 1952

D) D. A. Gould: Quick Dry Inks and Ballpoint Pens. Part I. American Ink Maker, *32* (2), pp. 34–37 and 65 (1954). Part II, ibid. *32* (3), 38–42 and 69 (1954)

E) J. Grant: A Laboratory Handbook of Pulp and Paper Manufacture, 2nd edit. London: Edward Arnold 1961

F) W. R. Harrison: Erasures. In: Methods of Forensic Science, vol. 3, Curry, A. S. (ed.). London – New York, Sydney: Interscience Publ. 1964 pp. 289–336

G) W. R. Harrison: Suspect Documents: Their Scientific Examination, 2nd edit. London: Sweet & Maxwell Ltd. 1966

H) E. Husemann: Cellulose. In: Ullmans Encyklopädie der Technischen Chemie, vol. 5, Foerst, W. (ed.). München – Berlin: Urban & Schwarzenberg 1954, pp. 156–165

I) R. Korn, F. Burgstaller (eds.): Handbuch der Werkstoffprüfung. Of Forensic Science, vol. 2, Lundquist, F. (ed.). London – New York: Interscience Publ. 1962, pp. 35–75

J) E. J. Labarre: Dictionary and Encyclopaedia of Paper and Papermaking. 2nd edit. Oxford: Swets 1952

K) H. Lobschat et al.: Schreib- und Zeichenmaterialien. In: Ullmans Encyklopädie der Technischen Chemie, vol. 15, Foerst W. (ed.). München – Berlin: Urban & Schwarzenberg 1964, pp. 343–357

L) E. Martin: An Analytical Technique for Determining the Origin of Unidentified Paper. In: Methods of Forensic Science, vol. 2, Lundquist, F. (ed.). London – New York: Interscience Publ. 1963, pp. 1–33

M) F. H. Norris: Paper and Papermaking. Oxford: Oxford University Press 1952

N) W. Schoch: Papier. In: Ullmans Encyklopädie der Technischen Chemie, vol. 13. Foerst, W. (ed.). München – Berlin: Urban & Schwarzenberg 1962, pp. 112–158

O) A. H. Witte: The Examination and Identification of Inks. In: Methods of Forensic Science, vol. 2, Lundquist, F. (ed.). London – New York: Interscience Publ. 1963, pp. 35–77

8. Specific References

1) J. A. W. Barnard, D. E. Polk, B. C. Giessen: Forensic Identification of Papers by Elemental Analysis Using Scanning Electron Microscopy. Scann. Electr. Microsc. Part II, 519 (1975)

2) D. B. Blanchard, S. H. Harrison: Trace Element Profiles and Ratios Determined by Instrumental Neutron Activation Analysis for Fine Paper Identification. J. Forens. Sci. *23*, 679 (1978)

3) A. Bober, R. C. Berkshire: The Use of Attenuated Total Reflectance (atr) in Determining the Nature of a Coating on a Paper. Appl. Spectrosc. *33*, 184 (1979)

4) Ch. Brown, P. L. Kirk: Horizontal Paper Chromatography in the Identification of Ballpoint Pen Inks. J. Crim. Law Criminol. Police Sci. *45*, 334 (1954)

5) C. L. Brown, P. L. Kirk: Identification of Typewriter Ribbons. J. Crim. Law Criminol. Police Sci. *46*, 882 (1956)

6) R. L. Brunelle et al.: Use of Neutron Activation Analysis for the Characterization of Paper. J. Ass. Off. Analyt. Chem. *54*, 920 (1971)

7) R. L. Brunelle, M. J. Pro: A Systematic Approach to Ink Identification. J. Ass. Off. Analyt. Chem. *55*, 823 (1972)

8) R. L. Brunelle et al.: Comparison of Typewriter Ribbon Inks by Thin-Layer Chromatography. J. Forens. Sci. *22*, 807 (1977)

9) R. L. Brunelle, A. A. Cantu: New Developments in the Dating of Inks. Report, No. 4 : 2, ICPC Interpol 4th Intern. Forens. Sci. Symp (1978)

10) D. A. Crown, J. V. P. Conway, P. L. Kirk: Differentiation of Blue Ballpoint Pen Inks. J. Crim. Law Criminol. Police Sci. *52*, 338 (1961)

11) D. A. Crown, R. L. Brunelle, A. A. Cantu: The Parameters of Ball Pen Ink Examinations. J. Forens. Sci. *21*, 917 (1976)

12) D. L. Duewer, B. R. Kowalski: Forensic Data Analysis by Pattern Recognition. Categorization of White Bond Papers by Elemental Composition. Anal. Chem. *47*, 526 (1975)
13) J. M. English: Dye Composition of Typewriter Inks as an Indication of Date of Typing. J. Police Sci. Admin. *6*, 74 (1978)
14) C. L. Garey, J. W. Swanson: Characterization of Filler and Coating Pigments in Paper by X-Ray Techniques. Techn. Assoc. Pulp Paper Ind. *43*, 813 (1960)
15) D. A. Gould: Quick Dry Inks and Ballpoint Pens. Parts I and II. American Ink Maker *32*, 34 65, *32*, 38 (1954)
16) J. Grant: The Rôle of Paper in Questioned Document Work. J. Forens. Sci. Soc. *13*, 91 (1973)
17) O. Hilton: Identification and Differentiation Between Colored Pencils. Forens. Sci. *6*, 221 (1975)
18) A. Kondo, S. Tanaka: Identification of Paper by X-Ray Fluorescence Analysis. Rep. Nation. Res. Inst. Police Sci. *31*, 299 (1978)
18 a) H. R. Lukens et al.: Forensic Neutron Activation Analysis of Paper. U. S. Atomic Energy Commiss. Rept. *1953*, 1
19) R. Mally: Der Kugelschreiber. Kriminalistik *10*, 55 (1956)
20) V. K. Mehrotra, S. K. Sidhana: Non-destructive Identification of Metallic Ions of Inks Used on Documents. Forens. Sci. *9*, 1 (1977)
21) G. R. Nakamura, S. C. Shimoda: Examination of Micro-Quantity of Ballpoint Inks from Documents by Thin-Layer Chromatography. J. Crim. Law. Criminol. Police Sci. *56*, 113 (1965)
22) R. A. Parham: On the Use of SEM/X-Ray Technology for Identification of Paper Components. Scann. Electr. Microsc. Part II, 511 (1975)
23) D. E. Polk, A. E. Attard, B. C. Giessen: Forensic Characterization of Papers. II. Determination of Batch Differences by Scanning Electron Microscopic Elemental Analysis of the Inorganic Components. J. Forens. Sci. *22*, 524 (1977)
24) H. L. Schlesinger, D. M. Settle: A Large-Scale Study of Paper by Neutron Activation Analysis. J. Forens. Sci. *16*, 309 (1971)
25) N. K. Sen, P. C. Ghosh: Dating Iron-Base Ink Writings on Documents. J. Forens. Sci. *16*, 511 (1971)
26) S. N. Tewari, N. Bhatt: Thin-Layer Chromatographic Analysis of Indian Fountain Pen Inks. Intern. Crim. Police Rev. *27*, 201 (1972)
27) J. Tholl (Cf. Ref. 7.) Police *1960*, 7
28) J. W. Thompson: The Identification of Inks by Electrophoresis. J. Forens. Sci. Soc. *7*, 199 (1967)
29) R. N. Totty et al.: X-Ray Microprobe Analysis of Coated Papers Used in Photocopying Processes. Forens. Sci. Intern. *13*, 31 (1979)
30) R. S. Verma, K. N. Prasad, G. J. Misra: Thin-Layer Chromatographic Analysis of Fibre-Tip Pen Inks. Forens. Sci. Intern. *13*, 65 (1979)
31) Unpublished experiments by J. Andrasko and R. Selse

9. Tables B. 43, 44

Table B. 43. The composition of typical wood (dry weight in per cent)

Cellulose	58–67
Lignin	20–34
Carbohydrates	11–25
Pectin	~2
Resins	a few percent
Waxes	
Tannins	
Coloring matters	

Table B. 44. The chromatographic analysis of ball-point pen inks by various methods, arranged in the chronological order of their introduction

Year	Method	Comments	Author(s)
1954	Horizontal paper chromatography	58 ballpoint inks tested	Brown and Kirk[a]
1955	Descending paper chromatography	25 ballpoint inks, 60 fountain pen inks and 32 stamp pad inks tested	Coldwell[b]
1958	Ascending paper chromatography	scraped-off ink placed into "pockets", formed in the chromatography paper	Martin[c]
1960	Ascending paper chromatography; sample placed into "pockets" cut into the paper	61 ballpoint inks tested and assigned to 17 groups	Bosch and Mueller[d]
1960	Thin-layer chromatography		Tholl[e]
1961	Circular paper chromatography	Extraction with pyridine; otherwise similar technique as Martin. The paper lists 25 references	Mathyer[f]
1965	Micro thin-layer chromatography; sample removed by hypodermic needle	Table showing some chemical properties of 55 different ink dye-stuffs	Nakamura and Shimoda[g]
1969	Micro thin-layer chromatography; sample cut out and placed between two glass plates	No advantage over previous procedure	Smalldon[h]
1970	Thin-layer chromatography; new type of micropunching device for ~1 μg of ink	No dramatic improvement	Naumann and Rentz[i]

[a] cf. Ref. [4]
[b] B. B. Coldwell: The Comparison of Inks and Writings by Paper Chromatography. Analyst *80*, 68 (1955)
[c] E. Martin: Identification of Ballpoint Pen Ink. Criminol. Rev. *12*, 18 (1958)
[d] K. Bosch, B. Mueller: Die Möglichkeiten einer Differenzierung von Kugelschreiberschriften durch Papierchromatographie. Arch. Kriminol. *126*, 75 (1960)
[e] cf. Ref. [27]
[f] J. Mathyer: Comparison of Inks by Paper Chromatography, Practical Methods. Intern. Crim. Police Rev. *148*, 130 (1961)
[g] cf. Ref. [21]
[h] K. W. Smalldon: The Comparison of Ink Dyestuffs Using Minimal Quantities of Writing. J. Forens. Sci. Soc. *9*, 151 (1969)
[i] E. Naumann, A. Rentz: Neue Mikro-Stanzmethode zur Entnahme eingetrockneter Schriften für die Dünnschicht-Chromatographie. Arch. Kriminol. *145*, 149 (1970)

Table B. 44 (continued)

1972	Spectrophotometry of the spots on thin-layer chromatography plates. Gas chromatography for solvents used	Review of various analytical methods with 13 references	Brunelle and Pro[j]
1974	Removal of part of the thin-layer for achieving better separations		Kuranz[k]
1977	High pressure liquid chromatography after punching out ink and extracting the punches with 2% formamide in methanol	Components with UV absorption only can be "seen", semiquantitative analysis possible; vehicle components can be analyzed	Colwell and Karger[l]

[j] cf. Ref. [7]

[k] R. L. Kuranz: Technique for the Separation of Ink Dyestuffs with Similar R_f Values. J. Forens. Sci. *19*, 852 (1974)

[l] L. F. Colwell Jr., B. L. Karger: Ballpoint Pen Ink Examination by High Pressure Liquid Chromatography. J. Ass. Off. Analyt. Chem. *60*, 613 (1977)

XI. Toxic Substances in Food

1. Introduction

Our laboratory receives about 30 cases per year with a request for an analysis for poisoned food or drinks. The forensic significance of such cases and collection of such food or drink samples will be discussed in Sect. C.I.12 but it can be stated already that positive findings are rare.

A search in the literature for suitable screening methods for the detection of deliberately added toxic substances gave negative results. Textbooks and abstract journals were consulted, and a computer search for the literature carried out. Countless articles on the determination of biocides, of naturally occurring toxins, of trace metals, bacteriostatic componds and artificial taste, smell and coloring agents were found, but nothing on the detection of deliberately added poisons. The one exception known to the authors in an article by J. Jackson [1] where he condenses the problem of analysis of "poisons" in various materials into 10 well written pages. In the authors' laboratory, a general scheme for the extraction of drugs from food was worked out [2]. The following paragraphs summarize some of the findings from these two sources.

2. The Analysis

The task for the analyst is greatly facilitated if there is some indication as to what kind of toxic substance might be present. Otherwise the work can be extensive. To quote Jackson: "With no information to guide him, the analyst must eliminate the various types of drugs and poisons step by step (and this is a slow process!)".

Fluids can be roughly divided in two groups according to their contents of solid material. Low solid concentrations are found in water, tea, coffee, wine, beer, liquor and soft drinks. High concentrations are met with in milk, syrups, soups and liqueurs. Solids may be subdivided into simple systems (salt, sugar, powders) and complex systems (vegetables, bread, butter, flour, pies, meats, creams etc.).

Extraction procedures are relatively straightforward in watery liquids but get more complex with increasing contents of solid matter.

For analytical purposes, toxic agents may be roughly grouped into volatile, extractable and mineral substances. In practice, volatile agents are not often met with, except for alcohols. A small representative sample of the food to be analyzed can be placed in a tight container with a punchable rubber casket and heated to about 80° C. About 1 ml of the headspace is injected into a gas chromatograph as a check for volatiles (a Porapac column is suitable).

Another part of the sample is prepared for solvent extraction. One scheme for such an extraction is rendered in Fig. B. 58. The scheme is based on practical experiments at our laboratory [2] but can be modified in various ways. An alternative method, using Soxhlet extraction, did not give satisfactory results. The solid residues from the centrifugation are kept for subsequent analyses for minerals. The water layers are evaporated *in vacuo* to a small volume and tested for inorganic or non-extractable compounds such as sugars. Finally, they are evaporated to dryness.

The "dry" residues from centrifugation or evaporated liquids respectively, are now ready for screening analyses for metallic toxic agents. They are dried at 120 °C and subsequently heated in a muffle furnace at about 400 °C until the residues are charred. Portions of these materials are now tested for toxic metals by emission spectroscopy, atomic absorption spectrometry, or electron induced X-ray emission.

The analysis of the acid and alkaline extracts proceeds, in principle, along the same lines as the corresponding analysis in the field of forensic toxicology. The main

Isolation of extractable drugs and poisons

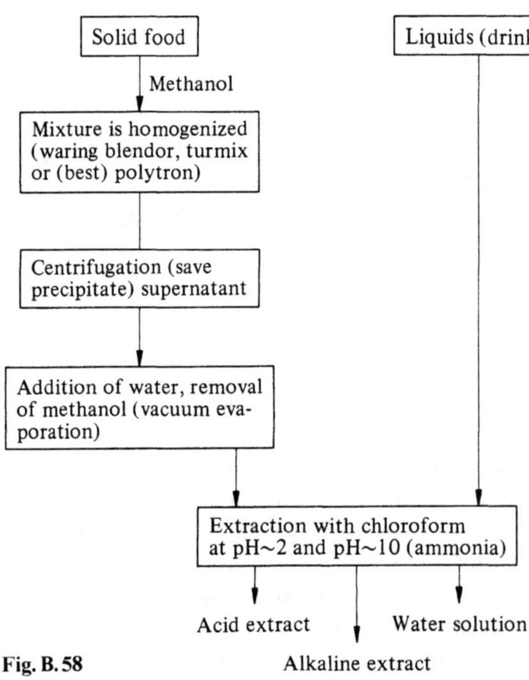

Fig. B. 58

difference is the usually much higher concentration of any drugs present in the extracts. Neutral and acidic drugs, together with the ion pairs of some classes of drugs, are found in the acid extract. Alkaline compounds are present in the alkaline extract. A few amphoteric compounds are only extracted within a relatively narrow pH-range; morphine e. g. requires a pH from about 8–9 (bicarbonate buffer) for efficient extraction.

The extracts can be screened by thin-layer chromatography, gas chromatography or liquid chromatography. In rare cases, ultraviolet spectrophotometry can also be used. Details about these screening methods are found in the literature on forensic toxicology.

3. Practical Examples

a. The Baby Food Terrorist

In 1974 an unknown person placed glass splinters and different toxic materials inte baby food jars in various supermarkets in Middle Sweden. One typical analysis will be related here.

A woman bought a jar of baby food on May 30, 1974. The contents looked unusual and she tasted some of the food. She had to vomit right away. On closer inspection she found a piece of paper, wrapped in cellophane with the message: "We can poison baby food. The children die". In addition, the police found a copper wire and a piece of glass. These were returned to the jar.

Some of the food was investigated by a pharmacologist who fed it to rats; the rats, however, showed no ill effects.

The jar was finally sent to the National Laboratory of Forensic Science for analysis. The label said: "Banana-apple dessert". A jar of the same make and type was purchased as a reference sample.

Occular inspection showed that the color was darker in the suspect jar than in the reference. Dark particles (metal shavings?) were observed in addition to a piece of glass (20 × 20 × 5 mm) and a copper wire (60 × 0.7 mm). The pH was 3.75 for both samples.

The dark particles were analyzed by emission spectroscopy and shown to consist of lead in a concentration of about 0.25 per cent (w/w) in the contents of the jar. An aliquot of the contents was extracted with deionized water, and the concentration of water-soluble lead determined as 0.01 per cent (w/w). Ten grams of the food were acidified and extracted with ether. The ether layer was evaporated and yielded a white, solid residue of 0.13 g. The residue was dissolved and the ultraviolet spectrum recorded at pH 10 (0.5 N NH_4OH) and at pH 2 (HCl). Spectra typical of barbituric acid derivatives were obtained. Using the absorption difference at 238 nm, the concentration of barbiturate was determined in six separate analyses and found to be between 30 and 75 mg for the whole jar. The large interval is due to the inhomogeneous distribution of the barbiturate in the food. The barbiturate was identified as amobarbital by gas chromatography (GC), followed by infrared spectrophotometry of accumulated GC-fractions.

b. Attempted Poisoning of His Wife

A woman arrived at a police station with a mug containing a mixture of fruit cocktail, ice cream and whipped cream. This dessert had been prepared by her husband but she had the impression that it had a strange taste. The police had previously to deal with a fight between the couple and therefore took the matter seriously.

The suspected food arrived at the laboratory 5 days after it had been brought to the police. It was to be analyzed for possible toxic substances. The dessert was blended and extracted with ether. The extract was analyzed by thin-layer and gas chromotography, ultraviolet spectrophotometry and mass spectrometry. The following compounds were identified with certainty:

dextropropoxiphene,
fenazone,
butenemal.

In addition, the presence of acetylsalicylic acid and diethylaminoethyl-phenothiazine-carboxyl hydrochloride was made probable.

These five compounds all form part of the commercial specialty Doleron Natt® (Astra) which is classed as a dangerous habit forming drug in Sweden.

4. References

1) J. Jackson: Isolation Procedures. In: Handbook of Analytical Toxicology. Sunshine, I. (ed.). Cleveland (Ohio): The Chemical Rubber Co. 1969, p. 391
2) A. C. Maehly, S. Fors.: Unpublished results 1976

XII. Restoration of Erased Markings

1. Introduction

Many kinds of machinery and other equipment are marked with various types of identification symbols (numbers, letters, trademarks). When such objects are stolen, the thief often tries to remove these markings to make identification impossible or at least very difficult. Examples of such markings are serial numbers on motor vehicles, firearms, TV-sets, jewelry and art objects with identification marks.

Markings can be made by writing (ink, pencil, fluorescent paste), by punching holes, by engraving, welding or punching indentations into a metal (or cardboard) surface – either from the front or the back of the material. Not all of these markings can be effectively removed. In many cases they will leave clear marks. On the other hand, the erased markings cannot always be easily restored.

Two groups of markings stand out both for their relative ease of obliteration and the good chances for restorations. These are the markings punched into a metal surface, leaving an indented mark; and markings produced by welding symbols onto a metal surface.

Wilson [2] has written a very clear review on the restoration of erased markings.

2. Indented Markings

When striking the surface of a metal object with hardened steel punches, strong forces are transferred to the punched region. These forces change the crystalline microstructure of the metal to a considerable depth. When the metal is ground down to the level of the lowest indentation, the thief is usually satisfied and stops further attack least he damages the object unnecessarily. Sometimes he paints the surface to hide the attack.

The forensic chemist can easily remove the paint. He then has to polish the surface in question. The regions of changed microstructure tend to be more prone to react with acids or alkali than the unchanged metal. A suitable chemical treatment can then dissolve these regions preferentially and reveal the original number. Another technique for restorations of indented marks has recently been worked out by Young [3] who treats the object with ultrasonically induced cavitation in water.

In relatively rare cases, the thief, after removal of all, or parts, of the original marking punches his own number or letters into its place. It is then advisable to make a cast of the region in question before proceeding with the restoration work. Punches have often some defects and can be "connected" with the punched objects.

In our laboratory the following procedure is used for restoration of markings in metal objects:

1. The surface is polished with a grinding and polishing machine (dentist's drill).
2. The polished region is wiped clean and washed with gasoline or ethanol and dried.
3. So-called "tops", or matches rolled in a piece of cotton are soaked with the etching liquid (cf. below) and placed on the metal object.
4. The surface is frequently inspected: the markings may appear after intervals from around a minute upt to several days. When this happens, the cotton is removed and the surface immediately washed with water and dried.
5. The region is now photographed. The best result is usually obtained after a slight application of a transparent oil to the metal surface.

Etching liquids used:

A. For steel: 40% conc. HCl, 30% H_2O, 25% ethanol (95%) and 5 g of $CuCl_2$.
B. For aluminum and other light metal alloys: 20% NaOH.

Chisum [1] has recommended an alternative procedure for aluminum using first a rinse with 1 N NaOH and then a treatment with 0.1 M $HgCl_2$ in 0.1 N HCl.

3. Welded-on Markings

The technique for welding markings onto the surface of a metal object is used for larger pieces of equipment. The heat produced causes local changes in the surface's microstructure under the welding material.

Wilson [2] describes at least one case where chemical etching successfully restored the outlines of the marking welded on a large number of skips used for rubbish collection.

4. Practical Examples

a. Number on Radio Transmitter Restored

A mobile radio transmitter had been stolen from an official government car. It was found in a suspect's home, but the stamped-in indented number had been filed down. After polishing and treatment with acid (cf. above) the number 00043 became clearly visible as shown in Fig. B. 59. As a result of this finding the suspect was convicted of theft and for illegal possession of a radio transmitter.

b. Identification of a Punch

This case does not illustrate a chemical process but exemplifies the possibility of identifying a punch used by the criminal.

In a suburb of Stockholm a large number of motorcycles and motorbikes were stolen within a short period. In all cases, the original numbers of the motors and the frames had been neatly filed off and replaced by other numbers. A search of a suspect's house revealed a set of number punches.

The laboratory received the set of punches and a motrocycle engine.

The questions put were:

1) Has the new number been punched with any of the punches in the set?
2) Can the original number be restored?

The motor had been filed down in the region where the original number was placed. According to the manufacturer, this number always has 7 digits. The "new number" however read "576314". The same number was punched using the "suspect set" into another metal surface and compared with the "new number". No differences in size, shape and details could be discovered on comparing all six numbers under a microscope. Figure B. 60 shows the number 4 stamped into the motors surface (left) and the reference surface (right). Several small defects are clearly seen in both cases.

The region with the "new number" was now polished down and etched with acid. However, the original number could not be restored.

Fig. B. 59. Restored number on the radio transmitter mentioned in example a

Fig. B. 60. The punch (no. 4) and a cast of the punched surface shown side by side in the comparison microscope. This figure refers to case b

The conclusion read:

Based on the investigation described above one can state that the number "576314" has with virtual certainty been punched with the suspect's number punches sent to the laboratory.

5. References

1) W. J. Chisum: A Catalytic Process for Restoration of Serial Numbers on Aluminum. J. Forens. Sci. Soc. *6*, 89 (1966)
2) P. B. Wilson: The Restoration of Erased Serial Identification Marks. Police J. *52*, 233 (1979)
3) S. G. Young: The Restoration of Obliterated Stamped Serial Numbers by Ultrasonically Induced Cavitation in Water. J. Forens. Sci. *19*, 820 (1974)

XIII. Miscellaneous

1. Introduction

Chapters B.I–B.XII have dealt with chemical forensic analyses of major importance and high frequency. Depending on the resources, the organization, and the traditions of the laboratories a number of other analyses can occur. The examples discussed below reflect accordingly those topics which the Swedish National Laboratory of Forensic Science handles and of which the authors have a certain amount of experience. There are, of course, many other types of forensic analyses carried out in other places and countries which cannot be covered here.

2. Dye Spots on Money Bills

In 1977, a new type of crime appeared in the major cities of Sweden. Many banks had introduced a new costumer service: the night service box. Customers could deposit money and checks at any time of the day and night in these boxes by inserting marked bags through an entrance slit into the box, usually built into the outer wall of the bank building. The system has been in use for many years in other countries.

Criminals soon learned to push dynamite or other explosives into the box and to detonate the charge by fuse, timer, or radio signal. If the charge was well designed, the whole box would be thrown out of the wall and the depositor's bags could be recovered more or less intact by the perpetrators.

The banks then distributed small bags filled with a dye solution to their customers, who were asked to insert these into their deposit bags. This countermeasure had the desired effect since the explosion efficiently spread the dye over the money bills which were apparently irreversibly marked. The laboratory assisted the banks in finding dyes which could only be removed with great difficulty. Two types of dyes (and later a third) were finally selected for this purpose. During the following years, several hundred cases involving stained bills were handled in the laboratory.

The analysis consisted in comparing physical and chemical properties of the dye stains with those of known reference dye samples by observing their fluorescence and by the use of thin-layer chromatography and ultraviolet spectrophotometry of suitable extracts. Later, some gangs developed a counter-counter measure by devising a chemical method for removing the stains. These "money bill laundries" would charge their customers the major part of their loot.

As a counter-counter-counter measure the laboratory developed analytical methods for tracing residues of the extraction liquids used at the "laundries" and for demonstrating faintly colored bands at the lower edges of many of the laundred bills. At the same time, new dyes are being investigated.

In an actual case the following bills, found in a suspect's home were sent to the laboratory:

A. Twenty-two 100 kr bills and two 50 kr bills.
B. Ten 100 kr bills.
C. Three 100 kr, one 50 kr, two 10 kr, and two 5 kr bills.

The police wanted to know:

1. If the coloring of and the damage to these bills were due to blasting of a night service box, and
2. If the bills had been "laundered".

The bills were examined in daylight and in UV-light. Colored parts were extracted and the extracts analyzed by ultraviolet spectrophotometry and thin-layer chromatography. Some of the bills which had a bleached appearance were extracted with water and tested for chloride ions (the "laundry" procedure includes treatment with hydrochloric acid).

The following results were obtained:

1. Ten of the 100 kr bills and both 50 kr bills in material A looked "bleached". Water extracts from these bills contained chloride ions which indicated that they possibly had been "laundered".
2. The other 12 bills in material A contained a brown variety of "bank box dye".
3. All bills in materials B and C contained a red variety of "bank box dye".
4. One bill in material B and one bill in material C showed transfers of printed text, caused by the pressure of the explosion (cf. Fig. B. 61).
5. Several bills from materials A–C showed damages from an explosion.

3. Herbicide Analysis

a. The Analysis

The analysis of samples from major pollution accidents in Sweden is at present carried out by a number of different laboratories. So far, the Forensic Science

a

Fig. B. 61 a, b. A transfer of print from one bill to another, caused by the pressure of the explosion. **a** The number 100 from the front side of the upper bill is partially transferred to the region where the number 100 is printed on the reverse side of the bill. Note also the typical fraying of the edges, due to the explosion

Laboratory is only rarely involved in these. On the other hand, it receives annually about 100 cases of minor pollution, usually caused by disputes between neighbors over trees or bushes that are felt to be in each other's way. In such cases, herbicides of all kinds are used to destroy the disputed plants.

The analytical difficulties for the forensic chemist are considerable in such cases since, as a rule, he has to deal with very small samples and has usually no clue to the kind of herbicide having been used.

The Swedish National Laboratory of Forensic Science is now in the process of systematizing this type of microanalysis by using a combination of extraction methods, screening analyses, and capillary gas chromatography.

b. Practical Examples

Herbicide spread by neighbor.

Two neighbors had been quarrelling for some time. One of them appeared at the local police station with small samples of red beets, carrots, radishes and sallad – claiming that the neighbors had attempted to poison these vegetables.

Fig. B. 61 b. The transferred details are outlined by small dots

At the laboratory each sample of vegetable and the soil adhering to each sample were extracted with distilled water. The extracts were filtered, evaporated to a small volume and decolorized with active charcoal. The solutions were analyzed for chlorate ions with o-toluidene and potassium iodide/starch, as well as for chloride ions with silver nitrate. All tests were positive.

The conclusion was worded thus:

"Chlorate was shown to be present in all vegetable samples and in the soil adhering to each sample."

Suspected poisoning of bushes by a herbicide.

A garden owner observed that some of his bushes had wilted and that there was a strange smell in their vicinity. He brought some wilted leaves (in a plastic bag) and some soil (in a glass jar) to the police who sent the samples to the laboratory.

The soil sample was extracted with water and 2-propanol/hexane. The water extract had neutral pH and gave negative tests for ions not normally found in the soil. The organic extract was evaporated to dryness, yielding a residue of about 5 mg/g of soil. The residue was redissolved and fractionated by column chromatography. Three fractions contained significant amounts of residues for further analysis. The analysis was carried out by gas chromatography/mass spectrometry and by nuclear magnetic resonance (NMR). The following compounds were identified in this way (their common names used in Sweden in brackets):

1) 2-(2,4-dichlorophenoxy) propionic acid (Diklorprop)
2) (4-chloro-2-methylphenoxy)acetic acid (MCPA) and
3) 4-hydroxy-3,5-diiodobenzonitrile (Ioxynil)

Semiquantitative analysis by NMR gave the following concentrations:

1) 69%, 2) 20% and 3) 11%

In Sweden there is a product on the market (Certrol Trippel) which contains the compounds listed above with contents of 60% for (1), 25% for (2), and 15% for (3).

The leaves were extracted in the same way as the soil. Due to the much lower amounts of herbicide obtained, the organic extract was analyzed only by thin-layer chromatography, parallel with standards of compounds 1, 2 and 3. The resulting chromatogram of the extract gave 3 spots with R_F values corresponding to the standards.

The conclusions were formulated thus:

"The soil- and leaf-samples contained the herbicides Diklorprop, MCPA, and Ioxynil. The commercial brand Certrol Trippel contains the potassium salts of these compounds in roughly the same proportions as those found in the crime samples".

4. Analysis of Coins

There are several categories of counterfeit coins. The simplest, and crudest, type of coin is a metal disk designed to be a cheap replacement for genuine coins for use in vending machines and the like. Sometimes, a thread is attached to the coin so that it can be used several times.

More sophisticated work goes into the change of a genuine, but common type of coin into a rare and therefore valuable specimen. One example is the addition of a dot to a 1936 Canadian coin. The investigation of such coins by electron optical means has been described by Rodgers et al. [8]. Another example is the "production" of a 1910-year Swedish 5 öre coin by removing suitable parts of the number 1916 or 1919*). This also can be discovered by optical observations which always show marks left by the perpetrator (cf. Fig. B. 62 a, b).

A more difficult task is the comparative investigation of total falsifications of valuable coins. As a rule, these are produced by casting and show small details with lesser precision than on genuine coins. However, the definite proof is best achieved by a comparative quantitative elemental analysis. The older methods required the sacrifice of a detectable part of both the questioned and a genuine coin. The method of choice now is an analysis using the scanning electron microscope (SEM) with an energy-dispersive (or wavelength-dispersive) X-ray analysis attachment. The comparative analysis of $ 20 gold coins by SEM-X-ray analysis has been described some

*) Five-öre coins from 1910 have a collectors value of up to 800 dollars, depending on their condition

a

b

Fig. B. 62. a A Swedish 5-öre coin from the year 1919, on which the lower part of the second 9 has been filed off, **b** Detail of the same coin as in Fig. B. 62 a. The file marks are clearly visible

time ago [2]. In the authors' laboratory several dozen suspect coins have been analyzed in this way for the benefit of coin specialists (cf. Fig.B. 63 a, b).

The most difficult investigation concerns counterfeit coins which have been struck. Recently, a number of Swedish and Norwegian gold coins with nominal values of 10 and 20 crowns respectively were received at the laboratory. Only minute differences between these coins and their genuine counterparts could be detected by electron microscopy. Also, the gold content was equal to or even higher than that of the genuine coins. However, measurements by X-ray diffraction showed that the coins must have been struck much more recently than the date of minting indicated [3].

Fig. B. 63 a. An electron-optical comparison of a genuine U. S. $ 20 gold coin (on the left) and two counterfeit coins (in the middle and on the right). *Top row:* The letter "O" from the text "In God we trust". The details on the genuine (struck) coin are neatly formed. On the counterfeit (cast) coins the details are blurred. *Bottom row:* Parts of the U. S. coat of arms from the same coins. Note the crude relief on the counterfeit coins

5. Lipstick Analysis

On rare occasions, lipstick smudges can have forensic significance. The smudges can be seen on an assailants clothes, or they may be left on cigarette butts, tissue paper etc.

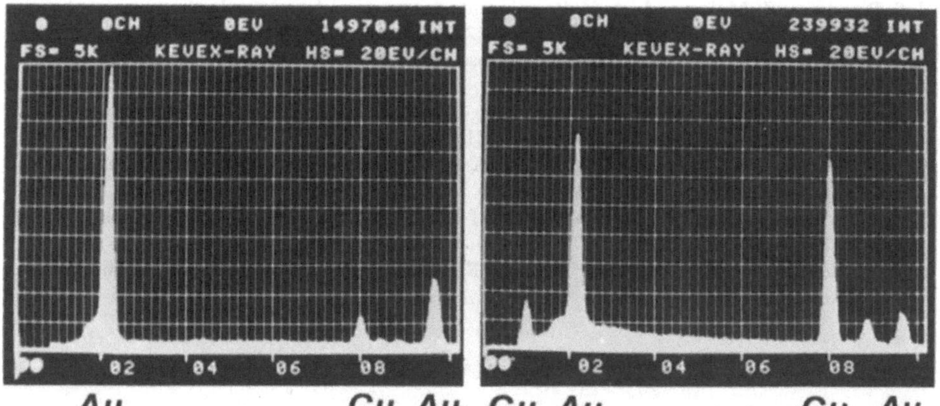

Fig. B. 63 b. The energy-dispersive X-ray analysis of the surface of a genuine coin (left) and a counterfeit coin (right). The trace of copper in the genuine coin is normal. The large signal at ~2.1 keV is the M_α-line for Au, the smaller one at ~9.7 keV the L_α-line for Au

Several methods for the analysis of lipsticks have been described in the literature, but most of these are concerned with quality control etc. in the laboratories of the cosmetic industry. Consequently, relatively large sample sizes are available.

The forensic chemist, however, has usually to deal with trace amounts of lipstick material (<1 mg). The authors have found nine publications concerning forensic lipstick analysis. In 1959, Pougheon and Moloster [7] used paper chromatography (PC). Two years later, Lucas and Eijgelaar [5] combined PC with UV-spectrophotometry. In 1972, Barker and Clarke [4] introduced the analysis of lipsticks by thin-layer chromatography (TLC). This technique was also described in two technical reports from the Metropolitan Police Forensic Science Laboratory [11, 12]. Another report from that laboratory deals with the gas chromatography of the waxes used in lipsticks [10].

Neutron activation analysis was applied to forensic lipstick investigation by Merli et al. [6]. Recently, Revland and Trinler [9] used high pressure liquid chromatography (HPLC) for the same purpose. Finally, Andrasko [1] published a paper on lipstick analysis where a combination of several methods is used: color, TLC, HPLC and energy-dispersive X-ray analysis.

6. References

1. J. Andrasko: Forensic Analysis of Lipsticks. Forens. Sci. Intern., in press (1981)
2. J. Andrasko, S. Bendtz, A. C. Maehly: Practical Experiences with Scanning Electron Microscopy in a Forensic Science Laboratory. Scanning Electron Microscopy II, 879 (1979)
3. J. Andrasko et al.: Analysis of Counterfeit Gold Coins by Scanning Electron Microscopy and X-ray Diffraction. Scanning Electron Microscopy I, 455, 410 (1979)
4. A. M. L. Barker, P. D. B. Clarke: Examination of Small Quantities of Lipsticks. J. Forens. Sci. Soc. *12*, 449 (1972)
5. D. M. Lucas, G. Eijgelaar: An Evaluation of a Technique for the Examination of Lipstick Stains. J. Forens. Sci. *6*, 354 (1961)
6. S. Merli et al.: Weitere Beiträge zur Identifizierung von Mikrospuren kosmetischen Materials mittels Neutronenaktivierungsanalyse. Arch. Kriminol. *161*, 137 (1978)
7. S. Pougheon, Z. Moloster: L'étude des colorants des rouges à lèvres par la chromatographie sur papier. Rev. Intern. Criminol. Police Techn. *12*, 298 (1958)
8. P. G. Rodgers et al.: Scanning Electron Microscopy of Forged Dots on 1963 Canadian Coins. J. Forens. Sci. *16*, 92 (1971)
9. D. J. Revland, W. A. Trinler: A Comparison of Lipstick Smears by High Performance Liquid Chromatography. J. Forens. Sci. Soc. *20*, 111 (1980)
10. L. W. Russel: Gas Chromatography of Waxes. Part 3. The Analysis of the Wax Components of Lipsticks. Technical Report No. 34, March 1976, Metropolitan Police Laboratory, London, England
11. A. E. Welch: Analysis of Lipsticks. Technical Report No. 13, June 1975, Metropolitan Police Laboratory, London, England.
12. A. E. Welch: Analysis of Lipsticks. Technical Report No. 39, October 1976. Metropolitan Police Laboratory, London, England

C. Auxiliary Activities

I. The Forensic Significance of Physical Evidence and its Collection

1. Introduction

The criminalistics branch of Forensic Science is concerned with the search of the crime scene and the belongings of suspects followed by the investigation and comparative analysis of the collected evidence material. The former activity may be designated as criminalistic field work, the latter as criminalistic laboratory work.

This book is mainly concerned with the laboratory aspects of criminalistics, with special emphasis on forensic chemistry. The authors do not claim to be experts on forensic field work and will therefore not go into this matter in any detail. There is a good deal of literature on this subject available, and the major publications have already been listed in Chap. A. II of this volume.

Nevertheless, it seems useful to discuss the forensic significance of the type of evidence discussed in some depth in part B so as to establish the necessary link between field work and laboratory work. Some remarks on the collection, preservation, and transportation of evidential material are also included. These remarks reflect mainly the wishes and opinions of the forensic chemist rather than the aims and tactics of the police officer engaged in field work.

The extent to which field work and laboratory work should be linked to or separated from each other has been a matter of much discussion, and various judiciary systems have arrived at very different solutions. In some police districts or regions, the same personnel handles the search of crime scenes and most of the laboratory work – whereas others have established a clear division of tasks. Both systems have their advantages and disadvantages, and some of these are discussed in

Chaps. A. II and C. V of this book.

Since the authors have only first-hand experience of the Swedish system it has to serve as the basis for the following discussion.

2. Narcotics And Dangerous Drugs

In principle, there are two types of offences against drug laws,

(i) the illegal manufacture of drugs, and
(ii) the smuggling, transportation, distribution and sale of illicit drugs.

The role of the forensic laboratory is decisive in these cases since the sentence depends wholly on the qualitative and quantitative analyses performed. Other evidence may have some influence on the deliberations of the court but this is the exception rather than the rule. Recognizing the great responsibility for a just decision, the forensic chemist must be absolutely convinced of the correctness of his analysis. In some countries, e. g. in Sweden, the sentences for serious drug offences can be of the same order of magnitude as those for murder (10 years imprisonment for serious drug offences vs, 10–15 years for murder), so that the evidence presented in court must have the same high standard.

In the case of smuggling, transportation, distribution and sale of illicit drugs, the police (or customs) officer has usually no other task than searching (in the clothes of a suspect, his living quarters, his car, boat or plane) packing and sending suspect items to a suitable laboratory. When raiding a suspect clandestine laboratory, the situation is very different. In this case the collection of evidence has to be done in collaboration with a forensic chemist.

Let us consider the former type of evidence handling first. The search for illegal drugs is usually based on information from police intelligence units, and only rarely from the general public. The search of a person is usually carried out "manually", and by police officers of the same sex as the suspect. Often, specially trained searching dogs are used to locate drugs. Among favorite hiding places can be mentioned: underwear, shoes and physiological cavities (mouth, anus, vagina). Drugs are also swallowed (usually in contraceptive sheeths), and in such cases only time can prove this. Samples may consist of tablets, pills, capsules, pieces of impregnated paper (LSD), samples of herbal material (cannabis, opium), powders, solutions, and syringes. During search, and especially when handling syringes precautions against hepatitis infection must be taken (disposable rubber or plastic gloves, special packaging for syringe needles, clear warning labels on such items for the analyst etc.).

The officer should enclose a short description of each sample, pack it, and label it clearly so as to avoid any chance of exchange or confusion. If the samples are sent by messenger or by mail (quite frequent in Sweden), tablets, and especially capsules, must be protected from being crushed. Placing them in a hard box of metal, wood or strong cardboard is usually sufficient. Even a matchbox is better than a paper or plastic bag.

Numerous makes of test-kits for the field-testing of illegal drugs are available on the market. Many police departments use these. In the authors' opinion these kits are of questionable or at least very limited value. They may be helpful at ports of entry or airports in order to enable the customs officer to make a fast decision about holding a busy passenger or not. In the majority of other cases the kits do not help much and may be harmful. The reasons are simple. For once, the police (or customs) officers are not chemists and cannot be expected to

(i) carry out the tests correctly each and every time, and
(ii) have the necessary critical scientific mind for interpretation of the results.

It is further known from laboratory testing of these kits, that false positive and false negative results can rarely be excluded. Thus, two unwanted situations can arise:

1) The test is positive. The suspect is arrested. The sample is sent for chemical analysis, which may give a negative result.
2) The test is negative. The suspect is released. The officer is not sure if this is a "false negative". He sends the sample for analysis, which may give a positive result.

A laboratory analysis can thus not be avoided anyway.

The raiding of a clandestine laboratory is a complicated matter. Suspected laboratories are usually located with the help of police intelligence work, often on an international scope. Once the suspicion has been supported by surveillance or reliable witness hearings a tactical plan has to be set up for raiding the place. No guidelines can be given in this treatise of how this is done. A major difficulty in this connection is timing. If the raid comes too early, only preparations for the synthesis may have been made (leading to moderate indictments). If it comes too late, some drugs may have left the laboratory and reached the illegal market. This can obviously lead to accusations of the police for knowingly allowing such activities to take place.

Once the time of the raid has been decided upon, the forensic chemist should be alerted. He should be allowed to enter the scene as soon as the laboratory is firmly in the hands of the police. The chemist's first role is that of a safety officer: He has to inhibit smoking in the whole area. Only he can advise how and in which order apparatus and equipment can be turned off and removed. Once the operation of the laboratory is safely halted, the chemist has to decide on which evidence to secure. There have been cases where ill-advised police personnel have transported the entire equipment plus all of the stored chemicals to the forensic science laboratory. This is not only an unnecessary and labor-consuming process, but also often a dangerous one.

In general, the final and intermediate products of the laboratory should be secured. Chemicals in labeled containers may be checked by securing a small sample. Apparatus and equipment can generally be left at the scene for later investigation. The chemist will, of course, make detailed notes of the equipment used since this gives very good indications of the synthetic methods used. Photographs (caution for flashes in explosive atmospheres!) are very helpful for reconstructing the chemical scenario.

The sampling, labeling and packaging of chemicals requires obviously care, accuracy and calm thinking.

A sufficient number of clean, tight and labeled containers must be taken from the laboratory to ensure proper and safe packaging and transport. It is advisable that two individuals check the identity of each item, e. g. the chemist and a police officer. After proper identification and labeling, the chemist should supervise the safe packing and transportaction of the evidence. If possible he should accompany the transport to the forensic science laboratory.

The Drug Enforcement Administration has some time ago published two articles on clandestine laboratories [6, 9], the International Criminal Police Organization (Interpol) has issued a booklet on this theme. Later, at the 8th International Meeting of the International Association of Forensic Sciences in 1978, a symposium on clandestine laboratories was held.

3. Explosives

The main purpose for the analysis of explosives is to establish the type and, if possible, the make of the explosive. In the Unites States there exists an agreement with the explosive manufacturing industry to add markers to their products. This facilitates considerably the identification of explosive residues at the scene of a crime.

Non-detonated devices as well as exploded remnants can give clues as to the materials used, such as batteries, wires, detonator caps etc. Sometimes toolmarks may be secured and used for comparison either with the suspected tools or with marks on other devices.

In securing the evidence in cases of criminal offenses involving explosives one is faced with two very different situations depending on whether the investigation takes place before or after the actual explosion.

In the former case one usually has to deal with intact devices and relatively large samples of explosive chemicals, in the latter case only traces of the explosive and, at best, remnants of the ignition devices may be found.

The disarming and further handling of "live" explosive devices is a highly specialized art and cannot be treated in this context. After disarming, the devices are sometimes sent to the laboratory for closer inspection. The questions posed are usually

(i) how is the device constructed?,
(ii) why did it not detonate?, and
(iii) what damage would have resulted had the explosion taken place?

The laboratory must assume that the device is safe to handle when it reaches the premises. Some sophisticated bomb arrangements are built to be ignited by X-rays, and many police forces have special bomb shelters where X-rays can be safely directed onto the bomb. This should be done before an X-ray examination is carried out at the laboratory – the logical first step in bomb examinations.

Residues from an exploded device can consist of ignition apparatus details and of traces of the explosive itself. The former should be carefully collected after suitable photography has been carried out, so that the exact location of each detail is apparent. Traces of explosives can sometimes be seen by the naked eye, but frequently this is not possible, and objects close to the center of the explosion must be collected for microscopic examination and/or treatment with solvents at the laboratory.

4. Polymers

The forensic significance of plastics has not been fully recognized in the forensic literature. Yet the increasing use of polymers in all sorts of industrial products cannot fail to influence the establishment of forensic evidence.

At present, there seem to be two regions of importance for "polymer evidence" so to speak. It is well-known that many polymers are highly inflammable and that they also can develop toxic and/or corrosive gases and fumes on burning. The identification of polymer residues will help in assessing these properties. Usually, relatively large amounts are available at the scene of a fire and their collection, dispatch and analysis present no great problem.

The other type of polymer evidence is part of the general problem of connecting items from a scene of a crime or a victim with items found on or in the possession of a suspect. The size of the samples can be very small indeed. No special rules for securing the evidence can be set up. As usual, larger pieces of polymers (visible to the naked eye) are picked up with the fingers or tweezers and placed in paper envelopes or clean glass jars (not in plastic containers, please!). Microtraces are removed by vacuum sweeping or the material carrying them is brought to the laboratory.

Just as in the case of paint chips, the use of adhesive tape is not recommended since the effect of solvents used for removing the polymers can be disastrous for the plastic itself.

5. Fibers, with Remarks on Dust

The forensic significance of fiber investigations can hardly be overestimated. Textile fibers are found everywhere in industrialized countries: in clothing, furniture, automobile seats, rugs, curtains, towels etc. Consequently, fibers are brought to and removed from most scenes of crime. Also, the offender can hardly protect himself from leaving contact traces even if he were conscious of the situation.

When considering the probative value of fiber transfers there are a number of problems to be faced. Thus, the frequency of a particular type of fiber or fabric is of great importance. Some types of clothing apparel are manufactured by the millions annually, others may be almost unique such as home-knitted clothes. Another consideration is the time of transfer: more recently transferred fibers would be expected to adhere to the surface of the recipient material whereas "old" fibers may be more deeply imbedded. The duration of an acquired transfer will vary greatly with the type of textile material (both of "donor" and of "receiver") and the kind of activity of the wearer.

The possibility exists that fibers transferred from a garment to another are such that have been picked up from other sources rather than from the fibers from a particular garment involved in a specific crime. This in turn means that *single* fibers may not be regarded as reliable evidence for a transfer, and that the evidential value of such fibers increases with the number or amount of these found.

The value of fiber transfer as proof for a crime increases dramatically in cases of "cross contamination". This phenomenon may occur especially in cases of violent crimes where the perpetrator and the victim can "exchange" fibers on, and from, their respective clothing.

Pounds and Smalldon have investigated some of the aspects of fiber transfer mentioned above in a series of publications in 1975 [20]. These authors give first a short review of earlier work on the forensic significance of fiber transfer. They then investigated the effects of pressure, type of recipient garment, number of contacts, and fiber length. They further studied the persistence of transferred fibers. One of their diagrams is rendered in Fig. C. 1. It shows that, under the conditions of their experiments, only about 1/3 of the fibers transferred remain on the recipient garment after 1 hour. After about 8 hours this amount has fallen to 10% or less, but approaches a few percent even after several days of wear. Results similar to those in Fig. C. 1 were obtained with other types of fibers and recipient fabric.

The choice of optimal methods for the collection of fiber evidence is a matter of considerable complexity and is being intensively discussed within the forensic community so to speak. It is generally agreed that there are, at present, six major methods for fiber collection available, i. e.

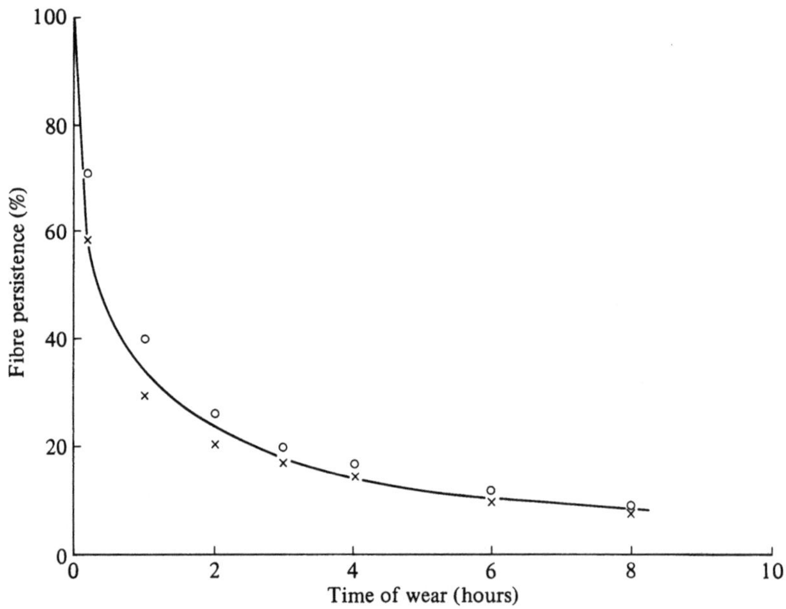

Fig. C. 1. Persistence of wool and acrylic fibers on a woolen jacket according to Pounds and Smalldon (20). The fabric to be transferred was labeled with fluorescein and then pinned onto a 14 cm × 14 cm polystyrene block and at least 100 fibers were transferred by rubbing against the jacket. The number of fibers was then counted under UV-light. The jacket was then worn and inspected after various intervals. The figure shows the percentage of fibers remaining on the jacket

1. The plucking of individual fibers under a magnifying glass with the help of tweezers or simply by hand.
2. Intensive shaking of the garment in question over a very clean paper.
3. Collection of fibers by suction through an appropriate filter, usually with the help of a vacuum cleaner.
4. Taping the garment, using more or less "sticky" plastic tape.
5. Brushing with an electrically charged item (nylon brush, shellac rod etc.).
6. Combing the garment, especially in cases were the cloth has a "wolly" surface, or in cases of furs. The comb is first fitted with a thin layer of pure cotton, on which fibers will fasten.

Prior to any investigation, wet clothes should be allowed to dry, if necessary in a drying cabinet. During this process, and indeed all the way from the collection of the clothes to the end of the analysis, great care must be taken to treat "crime scene material" separately from "reference material". In the laboratory the materials from these two sources should never enter the same room.

Another question of major importance is whether the search should be carried out on the spot or in the laboratory. In general, one can recommend that all wearing apparel is sent to the laboratory in separate, clean paper bags (plastic bags are less appropriate due to charging effects) whereas larger objects such as automobile seats, furniture, large rugs etc. are searched on the scene.

The first action should always be a search under a magnifying glass, using the fingers or tweezers to secure single fibers. Their exact location should be noted. The surface to be searched must be protected from contamination from the searcher (put paper sheets on rugs, work with bare arms etc.).

Suction through filters using a vaccum cleaner is an established method for collecting fibers as well as other small particles from garments as well as furniture, rugs, car seats etc. Each area to be vacuumed is collected on a fresh filter (tight nylon cloth or filter paper). The filters are put in small envelopes and properly marked. Suction is especially useful for collection of relatively large amounts of materials and for reaching small cavities such as pockets in garments, the inside of shoes or the corners and cracks in automobiles and certain types of furniture.

The vacuum technique seemed to have appeared in fiction in the 1880's (Conan Doyle) and in 1907 or 1908 (R. Austin Freeman) but the first practical application and technical description of this method is due to Türkel [24]. Around the same time, the Swedish criminalist Söderman and co-workers [10, 23] described a special vacuum cleaner for criminalistic purposes in detail.

The tape method was introduced by the Swiss criminalist Frei-Sulzer in 1951 [5]. Some time later, his countryman Martin developed the method further and published several papers on it [e.g. 13–16]. The method spread to southern Germany and is used in some other forensic science laboratories (e.g. in Poland) [18] but is not generally adopted. This is a pity since – while not a general panacea to all collection problems – it is a valuable complement to the other techniques discussed in this chapter.

The following points have to be considered when using the tape method:

1. Do not overload the tape; tapes looking like the one in Fig. C.2 cannot be analyzed.

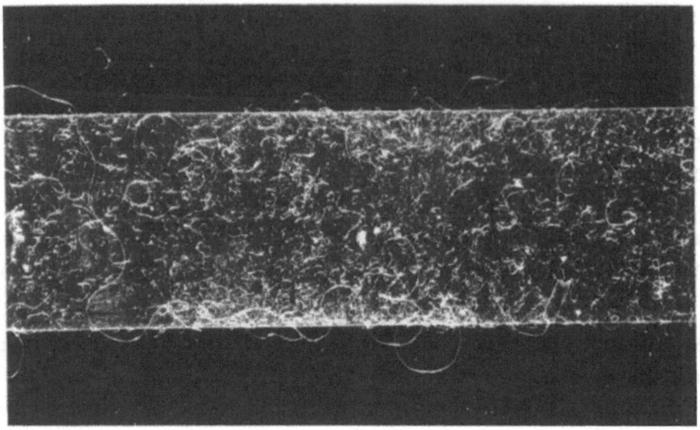

Fig. C.2. Example of an overloaded tape. Evidence collected in this way is worthless and cannot be investigated

2. Take reference tapings of the surface in question, but away from suspected areas of contact.
3. Avoid "doubling-up" of the tape. The method of protecting the evidence by folding the tape on itself, recommended by some authors, makes it difficult to recover single fibers for further analysis. Also, trapped air bubbles may disturb the investigation.
4. Alternative methods are placing the tape on microscope slides (specially cleaned slides commercially available) as recommended in Frei-Sulzer's original paper, or placing longer strips on clear plastic sheets.
5. Avoid the tape method for securing paint flakes – or polymer particles – so far a suitable solvent for removing such objects unchanged from the tape has not been found for all cases.
6. Fibers, welding sparks, and other inorganic traces can generally be recovered by using suitable solvents. These solvents must be found for the particular type of tape used. Usually xylene, petroleum ether, or butanol will work.
7. The type of tape used does not seem to be as critical as some authors, notably Martin, believe to be the case. The tape should be colorless, 1/2 to 1 inch, wide, and of intermediate "stickiness". The adhesive should be completely soluble in a solvent that does not attack the fiber material.
8. The tape method must be evaluated in the forensic laboratory and can then be used in the laboratory in certain cases. The officer in the field should not proceed using this technique before he is adequately briefed at the laboratory. Indiscriminate and uncritical use of tape can jeopardize the whole evidence.

The use of static electricity for collecting fibers and similar evidence was apparently first suggested by Heindl [7]. A similar method was described by Pounds [20] who used a nylon brush.

Pounds [20] compares the merits of the various collection methods discussed above. He determined the collection efficiency as well as the ratio of the amount of contact fibers to background fibers. The efficiency of collection, one hour after contact was 18% (!) for vacuuming, 28% for shaking, 55% for nylon brushing, 65% for low adhesive tape collection, and 85% for collection by high adhesive tape (fiber length 0–5mm). On the other hand, high adhesive tape collected 7 times more background fibers than low adhesive tape.

The choice of method still depends on the particular case. Table C.1 (p. 273) summarizes the properties of the five main collection methods.

Before closing this chapter we should like to recommend reading a very thorough treatise by the famous French criminalist Locard which he wrote in 1933 in the Archiv für Kriminologie. On 53 pages he discusses the types of "dust" and its forensic significance [12]. The treatise is a fine example of the one-man research typical of that area of criminalistics.

6. Paints, Varnishes And Lacquers

Paints (including varnishes and lacquers in this context) can provide valuable clues in criminal investigations. Two frequent examples are collisions of automobiles, either with each other or with obstacles, and traces of paint on the tools and clothing of burglars. Some less common cases have been described in Sect. B.III.5.

Collisions between two automobiles can lead to a mutual exchange of lacquer, a parallel to the mutual exchange of textile fibers between an assailant and his victim (cf. Sect. C.I.5). Such cases are especially favorable from the point of view of obtaining proof.

Smudges of paint on burglary tools have previously been disregarded since the amount of paint was usually too small for a meaningful analysis. However, modern methods of analysis (cf. Chap. B.V) can now be successfully used in many instances of small traces of paint found on tools. Their analysis constitutes a valuable complement to the comparison of tool marks. It is of course necessary to acquaint the investigators of the scenes of crime with this possibility so as to ensure the availability of all reference samples that had concievably been in contact with the tool.

Traces of paints are either found as "chips" of various sizes or as smears. The collection of paint evidence depends greatly on whether samples can be physically removed from their surroundings or not.

Chips and other loose fragments of paint are best collected with the help of tweezers or needles. Special attention has to be paid to the risk of losing the fragments through their tendency to jump around during this process. In the laboratory, vacuum cleaning is an efficient collecting technique. Adhesive tape should be avoided since it is often impossible to remove traces of paint from the tape without affecting the paint vehicle.

Smears of paint should not be treated at the scene of a crime or accident. Instead, the carrier of the smear should be secured and sent to the laboratory in tight containers. This is simple in the case of tools carrying paint smears. The relatively common situation of paint transfers between automobiles presents a more difficult problem. However, it is often possible to carefully cut away a piece of the automobile's original lacquer carrying the smear using a sharp scalpel. Placing a clean paper under the working area facilitates the search for pieces of paint that jumped off during this operation. This can be done in the following way: A clean, white sheet of paper (A 4 or similar size) is folded and taped onto the panel of the car below the paint smear (cf. Fig. C. 3). The paint flakes will collect on the paper, which then is folded and placed in an envelope.

When collecting reference samples, these should be taken close to the suspected area of impact in order to obtain a sample representing the same paint composition.

Obviously, great care must be taken in packing and labeling crime samples and reference samples appropriately, avoiding any cross-contamination.

7. Glass

In principle, all kinds of glass objects can become important in criminal investigations. In practice, however, certain kinds of glass occur much more frequently than others. The most common type of glass is window glass, either in shop windows or windows of homes. Next frequent are glass specimens from automobiles (windshield, side windows, headlights, stoplights). Bottle – and "glass-

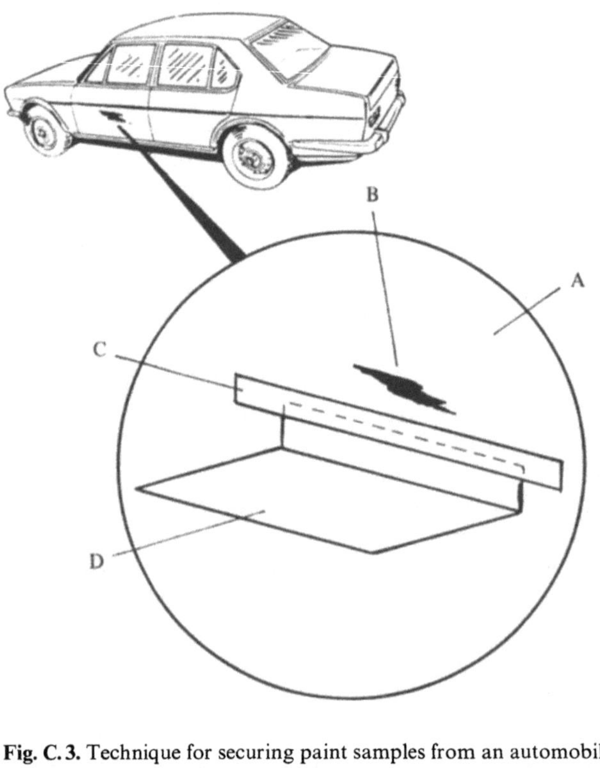

Fig. C. 3. Technique for securing paint samples from an automobile; acc. to Kjell Carlsson, Stockholm police, technical squad. *A* panel (in this case the door), *B* paint smear, *C* adhesive tape, *D* clean paper (A4 or similar size). After collection of the sample, the paper can be folded into a small package and placed into an envelope

glass" can sometimes be encountered, whereas optical glass, from spectacles or the like is rarely investigated.

A very common problem is the establishment of a link between a broken window and small pieces of glass found on the clothes, the shoes, hats or – less frequently – the hair of a suspect. The crime committed is usually burglary, but braking and entering in more serious crimes such as rape, armed robbery, manslaughter or murder do occur.

In the comparative analysis of window glass the forensic scientist meets with a number of difficulties. The purpose of the investigation is usually to prove or disprove the "identity" between glass splinters on or in the clothing, the automobile, or the dwelling of a suspect on the one hand and a specimen of the broken window on the other hand.

The first difficulty is the widespread occurrence of small glass particles in clothing and other locations (cars, apartments etc.).

A group of British workers [19] have examined 100 pairs of trousers and jackets at a large dry cleaning store for paint flakes and glass. A total of 551 fragments of glass larger than 0.1 mm were found in 63 of these suits. Two suits contained 46% of the glass fragments which leaves about 300 fragments in the other 61 suits.

The second difficulty is the introduction of large-scale industrial glass manufacturing processes where tens of thousands of bottles may be produced from the same melts. Melts of from a few tons up to 1000 tons are now common. In one of the modern plate glass processes, a large melt is continously feeding a production line where the flat glass floats on the surface of molten tin where it is gradually cooled.

A third difficulty lies in the variation in the physical and chemical properties over one and the same window pane. These variations may be very small but as the investigative methods become more and more precise and sensitive, the variations may become significant. The problem has been investigated e. g. by Dabbs and Pearson [3] and by Andrasko and Maehly [1].

Glass from automobiles can provide important clues in hit- and run accidents involving other vehicles or living beings (humans, dogs). Pieces of windshield glass can also, under certain circumstances, provide a clue to who has driven a certain automobile.

Occasionally, glass objects such as bottles, are used as weapons, and splinters on the victim can be compared with the remains of the glass weapon. Houshold glasses and spectacles can break, and pieces found on the clothing (or the car) of a subject.

The collection of glass splinters usually presents no special problems. Pieces visible to the naked eye are picked up by tweezers and placed in very well sealed and strong envelopes or better small plastic jars and clearly labelled. Vacuum cleaning gives very good yields of all microtraces but also picks up a large amount of "old" material of little interest to the case. The use of adhesive tape is an alternative but it is difficult to pick up glass samples from the inside of pockets where glass is likely to be found; however, the removal of glass from the tape presents no problem. Modern analytical methods can be used for the analysis of very small samples of glass, down to a few hundreds of a millimeter. Therefore, such particles need not be neglected.

A special problem is presented where multi-layered glass is encountered, as in safety-glass, automobile windows etc. The composition of the glass may vary a great deal from layer to layer.

8. Soil

The comparative investigation of traces of soil can, under favorable circumstances, yield corroborative evidence of crimes. If tracks from car tires or footwear are found on the scene of a crime there is a good chance for soil comparison analysis. However, there are less obvious cases where traces of soil also can provide clues. This can apply to situations where a criminal got soil on his clothes, tools, or upper parts of his footwear without leaving obvious inprints on the scene. In a way, such cases can be somewhat simpler to handle since there is a lesser risk of multiple soil layers left on these objects.

The incriminating samples of soil should not be removed from the object on which they are found. These objects should be handled as little as possible and be sent to the laboratory with great care. Transport can be a tricky matter since dry soil will easily fall out of tools and tires or even shoes or boots. The items are packed tightly, and if dry, possibly together with a humid but clean piece of cloth or paper. The laboratory investigator has to remove the soil samples carefully from the surface of the object in order to detect layering or the presence of other evidence such as plant material.

The task of collecting reference soil samples is relatively complicated. The value of soil evidence depends namely to a great extent on the correct collection of reference samples. The character and composition of soil can vary enormously between locations only a few meters apart, but the soil can also be relatively uniform for hundreds of meters or miles. Sometimes this is quite obvious (e. g. flower bed vs. lawn, open field vs. woods) but the mere appearance of the soil can also be misleading. It is therefore advisable to take soil samples at various distances from the suspected region as well as – in certain cases – from the neighborhood of a suspect's house. Another problem is the depth, or thickness of the samples. Generally, only the top layer (a few centimeters) should be removed.

Modern methods of soil analysis are aimed at traces such as chemical pollution substances or bacterial flora. Great care is therefore indicated to preserve these traces by using tight containers, and to avoid further pollution from automobile exhaust, cigarette smoke, dirty hands or bacteria.

The size of the sample depends on the methods used by the laboratory; incriminating soil deposits are worth taking as long as they are easily visible to the naked eye. Reference samples of 10–100 g are recommended in most cases.

Special cases of "soil" are traces of putty, cement, plaster, mortar etc. which may be found on burglars' shoes after entering buildings.

9. Firearm Discharge Residues

There are two kinds of firearm discharge residues of interest to the forensic chemist, i. e. the residues on the target, and the residues on the shooter's hands and clothes (cf. Chap. B. VIII). The former are of importance in deciding if a wound is caused by the entrance of a bullet and for the determination of the shooting distance if this distance does not exceed a meter or two. The latter residues, if found and properly identified, prove that the shooter had fired a gun. Positive findings on the skin prove a recent shooting, whereas the finding of residues on the clothes (especially sleeves) shows that a gun had been fired but does not allow to draw any firm conclusions as to the time of firing.

The significance of establishing the entrance hole of a bullet is obvious. The determination of the shooting distance can be essential for deciding if a shooting was suicidal or criminal. If the distance exceeds 30 or 40 cm suicide can be excluded; at smaller distances no conclusion can be reached based on this determination alone. At times, the knowledge of the shooting distance may also be needed for the reconstruction of a shooting incident.

If gunshot residues can be demonstrated on the hands of a person found shot with a gun close by, a suicide is virtually established. However, if the search for such particles is negative no absolute conclusion can be drawn since it is possible that the presence of the particles has been missed, either due to a removal of the residues prior to taping or due to a failure to find the particles in the SEM.

The collection of target residues does usually not pose any problems except in special situations. If the victim had been brought to a hospital it can be difficult to secure his/her clothing. Whenever possible, the police should rapidly inform the hospital team of the importance of preserving this evidence. Another difficulty arises if shots are fired at a naked part of the body. Medical examiners usually think of excising the region of the entrance hole for later examination at the forensic laboratory but they often cut out too small a region so that reference shots for shooting distance determination are hard to carry out.

The need for collecting residues on inanimate targets is rare. In such a case it is advisable to call the forensic chemist to the scene.

In Chap. B. VIII, the analysis of firearm discharge residues from the hand of a shooter has been described in some detail. At present, there are but two procedures to be recommended. In one, the hands are swabbed with cotton pieces impregnated with nitric acid. The swabs are then subjected to neutron activation analysis. In the other method, the web region of the hands are picked up on adhesive tape, and the resulting pieces of tape sent to the laboratory for investigation in a scanning electron microscope (SEM).

The collection of swabs is described in the literature cited in Chap. B. VIII. The authors have first-hand experience only with the taping procedure which will be described here.

The residue deposited on the shooter's hand consists of small irregular spheres made up of melted oxides and other compounds of metals which formed the detonator of the ammunition. These particles have diameters of up to 20 or 30 μm and are not visible at moderate magnification in an optical microscope. Moreover, these particles are very widely scattered over the surface of the hand. It is important to concentrate the particles to a relatively small part of the tape used. Ordinary office tape of 1/2 in. width is quite satisfactory if it is very "sticky". A piece of such tape about 6–10 cm length is used for taping. Only the middle region (about 1–2 cm) should be used to pick up any particles present. The same part of the tape is used all over the region of the hand indicated in Fig. B. 40. Another piece of tape is used for the same operation on the other hand of the shooter. Both hands should be taped in all cases irrespective of any knowledge of the right- or left-handedness of the shooter. The police officer should not start this work before having washed and scrubbed his own hands in the presence of a witness lest the counsel for defense should claim that the residue particles found on the tape stem from the investigator.

The tapes are examined in the SEM sticky side up. This means that this side must be protected from contamination and yet must not be covered. To this end, small plastic containers are used in our country: the tape is stretched over the open container sticky side down and the ends are pressed to the containers outside (cf. Fig. C. 4). The lock is pressed on, the container is properly labelled and sent to the laboratory in such way that the plastic container cannot be squashed.

The search of the tape for gunshot residue particles is rather tedious. If we imagine the tape magnified 200 times it will have a width of about 2.6 m and the particles would have a diameter of about 5 mm. On a length of a few meters 2 or 3 particles may be found. The search can last for hours (though our laboratory limits it to 2 hours) and the whole work: collection, packing, sending, receiving, coating with carbon, search, analysis, report writing will be in vain if the collection procedure has been faulty.

10. Fire Accelerants

While arson can, in principle, be committed by manipulating electric equipment or by tampering with mechanical devices, the most frequent means of setting fires is still the use of accelerants, either solid or liquid. If the presence of such accelerants, foreign to the normal environment of the original building can be proven, the probability of arson is very high. The real problem is linking a suspect to such a fire: this is difficult and usually unsuccessful. In rare cases, a connection between the accelerant discovered and a specific chemical agent in the possession of a suspect can be established.

As we have seen above (Chap. B. IX), there are mainly two chemical causes of fires: spontaneous ignition and chemical arson. The deliberate use of fire accelerants constitutes a more straightforward problem for the forensic chemist who is faced essentially with the isolation and the analysis of the accelerant, be it solid or liquid. Complications can occur when an ignition device has to be investigated. The investigation of spontaneous ignition, on the other hand, often poses considerable problems. The materials involved are often not known in detail, and reconstructions

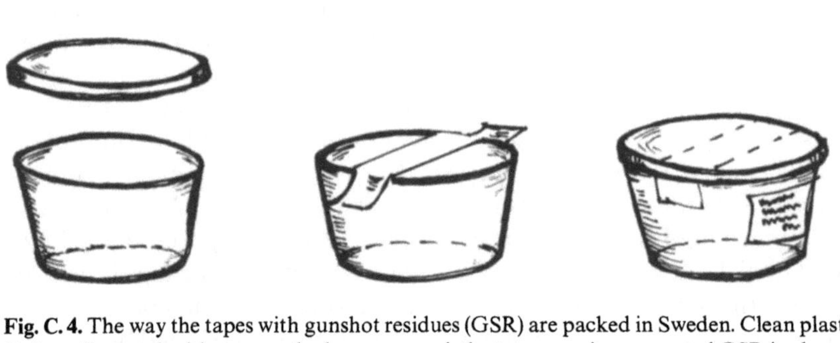

Fig. C. 4. The way the tapes with gunshot residues (GSR) are packed in Sweden. Clean plastic containers (4–8 cm diameter) with snap-on locks are opened, the type carrying suspected GSR is placed on the open container, sticky side *down*, in such a way that it is stretched tight, pressing down both ends, the lid is pressed on, and the container is carefully labeled

are difficult to carry out due to the many parameters involved such as amounts involved, insulation of the surroundings, access to oxygen, presence of solvents etc. In addition, the question whether the spontaneous ignition was accidental or staged cannot always be answered with certainty.

The search of a scene after a fire is a complex operation, described in many textbooks on crime scene investigations. We are here concerned only with proper sampling and packaging techniques from the point of view of the forensic chemist. Some of these have been discussed in Chap. B. IX.

The most efficient sampling technique is possibly the collection of vapors at the scene of the fire by aspiration through a suitably prepared absorbant. Such procedures are currently studied [2] but are not ready for general use by the police. Remains the securing of fire debris likely to have retained residues of accelerants.

The search should start near the origin of the fire (if this can be located). Liquid accelerants tend to run and seep downwards and may be found under rugs, floorboards, inside floor insulation, or even in the soil under a building. Porous material is much more likely to absorb liquid accelerants than material with hard surfaces. Wood (even heavily charred), rugs, soil, furniture upholstery and the like are suitable materials. The investigator's nose can be of help in locating remnants of accelerators, especically hydrocarbons. But the strong smells at a scene of a fire can mask this smell in many cases. There are commercially available testing instruments (cf. Chap. B. IX) which can be of use under favorable circumstances.

The material suspected of containing liquid chemicals have now to be packed, labelled and sent off. The most common packing materials are:

(i) glass jars with tight locks (rubber gasket or ground glass),
(ii) paint cans of metal, and
(iii) plastic bags.

Unfortunately, each of these containers has its drawbacks. Paint cans are usually not completely tight, and especially low boiling compounds can slowly escape. Fire debris are often wet, either from the fire hoses or from exposure to rain or snow. Experience has shown that most metal cans are quite rapidly corroded from the inside, possibly aided by acids formed in the debris. Glass jars, while tight, are hard to transport and relatively expensive (the local police wants to have them back); also, they have to be thoroughly cleaned after use. Plastic bags must be used with great care. Most polymers are unsuitable as packing material since they absorb hydrocarbons. A systematic search for a suitable bag material has been carried out by English forensic scientists. The found that nylon is sufficiently inert to hydrocarbons to be used in bags for packing of fire debris. Care must be taken to protect the bags from sharp details such as nails, wooden splinters, glass etc. Nylon bags are now widely used by the crime scene investigators in England. They have the additional advantage that headspace samples are easily taken by piercing the bag with the needle of a gas-tight syringe. The small hole can be closed with a bit of office tape.

Preliminary tests in the authors' laboratory [8] have confirmed the suitability of nylon bags for debris-containing hydrocarbons. However, the tests have shown that small amounts of alcohol seem to disappear quite rapidly. We do not know yet if this is due to leakage through the plastic wall, to adsorption or absorption. This is quite a problem in the Scandinavian countries where alcohols are frequently used as fire accelerators (cf. Table B. 41 (p. 214)). A possible solution may be the use of two-layer materials such as bags made of nylon-lined aluminum. Some of the advantages and drawbacks of the 3 types of containers discussed above are listed in Table C. 2 (p. 273).

Whatever the containers chosen, they should be rapidly sent to the laboratory where analysis should be started forthwith. If this is impossible, the containers must be kept in a cool place, if possible in a refrigerator.

The investigation of cases of spontaneous ignition can be a rather complicated problem. As a rule the suspected material has burnt up. The task of the police is then to establish the probable components of the material in question. All persons connected with the premises where the fire broke out should be questioned. The ignition components can sometimes be pinpointed, but often they are only approximately known. Examples of the former kind are paint or varnish from spray boxes, hay or sawdust, insufficiently washed garments etc. Examples of the latter kind are rags with turpentine or paint mixed with any kind of other refuse, spilled liquids on paper in waste paper baskets etc.

If the ingredients are well known, samples of equivalent materials should be sent to the laboratory. There, tests under a variety of conditions may prove or disprove a theory of spontaneous combustion. With unknown or insufficiently identified ingredients, experiments are often useless and the conclusions must be based on certain assumptions. These rarely lead to a clear and unequivocal type of evidence.

11. Questioned Documents

Questioned document investigations are mainly carried out along two quite different lines. One approach deals with the comparison of a handwritten text with specimens obtained from its suspected originator. These specimens can be obtained for this specific purpose, or can be collected among texts written earlier by the suspect. This process, comparative handwriting investigation, is necessarily subjective in nature, even when two or more forensic experts investigate the same case. The other task of a document examiner is the investigation of a document by technical, "objective" means such as photographic processes, chemical analysis, electrostatic phenomena, comparison macroscopy, electron microscopy etc.

The courts have often great difficulty in interpreting the opinion of the handwriting expert who cannot and should not use sweeping statements in his extremely difficult work. It is much easier to use the evidential value of a technical document examination. In the former case, the forensic expertise needs often be corroborated by other evidence, whereas the examination in the latter case can constitute definite proof in its own right.

A difficult question is the priority between fingerprint examinations and document investigations. Ever since the introduction of this technique by two Swedish scientists [16] the question of what to do first became pertinent: the ninhydrin treatment or the handwriting examination.

The treatment of documents with ninhydrin is a fast and reliable way of developing latent prints [22], and can rapidly lead to an identification. However, most solvents used for this purpose affect the majority of ball-point inks. This in turn may obliterate the finer details of the writing such as pen liftings, direction of the strokes and writing pressure. In addition, the ninhydrin treatment may disturb the chemical analysis of inks.

The alternative consists of sending the untreated document, properly protected by plastic envelopes, to the questioned document examiner. After completed examination, the document is then photographed and returned. The handwriting examination can be carried out with much better results – but valuable time in print identification may be lost.

There is a number of compromises available for solving this problem. One of them is the choice of a solvent, such as petroleum ether, or trichlorotrifluoroethane (Frigen 113) with a b.p. of 48° C. Such solvents do not affect the majority of ball point inks or affect them only slightly. Another technique consists of placing small pieces of paper over the most important parts of the handwriting before treating the document with ninhydrin. The document can also be photographed before applying ninhydrin, but a photograph (or worse, a photostat) does not allow an optimum investigation of handwriting specimens.

Documents to be examined must be properly packaged. Stiff cardboard should be used to prevent physical damage during mailing or other transportation. Documents needing fingerprint examinations should be placed in plastic "pockets" and properly labelled as such. Attention should be paid to avoiding artifacts due to impressions from stamping machines, staples or writing. In most climates, papers may need to be sealed into plastic. In Sweden, where much of the evidence is sent by mail, documents are usually sent by registered mail and insured.

12. Food Or Drinks Suspected of Containing Toxic Agents

Now and then, a person appears at a police station with samples of food or drink claiming that he or she were the victim of malicious poisoning attempts. Very often this suspicion is completely unfounded and exists only in the respective person's imagination. But actual poisoning attempts can happen, and each case must therefore be treated seriously. Recently, Klöppel and Adebahr [11] have described 23 cases of suspected poisoning in a region of Western Germany.

Depending on local conditions, the suspect material is sent to a Public Health Laboratory, a Toxicology Laboratory or a Forensic Science (Criminalistics) Laboratory. Irrespective of the destination, the officer responsible for the case should contact the investigating laboratory by phone and discuss the optimum way for packaging and transportation. All too often the evidence arrives at the laboratory in a sorry state: sour milk, dry fruit, molten chocolate, rotten vegetables etc. arrive in leaking, wet or evil smelling packages, sometimes with broken containers. If immediate transportation is not possible, the evidence is usually best placed in a refrigerator.

The evidential value of the analysis is somewhat doubtful in such cases where the "victim" had an opportunity to add foreign substances him/herself to the food in question.

Also, it is unwise to publish the analytical results uncritically. We remember a case where a woman erroneously suspected "person or persons unknown" to blow poison through the keyhole into her living room. The contents of her vacuum cleaner were analyzed, and traces of all sorts of elements found. She finally received the answer that 14 μg of arsenic had been found in the vacuum cleaner bag. This did nothing to alleviate the woman's fears.

When sending food or drink for analysis the police officer in charge has to observe two principle rules:

1) The package must be arranged so that no food, and especially no liquid can run out. It is not always easy to pack liquids tightly, especially if they contain artificial CO_2 or are fermenting. If necessary, the liquid has to be transferred to another *clean* container with a secure seal. In such cases, the original, now empty, container should also be submitted.
2) It is advisable to send food and drink in cooled packages since many of these items are liable to spoil at elevated temperatures which makes any comparison with equivalent food difficult. It is important not to send such materials over the week-end; the package may be kept in a refrigerator when necessary.

The laboratory is greatful if the police officer also can send food of exactly the same kind (if possible) for reference analysis. If this is not done, the laboratory has to purchase such items.

13. References

1) J. Andrasko, A. C. Maehly: The Discrimination Between Samples of Window Glass by Combining Physical and Chemical Techniques. J. Forens. Sci. *23*, 250 (1978)
2) J. Andrasko: unpublished experiments
3) M. D. G. Dabbs, E. F. Pearson: The Variation in Refractive Index and Density Across Two Sheets of Window Glass. J. Forens. Sci. Soc. *10*, 139 (1970)
4) R. A. Freeman: The Anthropologist at Large. In: John Thorndyke's Cases. London: Hodder & Stoughton 1927.
5) M. Frei-Sulzer: Sicherung von Mikrospuren mit Klebband. Kriminalistik *5*, 190 (1951)
6) J. W. Gunn, D. W. Johnson, W. P. Butler: Clandestine Drug Laboratories. J. Forens. Sci. *15*, 51 (1970)
7) R. Heindl: Eine billigere Methode, kriminalistisch bedeutsamen Staub zu sammeln. Arch. Kriminol. *93*, 169 (1933)
8) S. Jonson: unpublished experiments
9) D. W. Johnson, J. W. Gunn: Clandestine Drug Laboratories. N. F. P. A. Fire Command Magazine. November (1970)
10) A. Karsten: Eine Spezialstaubsaugerkonstruktion für kriminalistische Zwecke. Arch. Kriminol. *89*, 159 (1931)
11) A. Klöppel, G. Adebahr: Verdacht der Giftbeibringung. Erfahrung aus einem Zeitraum von 4 Jahren. Deutsch. med. Wochenschr. *104*, 266 (1979)
12) E. Locard, H. Söderman, R. Heindl: Beiträge zur kriminalistischen Staubuntersuchung. Arch. Kriminol. *92*, 148 (1933), *92*, 234 (1933), *93*, 63 (1933), *93*, 141 (1933)
13) E. Martin: The Comparison of Coloured Textile Fibres. Intern. Crim. Police Rev. No. *109*, 166 (1957)
14) E. P. Martin: New Types of Adhesive Strips and Protection of Microscopic Evidence. Intern. Crim. Police Rev. *21*, 200 (1966)
15) E. P. Martin: Proof in Criminal Cases by means of Microscopical Evidence. The Microscope *17*, 111 (1969)

16) E. P. Martin: Zur Frage des Beweiswertes von Wollfaserspuren im Strafprozeß.Arch. Kriminol. *152*, 172 (1973)
17) S. Odén, B. von Hofsten: Detection of Fingerprints by the Ninhydrin Reaction. Nature *173*, 449 (1954)
18) M. Owoc: Mikrospuren. Kriminalistik and forens. Wissensch. *29*, 23 (1977)
19) E. F. Pearson, R. W. May, M. D. G. Dabbs: Glass and Paint Fragments Found in Men's Outer Clothing – Report of a Survey. J. Forens. Sci. *16*, 283 (1971)
20) C. A. Pounds: The Recovery of Fibres from the Surface of Clothing for Forensic Examinations. J. Forens. Sci. Soc. *15*, 127 (1975)
21) C. A. Pounds, K. W. Smalldon: The Transfer of Fibres Between Clothing Materials During Simulated Contacts and Their Persistence During Wear. Part I. Fibre Transference. Part II. Fibre Persistence. Part III. A Preliminary Investigation of the Mechanisms Involved. J. Forens. Sci. Soc. *15*, 17, 29, 197 (1975)
22) O. Rispling: Ninhydrine Combats Bank Frauds. The Criminologist *6*, 30 (1971)
23) H. Söderman: Zur Frage des kriminaltechnischen Staubsaugeverfahrens. Arch. Kriminol. *93*, 156 (1933)
24) S. Türkel: Bemerkungen zur Technik der Staubextraktion aus Kleidungsstücken. In: Beiträge zur Krim. Symptomatologie. Stuttgart: L. Moser 1931, pp. 1–9

14. Tables C. 1, 2

Table C. 1. Advantages and disadvantages of different methods for collecting fiber evidence

Properties	Plucking	Shaking	Suction	Taping	Brushing	Combing
High overall efficiency	–	–	+	+	–	–
Takes up predominantly recently acquired fibers	+	+	–	+	?	+
Takes up much background material	–	–	+	–	–	+
Easy to remove fibers again	+	+	+	–	–	+
Easy to pinpoint fiber location	+	–	–	+	?	–
Easy to collect fibers from large surfaces	–	+	+	–	+	+

Table C. 2. Advantages and drawbacks for three types of evidence containers for debris from scenes of fire

Type of container	Glass jars	Metal cans	Plastic bags (nylon)
Sensitive to sharp or pointed objects (nails, glass)	No	No	Yes
Subject to rust on storage	No	Yes	No
Difficult to transport (e.g. by rail or mail)	Yes	No	No
Tight for hydrocarbon fuels	Yes	No	Yes, to a degree
Tight for alcohol-type accelerants	Yes	No	No, in the long run[a]
Repeated use	Yes	No	No
Price	High	Low	Moderate
Availability to police personnel	Limited	Good	Good, but require special sealing device

[a] Satisfactory for bags with metal foils

II. Reference Collections

1. Introduction

One of the most important activities of a forensic science laboratory is the setting up and constant updating of collections of reference materials and reference data. At the same time, this activity is very time-consuming and labor-intensive.

The International Police Organization (Interpol) has made an attempt of documenting reference collections presently existing in its member countries. Steinberg [1] has set up a list of standard reference collections of forensic science materials for the Law Enforcement Assistance Administration, U. S. Department of Justice. On a much smaller scale, the Scandinavian countries have informed each other of their respective collections. All these efforts are very useful, but nothing can replace collections available at immediate notice within the laboratory itself.

2. Type of Collections

There are, in principle, two kinds of collections, i. e. those involving physical objects (firearms, drugs, animal hair etc.) and those of data (cartridge markings, spectra etc.). In the following a short survey of some of these collections and their significance will be listed. Table C. 3 (p. 276 and 277) gives examples of physical collections, and Table C. 4 (p. 278) lists collections of various data. Many laboratories have set up collections not listed here.

The English laboratory system has set up a record of the blood group systems of all convicted persons. This record can then be used for comparison with traces of blood found on scenes of crime.

In the authors' laboratory (and probably many others), the rifling marks on bullets and the marks on cartridge cases found at the scenes of crime are centrally registered for eventual comparison with other crime ammunition and/or firearms. There is also a collection of wrapping papers from dynamite sticks: sometimes, a piece of wrapping paper can be fitted to another piece from a second crime scene – or to the remaining dynamite stick found in the possession of a suspect.

3. Reference

1) H. L. Steinberg: Standard reference collections of forensic science materials: Status and needs. Washington: Law Enforcement Assistance Administration, February 1977

4. Tables C. 3, 4

Table C. 3. Examples of physical reference collections

Type of collection	Use of collection
Firearms	1) Reconstruction of shooting incidents 2) Possibility of using parts as temporary replacement in firearms sent for investigation 3) Teaching and museal use 4) Help in the identification of sent-in weapons
Ammunition	1) Shooting of reference bullets and reference cartridge cases 2) Identification of bullets, cartridge cases and intact rounds
Detonators	1) Comparison of coding and physical appearance 2) Reconstructions
Explosives	1) Comparison of chemical and physical properties 2) Reconstruction of explosions (small scale only)
Locks	Study of lock mechanisms
Coins	Matching vs counterfeit coins
Tire marks	Identification of tire make and model
Electrical equipment	Comparison with burnt or otherwise damaged equipment
Safe filling	Comparison with traces of safe filling found on suspects etc.
Glass, paint chips	Statistical evaluation of physical and chemical properties
Solid dosage forms of drugs	1) Identification by physical characteristics 2) Possibility of comparative chemical analyses 3) Assistance to poison control centers
Pure drugs and other commercial chemicals (pesticides, desinfectants, solvents etc.)	1) Possibility of comparative chemical analyses 2) Study of yield after adding known amounts to samples

Table C. 3 (continued)

Type of collection	Use of collection
Plastics and polymers	1) Comparison of chemical properties 2) Analysis of products formed on burning
Lipsticks	Comparison of chemical properties and color
Herbarium	Comparison with unknown plants for identification
Seeds	Comparison of morphological and chemical properties
Spices	Physical and chemical comparison for identification
Wood	Comparison with unknown samples
Fibers	1) Physical and chemical comparisons for identification 2) Statistical evaluation of physical and chemical properties
Animal hair	Identification of unknown hairs
Cigarettes	Identification of whole cigarettes or parts thereof
Matches	Identification of whole matches or parts thereof
Drivers licenses	Comparison with counterfeit licenses
Passports	Comparison with counterfeit passports
Money bills	1) Comparison of counterfeit bills with genuine bills 2) Matching of counterfeit bills
Ball point cartridges	Optical and chemical identification and determination of earliest use

Table C. 4. Examples of reference collection of data

Type of collection	Use of collection
Specimens of typewriter writing	Identification of typewriters or parts thereof
Rifling data of firearms	Identification of firearms based on the marks found on bullets and cartridges
Properties of solid dosage forms	Aid in identification
Absorption spectra (IR and UV)	Identification of chemical compounds
Mass spectra	Identification of chemical compounds
Frequency data on glass, paint, fibers etc.	Statistical evaluation of evidence

III. The Forensic Expert

The forensic chemist must meet the same standards of professional quality and general ethics as forensic experts in other fields. The following discussion applies therefore to the forensic scientist in general.

It must further be realized that the role of the forensic expert in many respects depends on the judicial system used in his *) country. This is one of the reasons why "foreign" experts are rarely employed – another one being the language barriers. In some respects, this chapter is naturally colored by the experience of the authors in Swedish legal procedures.

Even within these limitations, enough fundamental principles remain to be emphasized.

By contrast to a witness, the expert not only has to report his observations but is required to draw conclusions from his findings.

Impartiality is probably the oldest and most obvious property demanded of an expert in court, be it in written or oral testimony. In some judicial systems this requirement is partly testifying "for the state", or "for the defense", respectively. A clearer status is that of the independent expert who is accepted by both parts.

A more or less hidden partiality can arise when the forensic scientist is employed by a police or district attorney organization. Many forensic science laboratories have frequent and friendly contacts with the prosecution and police. In such cases, a concious effort has to be made not to confuse good relations (which are valuable) with undue influence. If this is not clearly realized, unconscious or subconscious partiality may "sneak in". One obvious countermeasure are open contacts with defense lawyers, but this meets often with difficulties, sometimes due to one-sided or mutual distrust. An indirect threat to impartiality is the fact that the forensic scientist often has contacts initially only with the prosecution – simply because no counsel for the defense has been selected in the early stages of a criminal investigation.

Humbleness is a less often discussed quality of a forensic expert. Nothing is more devastating to the reputation of a scientist than over-confidence in his range of knowledge. This danger is probably greatest in the very young – and in the very much established expert. The enthusiasm of the newcomer can usually be held within proper limits by his superiors. The older, well-known and possibly famous expert is a much greater liability to the judicial process. Encouraged and possibly spoiled by success and flattery, he can be tempted to take on forensic problems

*) For the sake of brevity and simplicity, the word "his" will here stand for the clumsier "his or her"

which are outside (or just outside) of his training and experience. And who is going to stop him? Does he seek the opinion or advice of his younger co-workers, or of colleagues sometimes regarded as competitors? The chances are that he does not. Incorrect interpretations of his findings may go undetected for quite some time. The tragedy is complete when they are not.

The remedy is obviously the readiness to recognize one's limitations, and the consequences can essentially only be two: either the experts refuses to tackle the problem – or he refers it to a specialist. In the latter case, the ability for choosing the "best" specialist is also crucial. What if there are several individuals working in the special field in question? With regard to the not uncommon situation of distrust among scientists within the same specialty, it might be wise to contact a specialist in another part of the country or another country altogether (if possible, within the same language group).

Self-criticism is one of the weapons against accidents of the kind described above. It is obviously vastly preferable if the scientist discovers possible errors himself, rather than if such mistakes are found by others – quite apart from the possibly disastrous consequences of an error. One, rather torturous way for combatting the lack of self-criticism is the system of "reverse proof". Reports and opinions are put aside for a day or two, and are then read with the assumption that they are completely wrong. The process consists then in proving, to one's own *critical* satisfaction, step by step, that each statement is correct after all.

Readiness to be criticized by others is a valuable property of an expert but requires a good deal of self-restraint. Yet, each scientific expert should be deeply greatful to his co-workers or superiors if these find shortcomings in his reports *before* they leave the laboratory and become public documents. The procedure of double – (and – triple!) checking of records, results, and interpretations by several individuals within a forensic science laboratory is highly important for the proper functioning of justice. Moreover, it is of great educational value. Whereas vanity is the main cause for over-confidence (cf. above), pride is the main inhibitor for a smoothly operating system of multiple checking of data and their interpretation. Here again, the older expert is liable to sin: many senior scientist are more than reluctant to lose face by asking younger co-workers to check their reports. If this is done in a suitable manner however, the senior scientist will profit by it – even with respect to his leader image.

Compromises with time must be avoided. Forensic laboratories are often working under pressure to produce results within specified time limits. District attorneys especially are prone to demand speedy replies, often due to limited periods for detention or arrest of a suspect, or set dates for court sessions. It is not uncommon, especially in widely publicized legal cases, that the news media exert pressure on the legal authorities. Higher officials for the prosecution are then tempted to pass on this pressure to the forensic scientist, sometimes in a quest for publicity.

Important cases may deserve special efforts, mainly in the interest of the suspect who has a right to be exonerated (if innocent) as fast as this is safely possible. Similar considerations apply to individuals taken into custody. Such cases shall be given priority, either by holding back other cases, or by reassigning personnel, or both. But the work itself may by no means be accelerated or rushed. Also, all checking must be done deliberately and conscientiously, notwithstanding telephone calls or visits by so-called important people. It should be obvious that it is better to postpone a court procedure than to carry it out on erroneous premises. The responsibility for delaying a report in order to insure its quality rests usually with the director of the laboratory.

No discussion of the qualifications of the forensic expert would be complete without taking up the old question if this expert should be a strict specialist, or rather a so-called generalist.

Historically, there have been two areas of forensic science with claims for a "generalistic" approach. One is the continental model of forensic medicine, the other the more recent school of criminalistics in the United States. Forensic medicine, being the oldest discipline in the field of forensic science, has – for a long time – no difficulty in absorbing new fields of endeavour such as histochemistry, blood group serology, or human toxicology. It is only within the last two or three decades that the explosive development of forensic toxicology and serology has made it difficult or imposible for the expert in forensic medicine to master these fields in detail.

A similar development has taken place in criminalistics. Whereas the pioneers of this discipline, men like Gross, Reiss, Heindl and Locard, could still master the whole field, this has in recent years become much more difficult. The leading American criminalist of the period after the Second World War, Paul Kirk, still defended and supported the generalist approach, indeed he seems to have created the term. The California School, trained by Kirk, has applied this concept to practical work. In some laboratories, the criminalists rotate in a regular fashion from tool mark investigation to, say, blood stain analysis, arson tests, handwriting investigation etc., sometimes even including crime scene work.

Many of the larger forensic science laboratories, notably in Europe, but also in the Americas, Japan, India, and Australia have adopted a specialist system where groups of analytical chemists, biologists, tool mark and firearm examiners, document examiners etc. divide the work between themselves. This can mean that a single criminal investigation can be carried out by a large number of specialists.

Now, which approach is to be preferred? The answer must be: both. The criminalist working in a small organization will be forced by circumstances to study and penetrate into more subspecialities of the discipline than the scientist attached to a large laboratory. This means, however, that he has to remit on the average many more of his cases to specialists outside of the forensic community if he follows the rules set up above. Another category of criminalists who has to widen his field of knowledge and experience, is that of the laboratory director or head of a division inside of a large organization. In some cases, he has received a comprehensive training but this case is still quite rare. Usually, he starts off as a specialist, more often than not as a chemist, and has to make a series of serious studies of other fields in criminalistics as he advances in rank. Nevertheless, he really remains a specialist in his original field of basic training. Finally, there is the category of teachers in forensic science. Teaching and training of scientists in criminalistics is still in its infancy. Quite a number of colleges and a few universities offer courses in forensic science in the United States. Elsewhere, university education is still very rare. Comprehensive curriculae are not yet generally offered, and it is not clear how the teachers should be recruited. However, they need to have a general knowledge of the whole field of forensic science.

Whenever the size of a forensic science laboratory permits, the specialist approach seems to be more powerful, In this way, it can be assured that the individual criminalist has a thorough theoretical background for his work and, equally important, that he receives an intensive on the job training in his specialty.

This seems to be the right moment for bringing up the question of the relative merits of theoretical knowledge and practical experience in criminalistics. In some fields of forensic science a good academic background is essential and cannot be replaced by practical training. Apart from forensic medicine which is not discussed here, this applies to forensic biology, forensic chemistry, and to a certain degree forensic physics. The corresponding fields of work are morphological identification of flora and fauna, blood group serology and immunology, drug identification, toxicology, chemical analysis of various materials (cf. Part B of this book), ballistics, electrical investigations and development of instrumental methods. Also in document work an academic background is desirable since it ensures a measure of logical and critical thinking.

There are other activities, however, where an engineering degree is very helpful but where practical experience is the predominant requisite for quality work. This is true for firearm investigations, tool mark comparisons, the investigation of causes for fires, forensic photography – and again document work especially the comparative evaluation of handwriting.

Thus, a team of criminalists is ideally composed of academically trained scientists, engineers, and practical workers, including laboratory assistants with a 1–3 years theoretical training period.

If specialization has so many merits, why then bother with forensic scientists? Would it not be better to send the casework directly to specialists at universities, colleges or even industrial research specialists? There are two major reasons mitigating against such a system. One is the fact, that it is not enough, in forensic science, to obtain scientifically sound and carefully checked results. The forensic scientist must also evaluate, interpret, and explain these results in the context of the actual criminal investigation. This second task cannot (or only rarely) be left to the outside expert.

The second reason has to do with time limits. The forensic expert is hired for his work and is expected to deliver his reports within short time limits. On the other hand, one cannot expect a university professor to drop his research and teaching for wholly unpredictable periods in order to investigate a forensic case.

Both of these disadvantages must be accepted whenever even the forensic scientist feels to be incompetent in special cases. The uncertainties in interpretation and delays in legal procedures must, however, be kept within reasonable limits.

The most efficient way to guarantee a high level of forensic investigation in criminalistics seems to be a laboratory large enough to employ groups of specialists. However, the work of these groups needs to be coordinated by individuals with a general knowledge and the ability to synthesize the results – and especially the interpretation of the results from different individuals or groups within the organization.

IV. Sources of Information on Forensic Science

1. Introduction

It would be both difficult and impractical to discuss sources of information on forensic chemistry separately from those on the whole field of Forensic Science. Forensic Medicine lies on the borderline of this discussion and will not be covered systematically.

The reasons for this approach are simple. For one thing it has become obvious from part B of this book that chemical methods are used to a major or minor extent in most of the subspecialties of Forensic Science. Also, it is of importance for the forensic chemist to get acquainted with the whole field of criminalistics.

Since Forensic Science is an outspokenly multidisciplinary field of endeavour, scientific papers are found spread out over a large number of publications. This chapter will predominantly cover information on criminalistics and analytical chemistry applied to Forensic Science.

The following mechanisms for obtaining information will be considered:
Scientific journals.
Books and monographs.
Scientific meetings.
Major laboratories and
Information centers in Forensic Science.

2. Scientific Journals

Among the international journals in the field of criminalistics those of general interest are listed in Table C. 5 (p. 287 and 288). A more complete list is found in the Abstracts on Police Science. It cannot be the purpose of this book to recommend particular journals. It may be permitted to point out, however, that the Abstr. Police Sci. cover the majority of the relevant journals and are thus a useful reference. It can furthermore be noted that there are only European and American journals on the list, with the exception of one Japanese and one Russian periodical. The authors would be grateful for information on periodicals from other parts of the world. A great number of journals are being published in the field of analytical chemistry and other fields related to chemistry. Experience shows that the journals on Table C. 6 (p. 288) frequently carry articles on forensic chemistry. In addition, articles on forensic topics can be found in: Fire and Materials, Fire Research, Fire Journal, Human Genetics and many others.

Finally, it is useful to consult the large abstracting journals, such as:

A. A.	Analytical Abstracts,
B. A.	Biological Abstracts,
C. A.	Chemical Abstracts,
C. C.	Current Contents,
Exc. Med.	Excerpta Medica,
Index Med.	Index Medicus.

3. Books and Monographs

In recent years, a flood of books and monographs on topics in Forensic Science have been appearing. It is impossible to give a comprehensive or even extensive lists of these. The listing of this literature will therefore be focused on two types of publications, viz.

(i) those treatises that have stood the test of time and can be regarded as pioneer or classical works (cf. Table C. 7 (p. 288 and 289), and
(ii) recent publications of general interest to the forensic scientist working in criminalistics (cf. Table C. 8 (p. 289).

Roughly speaking, this means books published before the Second World War and after 1960 respectively.

4. Scientific Meetings

There are two kinds of regular international meetings in the field of Forensic Science. One of these are the Meetings of the International Association of Forensic Sciences. These were started in 1957 and are held every 3rd year. The past meetings convened in the following places:

1st Meet.	1957	Brussels and Ghent,
2nd Meet.	1960	New York,
3rd Meet.	1963	London,
4th Meet.	1966	Copenhagen,
5th Meet.	1969	Toronto,

6th Meet. 1972 Edinburgh,
7th Meet. 1975 Zürich,
8th Meet. 1978 Wichita, Kansas,
9th Meet. 1981 Bergen, Norway.
10th Meet. 1984, Oxford, England.

The following topics are discussed at these meetings: Criminalistics, Questioned Documents, Toxicology, Pathology and Legal Medicine and Odontology. Representatives of some of these disciplines are being organized in international organizations such as:

since 1938 The International Academy of Legal Medicine and Social Medicine.
since 1963 The International Association of Forensic Toxicologists.
since 1973 The International Society of Forensic Odonto-Stomatology.

The other regularly held meetings are those organized by INTERPOL, and held in Paris every 3rd year. These meetings have been held in the same years as those of the International Association, but this pattern has now been changed, so that the INTERPOL meetings will convene in 1980, 1983, 1986 etc. The themes of these meetings are generally restricted to criminalistics and consist mainly in reporting on research in collaborative projects involving two or three major laboratories, predominantly European ones.

In addition to these large gatherings, a large number of regional and national meetings are held – generally each year. In Forensic Chemistry, the most important congresses are probably those organized by

The American Academy of Forensic Sciences.
The California Association of Criminalistics.
The Forensic Science Society (United Kingdom).
The International Association of Forensic Toxicologists (TIAFT).
Die Deutsche Gesellschaft für Rechtsmedizin.

Information on meetings may be obtained from "The International Reference Organization in Forensic Sciences and Medicine (Inform)" Laboratory, St. Francis Hospital, Wichita, Kansas, USA – 67214.

5. Major Forensic Science Laboratories

The names and addresses of all major Forensic Science Laboratories, with the notable exceptions of the Soviet Union and the Popular Republic of China, have been collected by the Forensic Science Society and the resulting "World List" is available at the secretariat of the society*). The list includes laboratories and institutes of Forensic Medicine, Criminalistics, Toxicology, Questioned Documents and Odontology as well as private practitioners in these fields. Table A. 2 (p. 10–13) gives a summary of the distribution of these organizations in the world.

*) P. O. Box 41, Harrogate, North Yorkshire England, HG1 1QL

Some years ago (1968), the John Jay College in New York published a national survey of crime laboratories in the United States and Canada [1]. The survey covered, *inter alia*, the following topics:

training,
suggested educational program,
technical problems,
research needs.

Ninety two laboratories took part in the survey, and 21 of them were visited by the researchers. The largest laboratories in North America are the FBI-laboratory and the Forensic Laboratory of the Bureau of Alcohol, Tobacco and Firearms, U. S. Treasury Department – both in Washington, D. C., the New York City, Los Angeles County, Los Angeles City and Chicago Police laboratories, as well as the Forensic Science Centre in Toronto, Canada, all with well over 50 employees. In 1968, 3 laboratories had more than 30, 4 more than 20, 25 more than 10 employees. The remaining 55 laboratories employed less than 10 people each. The laboratories covered by the survey are listed in Table A. 2 (North America, p. 11 and 12).

Major centers of casework and research in Forensic Science outside of North America are The Metropolitan Forensic Science Laboratory in London, with well over 200 employees, the Home Office Central Research Establishment (HOCRE) in Aldermaston (GB), the Bundeskriminalamt in Wiesbaden (West Germany), the Central Forensic Science Laboratory in New Delhi (India), and the National Research Institute of Police Science in Tokyo (Japan).

6. Information Centers

At present, there are three major information centers in operation. The best organized and amply staffed one forms a part of the Home Office Central Research Establishment. About 15 scientists read and abstract the majority of journals containing articles on Forensic Science and adjoining fields. The HOCRE center has other information activities as well, such as collections of background information, computerized lists of reference spectra etc. The primary object is the dissemination of information inside the Forensic Science laboratory system in England, but there is an increasing amount of international contacts.

The second major center lies in Wichita, Kansas, USA and is known under the name of INFORM (International Reference Organization in Forensic Medicine and Sciences). This center aims at communications with forensic institutes over the whole world. It publishes a small journal with general information, an annual computerized abstracts list, and reports on special topics in the field of forensic h a certain focus on forensic medicine.

A third information center is run by the Bundeskriminalamt in Wiesbaden (West Germany).

7. Reference

1) "John Jay College National Survey", Project 013, The Office of Law Enforcement Assistance, U. S. Dept. of Justice, Washington D. C.

8. Tables C. 5–8

Table C. 5. International journals in the field of forensic science, especially in criminalistics, arranged according to the first year of publication

Year of 1st issue	Full title of journal	Abbreviation according to Chem. Abstr. or equivalent	Number of issues per year (1978)
1889	(Gross) Archiv für Kriminologie	Arch. Kriminol.	6
1893	Police Review	Police Rev.	52
1909	The Journal of Criminal Law and Criminology (before 1942: Journal of the American Institute of Criminal Law and Criminology, before 1972: The Journal of Criminal Law, Criminology and Police Science)	J. Criminal Law Criminol.	4
1918	Fingerprint and Identification Magazine	Fingerprint Identificat. Mag.	2–4 (ceased)
1922	Zeitschrift für Rechtsmedizin – Journal of Legal Medicine (before 1970: Zeitschrift für die gesamte gerichtliche Medizin)	Z. Rechtsmed. J. Leg. Med.	6
1928	Acta Criminologia et Medicinae Legalis Japonice	Acta Crim. Med. Leg. Jap.	6
1928	The Police Journal	Police J.	4
1929	Revue Internationale de Criminologie et de Police Technique	Rev. Intern. Criminol. Police Technique	4
1932	The FBI Law Enforcement Bulletin	FBI Law Enforcem. Bull	12
1932	The Medico-Legal Journal	Medico-Legal J.	4
1938	Royal Canadian Mounted Police Gazette	R. C. M. P. Gazette	11
1944	Kriminalistik	Kriminalistik	12
1947	The International Criminal Police Review	Intern. Crim. Police. Rev.	12
1948	Reports of National Research Institute of Police Science (Japan)	Repts. Nat. Res. Inst. Police Sci.	4
1949	The Bulletin of Narcotics	Bull. Narcotics	4
1949	Fire and Arson Investigation	Fire & Arson Invest.	4
1953	Forensic Science International (before 1972: The Journal of Forensic Medicine; from 1972–1978 Vol. 11 Forensic Science)	Forens. Sci. Intern.	6
1956	Journal of Forensic Sciences	J. Forens. Sci.	4
1958	Sudebna-meditsinskaja ekspertizo (U.S.S.R)		4
1960	Journal of the Forensic Science Society	J. Forens. Sci. Soc.	4
1960	Medicine Science and the Law	Med. Sci. Law	4
1963	Police Research Bulletin	Police Res. Bull.	2
1964	Blutalkohol	Blutalkohol	6
1966	The Criminologist	Criminologist	2–3

Table C.5 (continued)

Year of 1st issue	Full title of journal	Abbreviation according to Chem. Abstr. or equivalent	Number of issues per year (1978)
1966	The International Microform Journal of Legal Medicine	Intern. Microf. J. Leg. Med.	4
1968	Canadian Society of Forensic Science Journal	Canad. Soc. Forens. Sci. J.	4
1969	The Forensic Science Gazette	Forens. Sci. Gazette	4
1969	INFORM Newsletter		4
1969	Microgram (restricted)		12
1970	Zentralblatt für die gesamte Rechtsmedizin und ihre Grenzgebiete	Z. ges. Rechtsmed.	12
1970	Kriminalistik und Forensische Wissenschaften	Kriminal. forens. Wiss.	3–4
1973	Abstracts on Police Science	Abstr. Police Sci.	6
1973	Journal of Police Science and Administration	J. Police Sci. Admin.	6
1975	Mannheimer Hefte für Schriftvergleichung		2
1977	Journal of Analytical Toxicology	J. Anal. Toxicol.	6

Table C. 6. International journals in chemistry which frequently publish articles of importance to forensic chemistry, especially chemical criminalistics

Full title	C. A. abbreviation
The Analyst	Analyst
Analytical Chemistry	Anal. Chem.
Analytica Chimica Acta	Anal. Chim. Acta
Chromatographia	Chromatogr.
Fresenius' Zeitschrift für analytische Chemie	Z. Anal. Chem.
Journal of the Association of the Official Analytical Chemists	J. Assoc. Offic. Anal. Chemists
Journal of Chromatographic Science	J. Chromatogr. Sci.
Journal of Chromatography	J. Chromatogr.
Journal of Pharmaceutical Science	J. Pharm. Sci.
Organic Mass Spectrometry	Org. Mass. Spectrometry

Table C. 7. Classical works of historical interest

Alphonse Bertillon: Identification anthropométrique. Instructions signalétiques. Melun: Ministère de l'Intérieur 1885

M. A. Bischoff: La Police Scientifique. Paris: Librairie Payot 1938

F. Galton: Finger Prints. London and New York: MacMillan and Co. 1892

H. Gross: Handbuch für Untersuchungsrichter, Polizeibeamte, Gendarmen usw., Graz: Verlag von Leuchner & Lubensky 1893

E. Locard: L'Enquête criminelle et les Méthodes scientifiques. Paris: Ernest Flammarion (ed.) 1920

A. Lucas: Forensic Chemistry and Scientific Criminal Investigation. London: Edward Arnold & Co. 1921

Table C. 7 (continued)

A. S. Osborn: Questioned Documents, Albany, N. Y.: Boyd Printing Company 2nd ed. 1929
R.-A. Reiss: Manuel de Police Scientifique (Technique), Vol. I-IV, Lausanne: Librairie Payot & Cie and Paris, Félix Alcan 1911
H. Söderman, J. J. O'Connell: Modern Criminal Investigation New York and London: Funk & Wagnalls Company 1935
A. Svensson, O. Wendel: Crime Detection. Modern methods of criminal investigation. Amsterdam: Elsevier Publ. Comp. 1955

Table C. 8. Works published after 1960

Arbeitsmethoden der medizinischen und naturwissenschaftlichen Kriminalistik:

Vol. 7. K. Sellier: Schußentfernungsbestimmung. Lübeck: Max Schmidt-Römhild 1967
Vol. 14. K. D. Pohl: Naturwissenschaftlich-kriminalistische Spurenanalyse bei Verkehrsunfällen. Lübeck: Max Schmidt-Römhild 1975
Vol. 15. K. Sellier: Schußwaffen und Schußwirkungen. Forensische Ballistik, Wundballistik. Lübeck: Max Schmidt-Römhild 1977
E. B. Block: The Chemist of Crime. London: Cassell & Company Ltd. 1959
Ch. M. Bozza: Criminal Investigation. Chicago: Nelson-Hall Inc. 1978
P. F. Ceccaldi: La Criminalistique, Paris: Presses Universitaires de France 1962
Elsevier's Dictionary of Criminal Science, compiled by Johann Anton Adler. Amsterdam: Elsevier Publishing Company, 1960
Forensic Science. Scientific Investigation in Criminal Justice. J. L. Peterson (ed.). New York: AMS Press Inc. 1975
R. H. Fox, C. L. Cunningham: Crime Scene Search and Physical Evidence Handbook. Washington: U. S. Department of Justice 1973
M. Frei-Sulzer, A. Suter: Leitfaden zur Bearbeitung von Brandsachen und Explosionen. Hamburg: Kriminalistik. Verlag für kriminalistische Fachliteratur 1971
J. Gayet: Manuel de Police Scientifique. Paris: Librairie Payot 1961
H. Gross, F. Geerds: Handbuch der Kriminalistik, vol. I, 10th completely revised edition. Berlin: J. Schweitzer Verlag, 1977 vol. II, 1978
W. R. Harrison: Forgery Detection. A practical Guide. London: Sweet & Maxwell 1964
O. Hilton: Scientific Examination of Questioned Documents. Chicago: Callaghan & Company 1956
J. J. Horgan: Criminal Investigation. New York: McGraw-Hill Inc. 1974
S. Kind, M. Overman: Science Against Crime. London: Aldus Books Limited 1972
P. L. Kirk: Crime Investigation. Physical Evidence and the Police Laboratory. New York: Interscience Publ. Inc. 1966
Methods of Forensic Science. Vols. I-IV. Vols. I-II F. Lundquist (ed.) Vols. III-IV, A. S. Curry (ed.). New York and London: Interscience Publ. Inc. 1962–1965
Ch. E. O'Hara: Fundamentals of Criminal Investigation, 4th edit. Springfield: Charles C. C. Thomas Publisher 1976
J. W. Osterburg: The Crime Laboratory. Case studies of scientific criminal investigation. London: Bloomington and Indiana University Press 1968
R. Saferstein: Criminalistics. An introduction to forensic science. Englewood Cliffs, New Jersey: Prentice Hall Inc. 1977
J. D. Scott: Investigative Methods, Reston, Virginia: Reston Publ. Comp. 1978
H. J. Walls: Forensic Science. London: Sweet & Maxwell Limited 1968

V. The Organization of a Forensic Science Laboratory

1. Introduction

Ever since the establishment of the first Forensic Science Laboratories (cf. Chap. A. I), the tasks and the organization of these institutions have been under discussion. This is natural since such laboratories had yet to find their niche in the scientific field.

The discussion is still going on and concerns *inter alia* the following questions:

Who shall administrate the laboratory?
Should there be a few large units or many smaller laboratories?
How should the field work (crime scene investigation) and the laboratory work be connected?
What disciplines should be represented within the laboratory?
What are the educational requirements for the laboratory personnel?
How should research and case work be coordinated?
How shall education and training be organized for the benefit of the laboratory?
Which models for the internal organization of the laboratory are preferable?
What shall the laboratory building look like?
What instruments and equipment are needed?
What is the best way to achieve appropriate controls of the laboratory proficiency?
How shall external contacts to the "customers", the universities and other laboratories be organized?

Some of the more recent publications on these problems are listed at the end of this chapter as general references [A–M].

The following chapters are somewhat colored by the authors' personal opinions and their conclusions should not be regarded as the only valid ones. Each system has its merits and drawbacks, and each country, state etc has to make its own choice based on available resources.

2. Administration

The majority of Forensic Science Laboratories are administrated by police or prosecutor's organizations. However, various degrees of dependence on these bodies exist. Many of the smaller laboratories are more or less integrated with local police departments. In such cases there is often no clear-cut division between police personnel ("sworn officers" in US terminology) and civilian employees. Also, laboratory personnel often takes part in crime scene searches.

Larger laboratories tend to have looser ties with the local police forces and are often administrated by country or state police organizations or the district attorney's counterparts. Central laboratories, serving states or countries, are usually run by the Ministry of Justice (or Interior), either directly or via some national police organization. Table C.9 (p. 305 and 306) gives some examples of different administrative solutions. It is worthwhile to compare the administrative status of the criminalistics laboratories with those of forensic medicine. The medical examiner or other forensic pathologist is not, as a rule, incorporated into a police or prosecution administration. This has greatly contributed to the position of the forensic pathologist as an independent, unbiased expert. Forensic Medicine is usually part of a university organization or of a hospital. Nobody suspects the medical expert to be influenced by the prosecution.

It should be realized that the criminalist is at a disadvantage in this respect, and it would be useful to consider if a different, more independent status of the forensic science laboratories would not contribute to a higher ethical position of these institutions. One possible set-up could be modelled after the Swedish organization of Forensic Medicine. The six Institutes of Forensic Medicine are headed by a university professor who has two functions:

(i) he is responsible for teaching and research at a university department of Forensic Medicine (or of Pathology), and

(ii) he heads one of the government institutes for forensic medicine, charged with all case work.

This is one way for insuring an objective unbiased performance of the forensic services while being in constant touch with the academic research and teaching atmosphere.

The independence of forensic laboratories of police and prosecution would probably also mean that all of the staff would consist of civilians not connected to a police force. Experience at the author's laboratory has shown that this does not hinder a good and smooth cooperation with the police, based on mutual respect for each profession.

3. Geographical Distribution

One of the major decisions to be made when setting up or reorganizing a forensic science service is the degree of centralization of the laboratories.

The United States provide a good study region in this respect since all kinds of solutions can be found there. One group of laboratories has nation-wide organizations and includes inter alia:

1. The Federal Bureau of Investigation (FBI) Laboratory, Washington D. C.
2. The Bureau of Alcohol, Tobacco and Firearms, with a central laboratory in Washington D.C.
3. The laboratories of the US Post Office.
4. The US costums laboratories.
5. The Secret Service Laboratory.
6. The laboratory system of the Drug Enforcement Administration, DEA (Research Laboratory and regional laboratories in New York, Washington, Chicago, San Francisco, Dallas and Miami).
7. The US Army Criminal Investigation Laboratories.

The first step towards decentralization is the central State Crime Laboratory. A study report of the US Law Enforcement Assistance Administration (LEAA) from 1968 [1], found that 17 states lacked a central Crime Laboratory. This number has certainly decreased since then.

The 33 state laboratories serviced populations of 5000 (part of Alabama) up to 18 million (California). Out of the 14 county laboratories reported, the smallest served 330000 people and the largest 7 million (Los Angeles). Finally, 44 city laboratories each served 30000 (Highland Park, Michigan) up to 7 million (Los Angeles). The size of these laboratories can be judged by the number of employees who are entitled to witness in court (cf. Table C.10 (p. 306).

This array of laboratories (92) has more than doubled since but the pattern is still about the same, with many duplications leading to some complicated situations. A striking example is that of Los Angeles where the city, the county, the state and the US Government must divide the laboratory assignments among themselves.

What are now the arguments for and against centralization? The main advantages of relatively large, centralized laboratories are:

1. "Heavy" and expensive equipment can be justified because it is used often enough.
2. By the same token, specialized personnel can be trained to use this equipment efficiently.
3. The laboratory gains experience also with relatively rare cases.
4. The number of specialists (e.g. document examiners, drug analysts etc.) is large enough to form a backup when one or more quit the job, are sick etc.
5. Relatively large reference collections (guns, drugs, plants, ball point ink etc.) can be assembled in relation to the manpower available.
6. There will be a uniformity of quality and methodology for a large region which gives a fairer ground for weighing the evidence.
7. The large, centralized place can afford a reasonably complete library. Both books and journals tend to become very expensive.

Some of the drawbacks involve:

1. The central laboratory lacks close contacts with the local investigators and conditions.
2. There are considerable transport problems leading to losses in time and money.
3. It will be much harder for the laboratory personnel to take part in crime scene searches where their advice could be of considerable value.

Far-reaching decentralization gives close contacts between the local investigators and crime scenes on the one hand and laboratory personnel on the other. Unnecessary analyses can be avoided by direct discussion. Personnel can usually be reached off-hours for rush-cashes. The small staff acquires eventually some competence in many fields.

On the other hand, it will be hard to buy and maintain expensive instrumentation. When this is needed, local universities or larger laboratories have to be contacted. The staff will have to cover many disciplines and will not be able to penetrate a specialty very deeply. This will seriously hamper research and development.

One way to solve some of these problems is the establishment of an organized network of laboratories within a large region such as a country or a large state. The system of the English Home Office comes probably close to an optimal compromise. Eight regional laboratories plus the Metropolitan Police Forensic Science Laboratory (MPFSL) in London serve England (but not Scotland). These laboratories are large enough to have all-round experience, good equipment (some of them excellent) and the possibility to train specialists and yet are close enough to the local crime scene. In addition, the Home Office Central Research Establishment (HOCRE) does much of the more sophisticated research in the forensic field as will be discussed below.

Another approach is a system of satellite laboratories which are led by a central institute. This solution has been chosen in some American states. Here, the central laboratory has the heavy equipment, the good library, some research and development capability etc. whereas the satellites profit from their closeness to "the field". They can use the methodology developed at the center and can send their personnel there for training.

4. Connection between Crime Scene Work and Laboratory Investigations

By tradition, the investigation of a crime scene and the collection of physical evidence at the scene have been the task of police officers. In most cases there are specially trained policemen available for this work.

The laboratory investigations, on the other hand, have been progressively moved away from the sworn police personnel towards technicians and further towards academically trained scientists. This was a necessary consequence of the increasing sophistication of forensic analytical work.

At the same time it is obvious that all successful laboratory investigations depend on the proper selection, collection and dispatch of crime scene evidence.

Close cooperation between the crime scene investigator and the laboratory scientist is thus necessary. This can be achieved mainly in two ways, i. e. by rather extended study periods of the police specialists at the laboratory and by relatively frequent visits at crime (and accident) scenes of laboratory personnel on all levels. The police officers will acquire detailed information on the possible and the impossible achievements of a modern forensic science laboratory, whereas the scientist can give advice to the crime scene searcher and learns from him what is expected from the laboratory. New avenues to research and development may be opened that way.

An intensive cooperation between crime scene investigators and laboratory personnel has also an important psychological aspect since it increases mutual respect and appreciation between these groups.

Some countries, notably England, have started to train civilian personnel in crime scene search. This idea may prove very fruitful and would free many police officers for other tasks which they are better equipped to fulfill with their training.

5. Disciplines Represented in the Laboratory

The larger the laboratory, the stronger the trend towards specialization. The rapid development of some branches of criminalistics, notably drug analysis, blood group determinations, and the analysis of so-called microtraces, has made it virtually impossible for one individual to master all details. Also, for valid testimonies, it is necessary to have a deep insight in a specialty –good overall knowledge is not enough. This development threatens the existence of small local laboratories.

The survey by the John Jay College [1] included a question to the laboratories regarding the number of criminalists entitled to witness in court. The answers are represented in Table C. 10 (p. 306). There are probably fewer small laboratories in Europe than in USA.

Nearly all forensic science laboratories are concerned with the examination of questioned documents. This is understable since this specialty is not taught at universities and has little or no significance outside the forensic community. A similar situation exists in the field of firearm and tool mark examinations. The only other interested parties could be the Armed Forces.

The situation is more complicated regarding fingerprint examinations. The "lifting" of the prints is often carried out by laboratory personnel, but these do not always have access to regional or national files. An alternative is to leave this technique completely to the police and their crime scene investigators who are kept in constant training and acquire considerable experience in this field.

The investigation of trace evidence is one of the most important and at the same time most difficult tasks for the forensic scientist. It is not possible to divide this specialty clearly between forensic chemistry, forensic biology, and ballistics since it can concern all three of them. A team-work approach is therefore necessary in laboratories with highly specialized working units. Members of this team must have a good knowledge either of chemistry (especially analytical chemistry) or biology (especially morphology including pharmacognosy).

The analysis of explosives before and after ignition has become a frequent assignment for the forensic laboratory.

Forensic photography cannot be dispensed with in a well-staffed institute. Not only ordinary techniques are needed but forensic work requires the use of special methods such as macro-, micro-, fluorescence-, filter-, infrared- and stereo-photography.

During the last decade, the large increase in the abuse of illegal drugs had a considerable impact on most forensic science laboratories in the world. Their case loads have suddenly increased at an unprecedented rate and this has upset long-range planning of personal and material resources. Some countries, notably the United States, have started a special network of laboratories for drug analysis. Nevertheless, this relatively recent addition of analytical tasks still constitutes a major problem for most laboratories of forensic science.

Many of these also concern themselves with arson investigation. However, fires are also caused by electrical and mechanical failures – and those could or should be included in fire investigations.

There remain a number of disciplines which may or may not be incorporated into a forensic science laboratory.

One of these is forensic serology. It can find its home in medical examiners' offices, institutes of serology, laboratories for paternity examinations or even hospital departments for blood analysis. However, the examination of spots of blood, semen, saliva or sweat is often so closely linked to other forensic investigations that a separate analysis elsewhere poses considerable problems.

Another field of analysis often carried out elsewhere is forensic toxicology. This situation has an historical reason since forensic toxicology originally developed as a branch of forensic medicine. Nowadays it is very questionable if the forensic pathologist will have the possibility to penetrate into the complex details of the metabolic processes involved in present-day human toxicology. It seems, therefore, that special forensic toxicology laboratories will be needed, either as independent units or as parts of forensic science laboratories. The latter solution has a number of advantages such as the possibility to use common heavy equipment such as mass spectrometers, fourier transform infrared spectrophotometers, automatic multidetector gas chromatographs or sophisticated types of high pressure liquid chromatographs.

Legal medicine is usually regarded as a discipline separate from forensic science. Consequently, the forensic practitioners have generally their own institutes which are often connected to a hospital and/or a university faculty of medicine. Nevertheless, there are examples of close geographical and personal cooperation such as in the forensic science laboratory at Rijswijk (Netherlands).

In summary, it seems that the following branches of forensic science should belong to a well-organized laboratory:

questioned documents,
firearms,
toolmarks,
trace evidence,
forensic biology,
fire investigation,

explosives,
solid drug analysis,
photography.

Recommended fields are further:

serology, toxicology, blood alcohol analysis.

And possibly:

forensic pathology.

6. Educational Requirements for the Laboratory Personnel

The discussion in Sect. C.V.5 has shown the diversity of activities in a forensic science laboratory. The education and training of the individual members of the staff depends therefore on the specialty concerned. Forensic work always consists of a combination of theoretical knowledge and practical experience, but the proportion between these two components varies.

Unfortunately, an all-round education in forensic science is not available everywhere. The situation is better in the United States (where a number of colleges and universities provide courses and even give degrees in forensic science) than in the rest of the world.

According to a survey written by Lateef [2] in 1974 there were 14 universities and colleges in the United States having courses in forensic science. Many more offer such courses and give degrees in this field at this time.

In Europe there are, at present, only two schools for education in forensic science, namely "Institut de police scientifique et de criminologie" at Lausanne (Switzerland), and the University of Strathclyde (Scotland). The former is well known for its excellent courses in questioned document work, and tool marks whereas the latter's strength is chemistry and biology.

The majority of forensic scientists is recruited from universities and colleges, but requirements vary with the type of work concerned.

A successful career in forensic chemistry requires a very good knowledge of chemistry, especially analytical chemistry. A master of science degree should be a minimum requirement for a junior member of the staff. Group leaders and heads of departments should at least have a Ph. D. degree and ought to have some research experience.

A similar situation prevailes for the forensic biologist. The ability to identify botanical or zoological material, especially in microscopic samples cannot be acquired by practical experience alone but must be based on thorough academic studies. It is desirable that the forensic biologist also has a good theoretical knowledge of physiology and even biochemistry. Staff members should thus have a MS degree in biology. And again, group leaders must be expected to have a Ph. D. with research experience.

The situation is different in most other fields of forensic science. The examination of toolmarks and forensic ballistics require a basic engineering degree and an inborn ability to "see things" but cannot be carried out without a lengthy "on-the-job-training". Experience means an awful lot in these disciplines.

However, recent developments in methodology, and increasing demands on critical evaluation of results tend to change the traditional picture. Higher education is thus becoming increasingly desirable.

Even forensic photography is a combination of good basic education and much experience.

Finally, the questioned document (QD) examiner is placed in a special category. There is no formal education in document examination at any university with the exception of the university of Lausanne (cf, above). The future QD examiner can, of course, take a degree in forensic science but this is rather an exception. As a rule, the candidates are chosen from people with a good general education, if possible an academic degree. The candidates are then tested for their ability to recognize patterns and colors as well as their ease of expressing themselves clearly in the language of the country in question. The newcomer is the trained "on the job" for periods of 2–5 years before he/she will be allowed to take his/her own case work.

Choosing group and division chiefs as well as laboratory directors is a difficult task. In principle, there are two ways to do this: by promotion, or by recruiting from outside. Promotion has the obvious merit of acquiring a leader with practical experience, but these candidates sometimes lack a higher degree and research experience. For outsiders the situation is the reverse. Outsiders also often provide the laboratory with new ideas and "fresh blood". There is no patent medicine available for solving this problem, but the higher the position, the more important become the theoretical background and research experience. Also, administrative ability plays an increasing role in leading positions.

7. Coordination between Case Work and Research

The more sophisticated type of criminal is well aware of the possibilities of forensic investigations. Organized gangs would not leave fingerprints at the scene of the crime. They would destroy weapons, tools, shoes and other objects that had been in contact with the crime scene. They would employ specialists for producing counterfeit money or falsified documents etc. This situation is one of the reasons for the need of research in forensic science which should always be ahead of the adversary.

Another reason of greater importance is the need to improve and perfect the establishment of facts and thus to avoid the punishment of the innocent and to insure the conviction of the guilty, freeing others from suspicion.

Like in many other disciplines, not to search for new ideas means to slip backwards. Unfortunately, forensic science is still too little recognized as an important part of the judicial system. Policy makers in most countries are not aware of its possibilities or even its existence. As a consequence, the majority of forensic science laboratories drown in ever-increasing case-work and have very limited resources or none at all for research and development (RD). However, there are a few exceptions in this situation as we shall presently see.

For those who are aware of the importance of RD the major query is in which way it can be carried out under the present difficulties?

1. One possibility is to "sneak in " some development work between the case-work assignments. In this way no special funds have to be asked for but the case backlog is, of course, adversively affected leading to justified complaints from the prosecution. What is worse, such development work cannot be carried out systematically since it is constantly interrupted by "rush-cases". The researcher suffers from the "donkey-between-two-haystacks" syndrome and has a bad conscience in whatever he does. On the other hand, the direct coupling of case-work to development is an advantage.
2. Another way towards a successful RD programm is to model the research projects into thesis work (MS or PhD). The laboratory has thus no major expenses, and the young scientist is obviously eager to get a good result within reasonable time. The problem seems to lie in the sceptical attitude of many universities and colleges and the reluctance of the young scientist to make his thesis in such a narrow and specialized field. The latter factor is partially based on a misunderstanding since many, or most of the methodology of forensic science is used in many other fields. This is particularly true of forensic chemistry. The resulting research is usually of high quality. Programs of this kind are running in some English laboratories and probably elsewhere in the world.
3. A solution related to no. 2 is an agreement with a foreign student. The agreement would combine a prolonged study visit at a high-quality forensic science laboratory with a research program. Our laboratory has tried out such schemes in both directions, i. e. by sending a research fellow abroad, and by receiving a fellow from one of the developing countries. The main difficulty of this scheme is funding; fellowships in forensic science are deplorably rare.
4. Finally, RD positions may be created at the laboratory. At the present time, this is rather difficult due to the above mentioned negative attitude of most governments towards such projects. A research group can be integrated into the laboratory organization, or special research laboratories can be set up. The latter solution has been chosen e. g. by the Drug Enforcement Administration (DEA) in the US and notably by the English Home Office, which has created the largest research institute in this field in Aldermaston, the Home Office Central Research Establishment (HOCRE) with a staff of about 50 scientists and technicians.

The main difficulty with separate research institutes is the maintenance of strong links with the demands of the case-work. There should be a constant feedback and interplay between RD and the daily case-work problems.

The incorporation of research groups into a laboratory with case-work generally ensures a better, or at least more pragmatic approach to RD. Instead, other problems arise. Experience has shown that there can be rather hard feelings among the routine-bound criminalists towards the research people who are free of the grind of case-work. One way out of this is a rotation scheme, where individual staff members are freed from case-work for a year or so, and given the opportunity of developing new methods. They can then take their methods back to their case-work and test them in practice. Of course, there will be some grumbling about the choice of the individuals but that can hardly be avoided.

8. The Internal Organization of the Laboratory

The degree of specialization within the laboratory depends largely on its size as mentioned above. The division of the staff and technicians into working groups varies from place to place, but certain principles have generally emerged as far as the larger laboratoriers are concerned.

Usually questioned document examiners, firearm and toolmark specialists, and photographers are grouped into highly specialized teams. The forensic chemists outnumber as a rule all other criminalists and their groups assignments have the greatest variety. Often, they are assigned to a toxicology team, a general chemistry team, and a drug analysis group. Some laboratories have set up instrument divisions serving other teams with analytical problems. Biologists may be working in a team for serology and also in groups doing trace evidence work. Tables C. 11–13 (pp. 307–309) give some practical examples of internal organizations.

A division of the staff into highly specialized units usually ensures a high quality of work. However, it fails when the character of the incoming cases undergoes rapid fluctuations. These fluctuations may be due to new types of crime becoming suddenly "popular", or to different priority directives for the police.

It may be advisable, therefore, to train the forensic scientist in a second specialty so as to increase the flexibility of the laboratory. For example, all staff members may be tested for their ability to examine questioned documents. Biologists and chemists could learn some of each others methods. Firearm and tool mark examiners could from time to time switch work etc. This does not mean a full return to the "generalist" idea of the 1950's but constitutes a small and careful step in that direction. A by-product of such an arrangement may also be a welcome change of routine work for the staff members and especially for the technicians.

9. Architectural Solutions

A great variety of building principles and styles may be observed in the larger forensic laboratories in the world. Almost the only common principle is the demand for space. In contrast to administrative buildings, and even to other laboratories, the forensic scientists need large surfaces and much cabinet space. The reasons for this are:

1. Bulky objects are often brought to the laboratory. These can range from trucks to refrigerators, to motorcycles, to the contents of clandestine laboratories and to heaps of clothing.
2. The examination of these objects may take days or weeks, during which time new assignments and analyses may be added. It does not do to stowe away these objects in the meantime.
3. Cross contamination between objects from the crime scene or the victim and those from a suspect must be avoided at all costs. This means: separate rooms for each category of items.
4. Many items sent in have strong odors, and worse, are contaminated with bacteria and viruses. It is advisable, therefore, to provide separate rooms for the staff members for report writing, telephoning, small group discussions etc.

A total surface area of at least 50 m^2 but better 100 m^2 per employee is a reasonable requirement but will shock the administrators.

It is not easy to combine the demands of space and compactness in a building concept. Some architects have chosen "the vertical approach" and have built laboratories with relatively modest floor space but with a large number of floors stacked on top of each other. The former building for the Metropolitan Police Forensic Science Laboratory (MPFSL) was one example. Others are the Dutch "Gerechtelijke Laboratorium" in Rijswijk near The Hague and the Forensic Science Centre in Toronto.

Another solution is a dense packing of laboratories and offices into a sort of a landscape. Daylight is only available at the perimeter of the landscape, but communication is no doubt facilitated, and "dead space" in the form of corridors is almost absent. Noise, smalls, and biological contamination on the other hand can pose problems. This kind of architectural design is typical of many American laboratories, and has been used in Europe e. g. by the new MPFSL building and – on a smaller scale – at the police laboratory in Zurich (Switzerland). Drawings of such laboratories are available but are not published here for security reasons.

A third approach is the "horizontal solution". Where adequate space is available, the laboratory may be built only a few stories high but with relatively long corridors. These corridors would normally have rows of laboratories on one side and rows of offices on the other. Daylight is as abundant as in the vertical buildings, but the traffic is now less dependent on elevators and more on long hauls through the corridors. Heavy equipment can be moved (or installed) with relative ease and brought into the building by several entrances. One drawback is the greater vulnerability of the building to burglary or other unauthorized entry. Appropriate alarm systems can lessen this danger.

Examples for this type of architecture are relatively rare, probably due to high property prices. The authors' laboratory is built in this way and the new building for the central Belgian laboratory in Brussels is designed along similar lines.

Some activites in a typical forensic science laboratory need special attention. Much of the material to be analyzed can be contaminated by bacteria and viruses as mentioned above. It is therefore advisable to build safety laboratories where contaminated items can be handled with special precautions. The arrangements way include negative pressure areas with air locks and autoclaves for known high-risk material handling. Other rooms may be built in such a way that odors are removed by forced ventilation, table and floor surfaces can be desinfected by agressive agents, and hand-washing arrangements of the hands-off type are installed. Contaminated material is autoclaved or placed in double-walled containers for burning in special furnaces. The whole area should be placed in such a way that it can be declared "out of bound" in cases of massive contamination.

Less dramatic are other means to ensure efficient working conditions. A number of rooms need high ceilings such as a roomy garage for investigating large objects, a large photographic atelier and possibly a fire testroom. A relatively long room with special walls need to be provided for test shooting. Strong rooms are needed for safe keeping of illegal drugs, guns and ammunition. A large number of efficient hoods are an important integral part of the laboratories.

Cold rooms, and possibly a deep freeze room are necessary for storing toxicological specimens, for performing certain types of enzyme work and for long-term storage of sera, film material and other items likely to be spoiled at room temperature.

The protection of personnel from poisonous and infected material has been mentioned. The safety of the employees has to be considered in detail. This includes

fire-fighting material
(hoses, fire-extinguishers),
fire blankets,
emergency showers,
eye-washers,
gasmasks,
protective eye shields,
gloves,
first-aid boxes,
safe handling of compressed gases etc.

An important consideration is the protection of the laboratory from illegal entry and burglary. It is advisable to have only one entrance for normal in- and out-passage. This entrance is continously guarded during working hours. At all times off-work, the building must be locked and guarded, by automatic alarm systems and, if deemed necessary by guards with or without dogs. A direct alarm line to the nearest police station is mandatory. The local police force must get well acquainted with the localities and must know how to reach the laboratory director and some other staff members in case of a burglary, a fire, an explosion, a flood etc. Similar arrangements should be made with the fire department to which a direct telephone line ought to be established. Smoke detectors, sprinklers and possibly an internal fire alarm may be installed.

10. Equipment and Instrumentation

It is obviously impossible to give general advice regarding the purchase of apparatus and other equipment. Instead, a few examples will be given which hopefully may contribute to optimize the choice of instrumentation. A short list of basic equipment for a small (or new) laboratory is given (Table C. 14 p. 310). As the laboratory grows, further items are added (Table C. 15 (p. 310 and 311)), and finally advanced, and therefore expensive, equipment appears in Table C. 16 (p. 311). The lists are set up for laboratories handling criminalistics, toxicology, serology, blood alcohol analyses and questioned documents. It should be understood that they only are examples and have no general applicability.

11. Quality Control

Among the various types of laboratories carrying out analytical procedures there are at least two kinds which have particularly stringent responsibilities. One group comprises the clinical laboratories in the field of microbiology, tissue investigation, and chemistry. The other type is the forensic laboratory. In either of these fields faulty or inaccurate results can lead to serious consequences.

All forensic scientists recognize this situation but forensic science still has a long way to go until quality control procedures are an integral part of all laboratory work. There are various reasons for the slow progress in this direction. Some of these reasons are:

1. The forensic laboratories vary enormously in size, specialization and quality of work.
2. Most laboratories are still in a period of rapid expansion and reorganization.
3. Almost every laboratory experiences a rapid increase in case work which takes priority to most other activities.
4. Administrators are reluctant to provide funds and fear that quality control may be too expensive.
5. In smaller countries there are difficulties in finding cooperative partners; there is no organizational network for such activities.
6. Some laboratories fear that quality deficiencies may get known to the news media and damage the laboratory's reputation.
7. Many forensic scientists are convinced that their laboratory already has such a high level of professional ability that quality tests are not worth the effort and expense.

To the knowledge of the authors, there are only two major organized systems of quality control in progress. One of these is the "Crime Laboratory Proficiency Testing Research Program" conducted in the United States in 1974–1977 by the National Institute of Law Enforcement and Criminal Justice [3]. Some aspects of this publication have already been discussed in various chapters of Part B in this book.

The project was based on the voluntary participation of US and Canadian laboratories. To the pleasant surprise of the organizers no less than 240 forensic laboratories responded and participated in all or parts of the program. Sixty-six per cent of these laboratories employed less than 10 criminalists, 23% 10–19, and 11% more than 20 criminalists. Twenty-one different tasks were devised and the corresponding test samples distributed. Very stringent criteria for "correct answers" had been set up. This has to be kept in mind when perusing the data listed in Table C.17 (p. 311–313). These data are represented in a very simplified form, and the critical reader is advised to consult the original publication for details.

The other quality control system known to the authors is organized by the English Home Office Forensic Science laboratories. This service is carried out by the Quality Assurance Group (QAG) at the Central Research Establishment. The QA work is done on both "known trials" and "unknown trials". In the latter, samples with all associated paper work are routed via a police authority so that no-one in the laboratories knows that the case is a "QA trial".

There are, no doubt, other projects for quality control in progress elsewhere but certainly much remains to be done in this field.

12. External Contacts

A forensic science laboratory cannot function solely by receiving case-work from the prosecution and sending back reports. Contacts with other organizations and individuals are absolutely necessary to insure scientific quality, modern administrative organization and assurance of appropriate education and training.

In the field of research and development, contacts and collaboration with universities and places of higher technical education as well as with other forensic science laboratories are needed. This has been discussed in Sect. C. V. 7 above. In addition, developments in science and technology must be followed by studying journals, visiting symposia and carrying out study visits. These activities have been described in Chap. C. IV.

Another important prerequisite, more directly related to case-work, is the establishment of good relations with specialists in science, technology, and other fields of knowledge which cannot be adequately covered by the forensic laboratory. Specialists will at times be needed in all conceivable fields. A very few examples are listed in Table C. 18 (p. 313). The authors' laboratory has at present a list of about 140 external specialists. This list is being constantly updated. Similar lists are exchanged with the other Scandinavian countries.

Directors of forensic laboratories are usually not especially trained in administrative matters. It is therefore necessary to review the administrative prodecures frequently, if needed with the help of consultants.

Comparisons with other, similar laboratories' organization (domestic or foreign) are useful. These are so much more valuable as the local higher administrators ususally have little understanding of the special problems of a forensic science laboratory and are (or should be) grateful to receive information on administrative questions and solutions elsewhere.

Education and training are very dependent on outside contacts. Some colleges and universities have shown interest in education within the forensic disciplines (cf. Sect. C. V. 6) and others may become interested if properly approached. Locally, the university of Linköping, Sweden, has started some tentative activities. There is a diminutive "Center of forensic science" and a course in forensic chemistry. But Europe is generally far behind the United States in this respect.

Training proceeds essentially along three lines. One is the thorough, slow, and very essential on-the-job experience of the newcomer under the direction of his immediate superiors. The other is a planned system of training courses. These may be carried out in the laboratory but are usually organized at universities or by commercial companies. Typical examples are courses in instrumental techniques (chromatography, spectrophotometry, mass spectroscopy, scanning electron microscopy, light microscopy) or administrative activities.

Finally, the training is extended by visits to other institutes and to industry. Examples would be chemical and physical testing laboratories in special fields, firearm industry, factories producing lacquers, detonators, explosives, tools, plastic materials, electrical appliances, printed documents, typewriters etc etc.

13. General References (In Chronological Order)

A) R. F. Borkenstein: The Administration of a Forensic Science Laboratory. In: Methods of Forensic Science, Vol. 3. Curry, A. S. (ed.), London-New York-Sydney: Interscience 1964 pp. 115–1168
B) Anonymous: Crime Laboratories – Three Study Reports. Law Enforcement Assistance Administration (LEAA). Grant no, 013. Washington 1968, pp. 117
C) L. W. Bradford, A. H. Samuel: Resarch and Development Needs in Criminalistics. Proc. 3rd Nat. Symp. Law Enforcement, Science and Technology: Chicago Ill. April 1970, pp. 465–476
D) B. Parker, V. Gurgin: The Role of Criminalistics in the World of the Future. SRI Project PYU-1359. Menlo Park, Calif.: Stanford Research Institute 1972, pp. 209
E) M. C. Hall: A Proposal for a National Forensic Science Research Institute for Australia. Paper presented at the "Symposium on Law and Justice in the Australian Capital Territory", 24 & 25 August 1973, pp. 19
F) Anonymous: The Bundeskriminalamt (BKA). Intern. Crim. Police Rev. *29*, 26 (1975)
G) A. F. Carter: The West German "Bundeskriminalamt". The Polize J. *49*, 199 (1976)
H) K. Everett, D. Hughes: A Guide to Laboratory Design. Borough Green, Kent: Butterworths 1975
I) K. W. Smalldon: Chemical Research at CRE (The Home Office Central Research Establishment, authors' remarks). Police Res. Bull. *27*, 10 (1976)
J) H. J. Funk: The New Centre of Forensic Sciences. J. Canad. Soc. Forens. Sci. *9*, 20 (1976)
K) L. G. Clark: Planning a New Laboratory. Augmented version of a paper presented at a symposium on "The Laboratory of the Future", Univ. of N.S.W., May 27, 1975. Health Commission of New South Wales 1976, pp. 45
L) J. L. Peterson, P. R. De Forest: The Status of Forensic Science Degree Programs in the United States. J. Forens. Sci. *22*, 17 (1977)
M) F. J. Kreysa: Laboratory Support of Criminal Enforcement at Treasury's Bureau of Alcohol, Tobacco and Firearms. J. Police Sci. & Adm. *5*, 223 (1977)
N) K. W. Goddard: A Role for the Small Local Crime Laboratory. Police Chief *45*, 18, 20, 22 (1978)
O) J. Cadman: How to Get the Best from Your Crime Laboratory. Police Chief *45*, 64 (1978)
P) N. T. Lappas: Forensic Science Laboratories in the United States: A Survey. J. Forens. Sci. Soc. *18*, 171 (1978)

14. Specific References

1) A. Joseph: Study of Needs and The Development of Curricula in the Field of Forensic Science (A Survey of Crime Laboratories). Study Report, Grant No. 013. Washington D. C.: Office of Law Enforcement Assistance, U. S. Department of Justice 1968
2) A. B. Lateef: A Proposed Curriculum in Forensic Science. J. Canad. Soc. Forens. Sci. *7*, 26 (1974)
3) J. L. Peterson, et al.: Crime Laboratory Proficiency Testing Research Program. Washington D. C.: National Institute of Law Enforcement and Criminal Justice, U. S. Department of Justice 1978

15. Tables C. 9–18

Table C.9. The administration of some Forensic Science Laboratories

Laboratory's name and location	Number of employees	Administrative body	Region served by the laboratory
Bureau of Alcohol, Tobacco and Firearms, *Washington*, USA (also 4 field laboratories in Pa, Ga., Ohio, Calif.)	31 (55)	Department of the Treasury, Internal Revenue Service	In principle the whole United States

Table C.9 (continued)

Laboratory's name and location	Number of employees	Administrative body	Region served by the laboratory
Forensic Science Laboratory (Gerechtelijk Laboratorium), *Rijswijk* (The Netherlands)	75	The Dutch Ministry of Justice	The Netherlands (14 million inhab.)
Home Office Forensic Science Laboratory, *Birmingham*, West Midlands, United Kingdom	86	The English Home Office	4.2 million inhab.
The Laboratory, Central Criminal Police (Keskusrikospoliisi), *Helsinki* (Finland)	20	The Finnish Central Criminal Police	Finland (4.8 million inhab.)
Metropolitan Dade County Public Safety Department, *Miami* (Florida)	60	Dade County Public Safety Department (a police agency)	Dade County plus two other counties (1.7 million inhab.)
San Bernardino County Sheriff's Office, Criminalistics Laboratory, *San Bernardino*, California	15	Sheriff's Office, S. B. County	S. B. County (0.8 million inhab.)
Scientific Laboratory, City Police of Zurich (Wissenschaftlicher Dienst der Stadtpolizei Zürich) *Zurich*, Switzerland	30	Zurich City Police	The German and Italian speaking parts of Switzerland (4.5 million inhab.)

Table C. 10. The number of criminalists per laboratory entitled to witness in court (91 laboratories 1968). According to Ref. (3).

Number of criminalists in a given laboratory	Number of laboratories
1	3
2	10
3	16
4	12
5– 9	25
10–19	16
20–29	4
32	1
54	1
56	1
85	1
92	1
–	91

- Director + staff officer
- General administration
  - Administration
  - Library
  - Information services
  - Firearms
  - Photography
  - Questioned documents
- Chemistry I
  - General chemistry
  - Physics and metallurgy
  - Fire investigation unit
- Chemistry II
  - Drugs and toxicology
  - Alcohol
  - Special services
  - Scanning electron microscopy
- Biology
  - General biology
  - Grouping: blood, saliva and semen

Staff:

Director	1
Deputy directors	4
Scientific officers (all ranks)	213
Technical staff	20
Clerical staff	19
Total	257

Table C.11. The internal organization of the Metropolitan Forensic Science Laboratory in London (1978)

- **Director** — Director + deputy director
- **General division** — Head
  - Secretariat (6 empl.)
  - Housekeeping (7 empl.)
  - Receiving of evidence (4 empl.)
- **Service division** — Head
  - Technical services (3 empl.)
  - Photography (4 empl.)
- **Main division of forensic science** — Head
  - General criminalistics (6 empl.)
  - Physico–chemical investigations (4 empl.)
  - Biological investigations (6 empl.)
  - Chemo–technical investigations (3 empl.)
  - Technical criminalistics (8 empl.)
- **Section for toxicological, pharmaceutical and clinical chemistry** (8 empl.)
- **Section for the analysis of blood and urine from drivers for alcohol and drugs** (5 empl.)
- **Section for questioned documents** (5 empl.)
- **Section for research and development** (3 empl.)

The staff consists of:

17 Ph.D.'s
9 engineers (technical degrees)
25 laboratory technicians
26 technical and clerical assistants

77 total

Table C. 12. The internal structure of the Dutch National Laboratory of Forensic Science (Gerechtelijk Laboratorium van het Ministerie van Justitie) in Rijswijk in 1980

- Director
- Administration (15 empl.)
- Chemical and biological division — Head
  - Chemical section — Head
    - Drugs I (8 empl.)
    - Drugs II (8 empl.)
    - Technical chemistry (5 empl.)
    - Special analyses (3 empl.)
  - Biological section — Head
    - Morphology (5 empl.)
    - Serology (5 empl.)
- Division for physical technology — Head
  - Firearm section (5 empl.)
  - General technology section (4 empl.)
  - Fire investigation section (7 empl.)
  - Photography section (4 empl.)
  - Workshop (2 empl.)
- Questioned document division — Head
  - Handwriting section (6 empl.)
  - Document section (2 empl.)
- Research and development group (2 empl.)

Staff:

9 Ph.D.'s
18 MS
11 engineers
32 technicians
15 clerical staff

85 total

Table C.13. The internal organization of the Swedish National Laboratory of Forensic Science in Linköping (1980)

Table C. 14. Basic equipment for the small forensic science laboratory

Microscopes
Stereomacroscopes with light-guide illumination
Comparison microscope with camera attachment
Device for recovery of bullets (water tank or box with cotton)
Recording ultraviolet spectrophotometer
Recording infrared spectrophotometer
Gas chromatographs
Melting point apparatus
Electrophoresis unit
Centrifuges
Shaking machine
pH meter
Good system camera
Enlarging apparatus
Refractometer
Thin-layer equipment
Water baths
Hot plates
Stirring motors
Blendors
Balances
Refrigerators
Deep freeze
UV-lamps
Handheld calculators
Copying machine
Electrostatic document examination apparatus
Apparatus for distillation or deionization of water

Table C. 15. Additional equipment for the growing laboratory

Microcomputer(s)
Emission spectrograph (possibly with an inductively coupled plasma burner)
High pressure liquid chromatograph(s)
Atomic absorption spectrophotometer
X-ray equipment
Autoclave
Equipment for isoelectric focusing
Muzzle velocity meter
Infrared sensitive TV equipment
Microbalance
Microtome
Differential thermal analyzer
Developing machine for color work
Large document camera
Teleprinter
Workshop machines
Vacuum evaporator
Additional gas chromatographs
Additional UV spectrophotometers
Additional IR spectrophotometer
Additional comparison microscope(s)

Table C.15 (continued)

Additional electrophoresis units
Additional pH-meters
Additional cameras (e. g. macro)
Additional enlarging apparatus
Additional refrigerators and freezers
Additional centrifuges
Additional blendors and homogenizers
Additional water baths
Additional hot plates
Additional stirring motors

Table C. 16. Advanced equipment for the large laboratory

Mass spectrometer (with gas chromatograph and computer)
Scanning electron microscope (with induced X-ray capability)
Microspectrophotometer
Nuclear magnetic resonance equipment
Video equipment
Spectrophotofluorimeter
Advanced multipurpose microscope
X-ray diffraction unit
Minicomputer with terminals

Table C. 17. Test samples sent to the participating laboratories (cf. Ref. [3])

No.	Sample type	Description	No. of labs reporting	Correct answers	Unacceptable answers
1	Controlled drug (identification)	74% sodium barbital	205	189	16
2	Firearms (identification of types possible)	2 projectiles 2 cartridge cases	124	89	35
3	Blood (ABO-testing)	Dried blood, type B, from 4 drops	158	152	6
4	Glass (matching of crime sample with reference sample)	2 small sheets of glass	129	123	6
5	Automobile paint (matching crime sample with 2 reference samples)	Small pieces of metal with original car lacquer	121	97	24
6	Mixed street drug (components)	3% heroin 3% cocaine 3% procaine 91% lactose	181	178	3

Table C.17 (continued)

No.	Sample type	Description	No. of labs reporting	Correct answers	Unacceptable answers
7	Bullets and cartridge cases (which were fired from same gun?)	Crime scene 1: 1 bullet + 1 case Crime scene 2: 2 bullets + 1 case	132	125	7
8	Blood (if blood, if human) and grouping	Blood stain	132	38	94 (at least one grouping in error)
9	Glass (matching of two crime samples with one reference sample)	3 glass pieces from headlight lens	112	77	35
10	Building paint (matching one crime sample with two suspect samples)	3 chips of monolayered paint scraped from a glass sheet	111	54	57
11	Soil (matching one crime sample with two suspect samples)	3 samples sieved through 80-mesh sieve	93	60	33
12	Fibers (matching one crime sample with two suspect samples)	Small samples pulled from 3 carpets	120	118	2
13	Physiological fluids (identification of two spots)	Sperm spot (A) and saliva spot (B)	129 129	126 127	3 (A) 2 (B)
14	Arson (matching a cloth containing an inflammable liquid with a clean cloth and a liquid)	Two pieces cut from a 8 sq. in. cloth, one of which containing 2 ml of gasoline + 8 ml of gasoline	118	84	34
15	Mixed street drug (all qualitative and quantitative information)	1% methamphetamine HCl 1% ephedrine sulfate 49% lactose 49% sodium carbonate	143	117	26
16	Building paint (repeat of test no. 10)	Same samples as for test no. 10	103	68	35
17	Metal (differentiate between 3 samples)	Sample A: 94 B 17 steel Samples B and C: Open hearth steel	68	53	15

Table C. 17 (continued)

No.	Sample type	Description	No. of labs reporting	Correct answers	Unacceptable answers
18	Hair (species determination)	5 hairs: A dog, B cat, C deer, D cow, E mink	90	45	45 (A)
			90	65	25 (B)
			90	41	49 (C)
			90	29	61 (D)
			90	58	32 (E)
19	Wood (differentiate between 3 samples)	Fir, Pine, Maple	65	51	14
20	Questioned documents (identification of A, handwriting and B, typewriter)	1) letter 2) 4 reference handwriting samples 3) 3 reference typewriting samples	74	70	4 (A)
			74	60	14 (B)
21	Assignment of bullets to same firearm	2 bullets fired from one pistol, 1 bullet fired from another pistol	88	76	

Table C. 18. A few examples of specialists who may have to assist the forensic science laboratory

Specialist or special institution	Typical example of case
Osteologist	Finding of skeletal remains (animal, man, identification)
Forensic odontologist	Identification of skulls, bitemarks
Entomologist	Identification of insects on bodies, clothes etc.
Metallurgist	Investigation of fractures, oxidation of metals, determination of alloys
Mineralogist, geologist	Analysis of traces of minerals and their geographical location
Textile research institute	Statistical data on frequency of fibers, special types of fiber analysis
Foreign embassies	Queries regarding passports, driving licences, official stamps etc.
Pharmacologist	Testing of suspected poisons by animal experiments, testimony on possible effect of toxic substances
Industries	Queries regarding the physical or chemical properties of various products
Pollution laboratories	Sampling and analysis of certain rare types of chemical pollutants
Microbiologist	Testing for bacterial or viral contamination

Index

MIX
Papier aus verantwortungsvollen Quellen
Paper from responsible sources
FSC® C105338

If you have any concerns about our products,
you can contact us on
ProductSafety@springernature.com

In case Publisher is established outside the EU,
the EU authorized representative is:
Springer Nature Customer Service Center GmbH
Europaplatz 3, 69115 Heidelberg, Germany

Printed by Libri Plureos GmbH
in Hamburg, Germany